AF601477

Electrowetting

Electrowetting

Fundamental Principles and Practical Applications

Frieder Mugele and Jason Heikenfeld

Authors

Prof. Frieder Mugele
University of Twente
Physics of Complex Fluids
Drienerlolaan 5
7522 NB Enschede
Netherlands

Prof. Jason Heikenfeld
University of Cincinnati
Novel Devices Laboratory
Ohio
United States

Cover fotolia_©altmis

Library of Congress Card No.: applied for

British Library Cataloguing-in-Publication Data
A catalogue record for this book is available from the British Library.

Bibliographic information published by the Deutsche Nationalbibliothek
The Deutsche Nationalbibliothek lists this publication in the Deutsche Nationalbibliografie; detailed bibliographic data are available on the Internet at <http://dnb.d-nb.de>.

Print ISBN: 978-3-527-41229-7
ePDF ISBN: 978-3-527-41240-2
ePub ISBN: 978-3-527-41241-9
oBook ISBN: 978-3-527-41239-6

Cover Design Adam-Design, Weinheim, Germany
Typesetting SPi Global, Chennai, India
Printing and Binding CPI Group (UK) Ltd, Croydon, CR0 4YY

Printed on acid-free paper

C9783527412297_130524

Contents

Preface

Electrowetting went through a dramatic development within the last 20 years. The idea of introducing thin films of hydrophobic polymers as insulators between an electrode on the substrate and the active fluid, which was promoted in the mid-1990s by Bruno Berge and his coworkers, laid the grounds for the explosive scientific and technological progress that we have seen ever since. It made electrowetting evolve from a somewhat curious phenomenon in physical chemistry to arguably the most versatile and widely used active tool for manipulating wettability. Since the late 2000s, 150–200 publications on electrowetting appear every year in scientific journals in physics, chemistry, and engineering. Yet, electrowetting has long expanded beyond basic science. By now, the number of technological publications exceeds the purely scientific ones by far, and every year new groups and companies enter the field. Technological applications of electrowetting started out in the early 2000s with electrically tunable lenses, lab-on-a-chip systems, and display technology. Since then many technological concepts appeared within these application areas and beyond. Notable extensions include energy harvesting, electrochemical storage, micromanipulation, tools for analytical chemistry, and advanced optical systems.

From a purely fundamental perspective, the relevant physical concepts to understand electrowetting were established by 2005 when one of us published the first major review in the field. Excellent review papers have appeared ever since covering the recent progress. Yet, as electrowetting is used in more and more application areas, more and more researchers and engineers enter the field from very diverse backgrounds, such as mechanical or electrical engineering, biology, chemistry, biomedical technology, and so forth. They want to exploit the full potential of electrowetting and get the best out of it for their specific problems. But it is very challenging to acquire the relevant knowledge from the many different branches of physics and physical chemistry that are required to perform successful electrowetting experiments, let alone to develop reliable devices. Bits and pieces of knowledge in wetting, capillarity, electrostatics, fluid dynamics, molecular and colloidal adsorption, materials science, and best practices are spread all over the electrowetting literature. A consistent overview, however, is missing.

Therefore, we felt that it would be worthwhile to provide such an overview explaining both the relevant physical principles and the practical materials knowledge that has accumulated throughout the two decades of electrowetting research since its "outbreak" in the late 1990s. As early as 2012, during the Seventh International Electrowetting Conference in Athens, we decided to embark on the endeavor to write this book. It turned out to be a rather long journey. We hope that the present book will be useful not only for those entering the field of electrowetting but also for already experienced researchers in the field as a reference and as a stimulation to deepen their understanding. We included various derivations of electric field distributions, electrostatic energies, ion distributions, flow fields in thin films, and droplets in quite some detail. This may look a bit technical at times, and we point the reader to sections that can be skipped if one is primarily interested in the final results but not the derivation. However, the specific examples that we derive in detail are very generic and can be encountered in many applications of electrowetting in daily practice. If ever the reader runs into such a problem, the derivations given throughout the book would at least provide the relevant physical ingredients along with an idea of the type of calculation that needs to be done and sometimes probably already an approximate solution of the problem at hand. While writing, we were ourselves frequently amazed how many of the fancy phenomena that appeared in the literature throughout the years actually rolled out very naturally once we took the time to lean back, absorb the basic physical principles, and think for a moment about the specific problem. We hope that reading the book will enable more people to experience this enlightenment and help them combine the different pieces of physics and chemistry in a successful and creative manner to solve their practical problems. At the same time, we also hope that the present text may be used as lecture material for specialized courses in academic environment. We hope that the detailed derivations as well as the problems that we included at the end of each chapter will help students understand this exciting field.

Enschede, February 2018 — *Frieder Mugele*
Cincinnati, February 2018 — *Jason Heikenfeld*

1

Introduction to Capillarity and Wetting Phenomena

The goal of electrowetting (EW) is to manipulate small amounts of liquid on solid surfaces by tuning the wettability using electric fields. Many phenomena encountered in EW experiments are actually not special to *electro* wetting and *electro* capillarity. They are simply wetting and capillary phenomena that can be observed in many circumstances with surfaces of more or less complex geometry and more or less complex distributions of wettability. EW is rather unique in its ability to change contact angles over a very wide range in a very fast and usually very reproducible manner. Therefore, EW has enabled an unprecedented degree of control of drop shapes and dynamics – and along with it a plethora of possible applications. Nevertheless, many basic observations in EW are still variants of classical wetting and capillary phenomena and hence subject to the general laws of capillarity. To understand the phenomena that we encounter in EW experiments and to be able to fully exploit the potential of EW technology, we therefore need a good grasp of the physical principles of wetting and capillarity. The purpose of this first chapter is to introduce the reader to these basic concepts. We will discuss in Section 1.1 the microscopic origin of surface tension and interfacial energies starting from molecular interaction forces. In Sections 1.2 and 1.3, we introduce the two basic laws governing the mechanical equilibrium of liquid microstructures, the Young–Laplace law and the Young–Dupré equation. In Section 1.4, gravity is added as an additional external body force. Section 1.5 is devoted to a concise but somewhat formal mathematical derivation of the fundamental equations that were discussed in Sections 1.2–1.4. Less mathematically interested readers can skip that section without missing important physical results. In Section 1.6, we address aspects of wetting on the nanoscale and introduce concepts such as the disjoining pressure and the effective interface potential that are relevant, among others, in the vicinity of the three-phase contact line and for the stability of nanometric films. In Section 1.7, we discuss some consequences of surface heterogeneity such as contact angle hysteresis and superhydrophobicity. In order to provide the reader with an intuitive understanding of the theoretical concepts and formulas, we will discuss a variety of classical capillary phenomena frequently switching between force balance arguments and considerations based on energy minimization.

A deeper discussion of many aspects addressed here can be found in the excellent textbook by deGennes et al. [1] that we will refer to multiple times.

Electrowetting: Fundamental Principles and Practical Applications,
First Edition. Frieder Mugele and Jason Heikenfeld.

1.1 Surface Tension and Surface Free Energy

Capillary and wetting phenomena are important on small scales. Small scales always imply large surface-to-volume ratios: the smaller the system, the larger the fraction of atoms or molecules that is located at interfaces. This is *the* fundamental reason why interfacial effects are crucial in all branches of micro- and nanotechnology, including micro- and nanofluidics. Being *at* the surface means being within the range of molecular interaction forces of the geometric interface. Using a typical value of, say, $\Delta r = 1$ nm we can simply estimate the fraction of molecules in a small drop that is affected by the interface: The volume of a drop or radius r is $V = 4\pi r^3/3$. The volume of a shell of thickness Δr is $V_s = 4\pi r^2 \Delta r$. Hence the ratio is $\Delta V_s/V = \Delta r/3r$. For a millimeter-sized drop, one molecule in three million is thus *at* the surface. For a micrometer-sized drop, the ratio is one in three thousand. These surface molecules, not the ones inside the bulk of the drop, determine the equilibrium shape of the drop.

1.1.1 The Microscopic Origin of Surface Energies

Throughout this book, we will consider liquids as continuous media characterized by material properties such as density, viscosity, and cohesive energy. Interfaces between two different phases such as liquid and vapor or liquid and solid are characterized by an interfacial energy or tension γ. For the specific interface between a liquid and its own vapor, it is common to speak of surface energy or surface tension. Both expressions, interfacial energies and tensions, mean the same thing, and we will use them interchangeably throughout this book. A surface energy is an excess free energy per unit area of the surface. It is measured in Joule per square meter or more frequently in milli Joule per square meter because the latter turns out to be more convenient for most common liquids. A surface tension is a tensile force per unit length acting along the surface. It is measured in Newton per meter, or in milli Newton per meter, which is dimensionally equivalent an energy per unit area. (In older books, you will sometimes find surface tensions reported in dyn/cm, which is numerically equivalent to mN/m.) These two complementary perspectives have their origin in complementary experimental observations and conceptual approaches. As often in mechanics, a given problem can be considered either from the perspective of energy minimization or from the perspective of force balance. Both views are perfectly equivalent. As classical theoretical mechanics tells us, Newton's equations of motion are the differential equations that any solution of a mechanical problem has to fulfill in order to minimize the Lagrangian and thus, in equilibrium, the energy of the system. Whether energy minimization or force balance is more convenient or more intuitive to solve a specific problem depends on the problem at hand – and to some extent on personal taste.

Why is there an excess energy associated with an interface? To understand this point, it is convenient to deviate for once from our general continuum picture and to instead consider the individual molecules and their mutual interaction. Let us first look at a reference molecule somewhere in the bulk of the liquid drawn

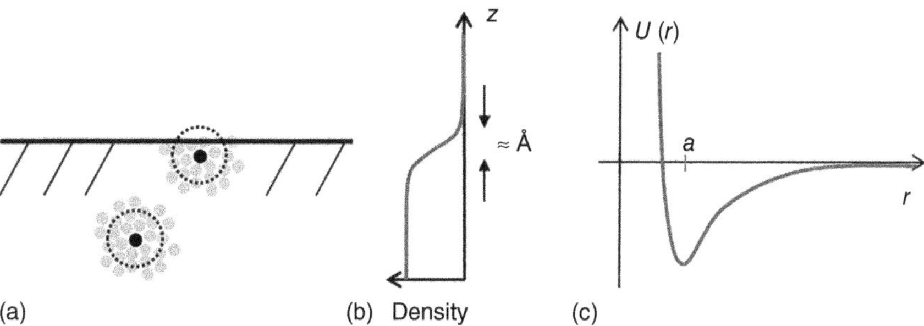

Figure 1.1 (a) Schematic illustration of one specific molecule (black) in the bulk liquid interacting with neighbors (gray) within the range of interactions of the molecular forces (dotted circle) and a second molecule close to the interface that is missing binding partners on the opposite side. (b) Density profile at the liquid surface with a gradual transition from the bulk liquid density to the vapor density. (c) Typical molecular interaction potential between two molecules with a minimum close to the molecular diameter *a*.

in black in the bottom of Figure 1.1a. The reference molecule interacts with its neighbors via some molecular interaction potential. The details of this potential are characteristic for each specific liquid. While all relevant molecular interactions are fundamentally electromagnetic in nature, there are different flavors such as direct Coulomb interactions between charged ions, charge–dipole interactions, dipole–dipole interaction, van der Waals interaction, etc. A detailed discussion can be found in classical textbooks of surface forces, such as the ones by Butt and Kappl [2] and by Israelachvili [3]. Notwithstanding all the details, the generic shape of the interaction potentials usually looks as sketched in Figure 1.1c. There is a strong repulsive barrier at short distances that prevents the molecules from overlapping, and there is an attractive force with a range of typically a few molecular diameters that depends on the specific interaction. The minimum of the potential determines the average separation of the molecules and thus the density of the fluid. In practice liquids are very dense, and molecules continuously bounce into the repulsive potential barrier. The target molecule interacts with all its neighbors (drawn in gray) within the range of the interaction potential, as indicated by the dashed circle in Figure 1.1a. The sum of the interaction energy with all the neighbors within this range determines the cohesive energy E_{coh} of the liquid. This is the reference energy of a molecule in the bulk. If we now consider a second molecule close to the surface, it is obvious that that molecule lacks binding partners above the liquid surface. As a consequence, it has less binding energy than the reference molecule in the bulk. This lack of binding energy constitutes an excess energy as compared with the reference state. If we denote the area per molecule at the interface as a^2, we obtain an estimate for the surface energy based on a very generic atomistic picture:

$$\gamma \approx \frac{E_{coh}}{2a^2} \tag{1.1}$$

We thus expect the surface tension to scale with cohesive energy of the liquid. Table 1.1 gives an overview of the surfaces tensions as well as other properties of liquids commonly used in EW. Generally, more polar molecules (stronger dipole,

Table 1.1 Physical properties of some common liquids.

	Surface tension ($mJ\,m^{-2}$)	Boiling temperature (°C)	Density ($kg\,m^{-3}$)	Viscosity (mPa s)	Vapor pressure (kPa)
Water (25 °C)	72	100	997	0.89	
Water (100 °C)	58	100	972	0.354 (80°)	100
Ethanol	22	78	789	1.074	6.1
Hexanol	25.7	158	814		0.1
Hexane	18.4	68	655	0.29	17.6
Decane	23.8	174	730	0.92	0.195
Hexadecane	27.5	287	770	0	< 0.01
Glycerol	64	290*	1260		$< 10^{-4}$
Ethylene glycol	47.7	197	1110		0.005
Silicone oil (MW 100–100 000)	≈ 20		≈ 900	$1–10^6$	$1–10^{-4}$
Mercury	485	357	13 540		$<2 \times 10^{-4}$

stronger Coulomb forces) exhibit stronger attractive forces. Hence, they typically display a higher cohesive energy, higher boiling point, and higher surface tension. Water with its very large dipole moment is an example of a liquid with a particularly high cohesive energy in view of its low molecular weight. Even stronger cohesive forces are found for liquid metals, for which the cohesion is caused metallic bonds and thus by delocalized electrons.

Our atomistic picture and the estimate of the surface energy are obviously rather rough. Nevertheless, it illustrates a number of important points. For instance, we see from the finite range of the molecular interaction forces that the surface tension is not *located* strictly at the surface. Rather, it arises within the first few layers of molecules at the interface. More rigorous theories for calculating surface tensions in liquid state theory include such effects (see, e.g. [4]).

Another aspect worth noting is that the molecules in a liquid are obviously in vigorous permanent thermal motion. Molecular neighbors change on a picosecond time scale in conventional fluids. What matters for the surface tension is the average interaction energy, which is nevertheless a well-defined sharp value because it reflects an average over macroscopic numbers of molecules and molecular collision events. The relevant type of energy under such conditions is a *free* energy. It includes the effect of random thermal motion in the form of entropy. Surface tensions are excess *free* energies – not pure binding energies as our simplistic picture suggested. For a liquid surface, thermally activated capillary waves carry the entropy similar to phonons in a crystal. The amplitude of these waves is very small, typically not more than a molecular diameter. Yet, the population and amplitude of these capillary waves increase with increasing temperature. As a consequence, surface energies of liquids typically decrease with a temperature coefficient of the order of $-0.1\ mJ\,m^{-2}\,K^{-1}$. For water, this means that the surface tension decreases from $73\ mJ\,m^{-2}$ at 25 °C to $58\ mJ\,m^{-2}$ at 100 °C.

Another consequence of the thermal agitation and the discreteness of the molecules is that the position of the liquid surface is not perfectly sharp. Rather than decreasing abruptly from the liquid density to the vapor density, the interface is characterized by a transition that is smeared out over a small but finite region, as sketched in Figure 1.1b. Except for the vicinity of a critical point, the width of this transition region is of the order of one molecular diameter. Nevertheless, this gradual decrease of the density provides us with an intuitive way of understanding that surface energies act as tensions, i.e. as tensile forces tangential to the surface: In the bulk, the average distance between the molecules is given by the minimum of the interaction potential $U(r)$ (see Figure 1.1c). As mentioned above, this distance determines the density of the bulk liquid. Within the transition zone from liquid to vapor, the average distance between adjacent molecules gradually increases. For intermolecular separations slightly larger than a in Figure 1.1c but still within the range of the potential, adjacent molecules experience net attractive forces. Averaged over the interfacial layer, this leads to a net tensile force within the range of the density gradient. This force tries to reduce the surface area like the tension of a rubber balloon.

So far, we have only been talking about surface tensions, i.e. about the interface between liquid and vapor. Our reasoning can be easily extended to estimate the interfacial energy of liquid–liquid interfaces. Assume that the vapor phase on the top of the surface in Figure 1.1a is replaced by a second fluid phase. Within the range of the molecular interaction potential, molecules of liquid A, lack binding partners of the same species on the opposite side of the interface. The same applies vice versa to the B molecules on other side of the interface. The fact that the liquid A and B don't mix implies that A and B molecules are each more strongly attracted to their own species than to the other one. As a consequence, the presence of the interface gives rise to an excess energy, i.e. a positive interfacial tension γ_{AB}. Now, what happens if A and B molecules attract each other more strongly than A molecules and B molecules among themselves? In this case, the presence of an A–B interface would actually reduce the free energy of the system, i.e. the A–B interfacial energy would be negative. As a consequence, the system gains energy by increasing the interfacial area; it spontaneously emulsifies and eventually mixes on molecular scales. Such systems exist. Yet, they do not form two thermodynamically stable separate phases.

In a similar manner, we can extend the concept of interfacial energy to interfaces between solids and liquids and solids and vapor. We will denote the corresponding interfacial tensions by γ_{sl} and γ_{sv}, respectively. For liquid–vapor surface tensions, we drop the subscript for simplicity and simply write the symbol γ. For the remainder of this book, we will return to the continuum picture, and we will consider interfaces between adjacent phases as perfectly sharp.

1.1.2 Macroscopic Definition of Surface Energy and Surface Tension

Thermodynamically, the macroscopic surface energy is defined as the amount of mechanical work δW required to deform a given amount of liquid in such a way that the total surface area changes by dA:

$$\delta W = \gamma dA \tag{1.2}$$

Given our microscopic picture of surface energy, this definition makes perfect sense. All the molecules at the surface have an excess energy compared with ones in the bulk. Creating more surface area while keeping the number of molecules (and the density) constant implies transferring molecules from the bulk to the surface. Hence, we need to perform work against the molecular interaction forces. That mechanical work is stored as additional surface energy in the total Gibbs free energy G of the system. Thus we can also write

$$\gamma = \left.\frac{\partial G}{\partial A}\right|_{N,P,T} \tag{1.3}$$

The existence of a finite macroscopic surface tension γ also explains why drops assume a spherical shape in the absence of gravity or other external forces: Drops are free to adjust their shape in order to minimize their energy. Hence, they adapt their shape such that the surface area is minimal while keeping the volume constant. Since a sphere is the geometric object with the smallest surface-to-volume ratio, it is the equilibrium shape of free drops. Obviously, this is only true for free drops. As soon as the drop is subjected to external forces or to boundary conditions, e.g. due to contact with a solid surface, the equilibrium shape is in general no longer spherical.

The macroscopic definition of surface tension as a mechanical force per unit length is perhaps even more intuitive. Unlike the mechanical work required to deform a drop, the tensile forces exerted by liquid surfaces on solid objects can be visualized and measured very easily: Consider a U-shaped solid frame, made, for instance, of an open-bent paper clip, as sketched in Figure 1.2a. We place a straw across the two legs to close the U and span a soap film between the three sides of the U and the straw. As soon as we release the straw, it gets pulled toward the left by the soap membrane (see Figure 1.2). Not surprisingly, the force that pulls the straw is proportional to the length l of the straw. Except for this geometric factor, the force is determined by the properties of the liquid, i.e. in this case the soap membrane. The coefficient relating the force F to the length l of the straw is the macroscopic surface tension:

$$F = 2\gamma l \tag{1.4}$$

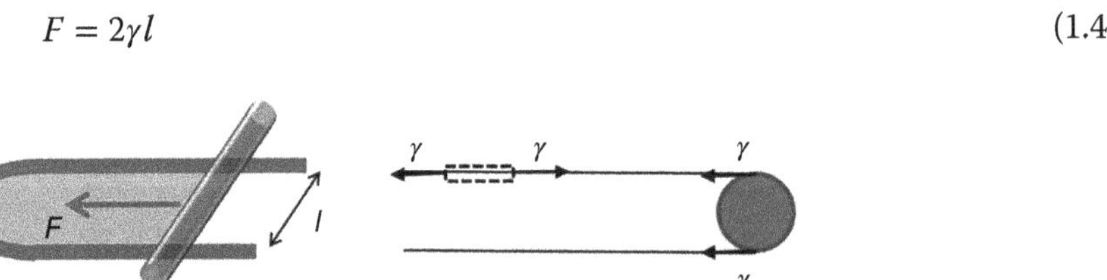

Figure 1.2 (a) Schematic of a rod placed perpendicular across a U-shaped object is pulled toward the left by the surface tension of a thin soap membrane (gray). (b) Zoomed view of the top and bottom surface of the soap membrane and the tensile interfacial tension forces acting on a control volume (dashed) within the interfaces and on the solid boundary holding the membrane.

The factor 2 in the equation accounts for the two liquid surfaces on the top and bottom side of the soap membrane. Both pull on the straw with the same force. Again, we can relate the macroscopic thermodynamic definition to our atomistic picture of the surface tension: Tensile forces are present everywhere within the liquid surface. For an arbitrary element of the surface, tensile forces pulling from different directions balance each other, as sketched in Figure 1.2b. Wherever the liquid surface merges into a solid border of our U-shaped paper clip or into the straw, these tensile stresses must be balanced by the solid surface according to Newton's third law. The U-shaped paper clip becomes slightly deformed, and the mobile straw accelerates. In fact, the measurement of mechanical forces acting on solid objects is one of the most direct and widely spread methods to determine the surface tension of liquids. Tensile forces due to surface tension form an important part of capillary forces that hold sandcastles together or stick our wet hair together into bundles.

1.2 Young–Laplace Equation: The Basic Law of Capillarity

1.2.1 Laplace's Equation and the Pressure Jump Across Liquid Surfaces

As already mentioned, the existence of surface tension has profound consequences for the equilibrium shape of liquid structures. One of the most basic consequences is the existence of a pressure jump across curved liquid surfaces (and liquid–liquid interfaces). The origin of this pressure jump can be illustrated by considering a drop that is connected via a thin needle to a reservoir with a piston, as shown in Figure 1.3. If we move the piston forward and push liquid from the reservoir into the drop, we increase the surface area of the drop. If the drop radius is increased by dr, the surface area has increased by $dA = 8\pi R\, dR$.

Consequently, we have increased the Gibbs free energy of the drop by increasing its surface energy by $dG = \gamma\, dA$. To achieve this increase in surface energy, mechanical work $dW = F\, dx$ has to be performed to move the piston. Since the force on the piston is product of the area A_p of the piston times the difference between the ambient pressure P_0 on the left and the drop pressure P_{drop} on the right (see Figure 1.3), we can equate $dW = (P_{drop} - P_0)A_p dx = \Delta P_L dV$. Equating mechanical work and increase in surface energy, we find

$$\Delta P_L = \gamma \frac{dA}{dV} = \frac{2\gamma}{R} \tag{1.5}$$

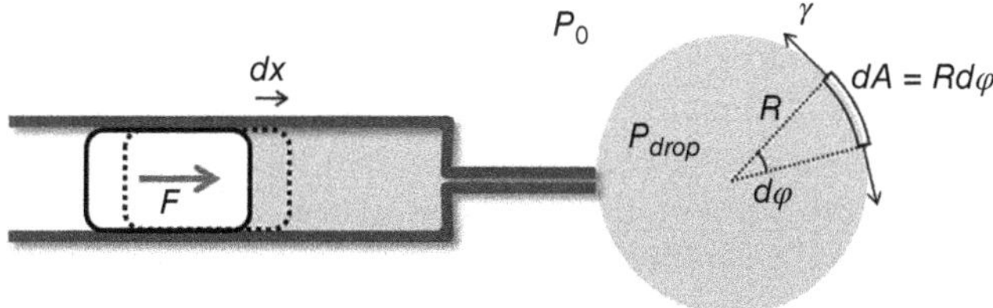

Figure 1.3 Illustration of the derivation of Laplace pressure.

For the right-hand part of the equation, we made explicit use of the spherical shape of the drop by inserting $dV = 4\pi R^2 dR$. The pressure inside the drop thus exceeds the pressure in the ambient atmosphere by an amount that is given by the surface tension times $2/R = 2\kappa$. $\kappa = 1/R$ is called the (mean) curvature of the surface. (We will see below why the expression *mean* curvature makes sense.) This excess pressure, which is a direct consequence of the existence of surface tension, is generally known as Laplace pressure, paying tribute to the French mathematician and physicist Pierre-Simon Laplace, who studied this phenomenon in the early nineteenth century. In parallel Thomas Young studied the same phenomenon in England. Equation (1.5) and its generalized version that we will discuss below are therefore known as Young–Laplace equation.

In addition to the energy picture, it is instructive to consider the same problem from the perspective of force balance. We consider a little volume element $dV = dR\, dA$ that is enclosed by area elements dA just inside and just outside the liquid surface spaced by a radial distance dR, as shown in Figure 1.3. The forces acting on this volume element consist of the surface tension force that pulls tangentially along the surface and the pressure forces on the inside and on the outside of the liquid. The components along the ϕ-directions cancel for symmetry reasons as in the case of the flat surface considered in Figure 1.2b. Due to the finite curvature of the drop, however, there is also a net surface tension force $d\vec{f}_\gamma = -\gamma\, d\varphi\, \vec{e}_r$ oriented radially inward. (For simplicity, we consider the system as two-dimensional.) The pressure P_{drop} on the liquid side is acting along the arc $Rd\phi$ and results in a net force pushing radially outward. Similarly, the ambient pressure P_0 leads to an inward-oriented force, leading together to a net pressure force $d\vec{f}_p = (P_{drop} - P_0)\, Rd\phi\, \vec{e}_r$. Balancing pressure and surface tension forces, we recover Eq. (1.5). It is important to note that the resulting excess pressure $\Delta P_L = P_{drop} - P_0$ in the fluid is a regular isotropic pressure that is felt everywhere inside the fluid notwithstanding the fact that it arises from the directional surface tension forces, which act along the interface.

The Laplace pressure inside a drop can become quite considerable for small drops. As a rule of thumb, it is useful to remember that a water drop with a radius of 1 μm has a Laplace pressure of approximately 1 bar – as much as the hydrostatic pressure due to a water column with a height of 10 m. For even smaller drops, as they are frequently produced and handled in nanofluidic devices, Laplace pressures become even higher.

The form of the Young–Laplace equation that we wrote down in Eq. (1.5) is actually only a special form for spherical drops of the full Young–Laplace equation. In general, liquids can assume more complex nonspherical shapes depending on the boundary conditions that we impose. Figure 1.4a shows a drop squeezed between two solid cylinders at two different distances. Obviously, a single radius of curvature is not sufficient to characterize such three-dimensional surfaces. In such cases, liquid surfaces are described by two different radii of curvature, R_1 and R_2. At any given point on the surface, R_1 and R_2 are defined by intersecting the liquid surface with two perpendicular planes, both containing

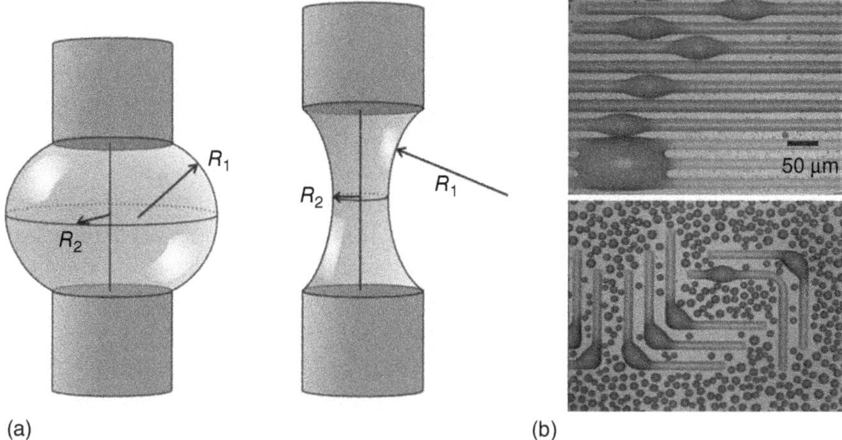

Figure 1.4 Complex three-dimensional liquid microstructures. (a) Schematic of a liquid drop squeezed between two solid cylinders at variable distance showing the two perpendicular radii of curvature. (b) Microstructures of water condensed onto parallel (top) and elbow-shaped (bottom) hydrophilic stripes on an otherwise hydrophobic surface. Source: Adapted from Ref. [5]. See also Ref. [6].

the local surface normal. The resulting values of R_1 and R_2 will in general depend on the azimuthal orientation of the two surfaces. As we rotate the two surfaces around the surface normal, there will be a unique orientation, for which R_1 and R_2 assume a maximum and a minimum value, respectively. These maximum and minimum values are known as *principal* radii of curvature. The *mean curvature*, the specific combination $\kappa = (1/R_1 + 1/R_2)/2$, *is* a well-defined quantity, which determines the pressure jump across the liquid surface. With these definitions, the general version of the Young–Laplace law reads

$$\Delta P_L = 2\gamma\kappa \tag{1.6}$$

In this generalized version, the Young–Laplace law has somewhat deeper implications than discussed so far. Of course, it still provides the local value of the pressure jump upon crossing a liquid surface of curvature κ. But in addition, the law also states that κ is constant everywhere along any connected liquid structure in mechanical equilibrium in the absence of other external forces. No matter how complex it may look, every equilibrated liquid surface is a surface of constant mean curvature. Any position dependence of κ along the surface would imply pressure gradients within the fluid that would immediately set the fluid in motion. The resulting fluid flow would continue until all initial pressure gradients are balanced in a final state of constant mean curvature.

A few more comments are in order: First, for a sphere, R_1 and R_2 are obviously both equal and independent of the position by symmetry. In this case, the expression for the mean curvature reduces to $\kappa = 1/R$, as already stated above, and the full Young–Laplace equation reduces to its simplified form, Eq. (1.5).

Another important limiting case is the situation $R_1, R_2 \to \infty$. This corresponds to a flat surface. In this case, the pressures in the liquid phase and in the ambient phase are identical, i.e. $\Delta P_L = 0$. Second, looking at the right-hand side (RHS) of Figure 1.4a, we notice that the center of the local circle of curvature can be either inside (R_2) the liquid or outside (R_1). The former corresponds to the situation of a sphere. It gives rise to a positive excess pressure inside the drop, and hence we assign the positive sign to R_2 under these conditions. The latter corresponds to a negative radius of curvature R_1 and hence gives rise to a reduction of the pressure inside the liquid as compared with the ambient phase. For liquid drops in ambient vapor, the sign convention is thus straightforward: A radius of curvature is counted positive if the center of the circle is inside the liquid phase and it is counted negative in the opposite case. If we only know it is a mathematical function $h(x, y)$ describing the shape of an interface, it is not a priori clear what is *inside* and *outside*. It is therefore always very important to verify that the sign of the pressure drop in any calculation is in accordance with the physical problem at hand. For two immiscible liquids such as oil and water, the situation is in fact ambiguous, and ΔP_L may either be defined as $P_{water} - P_{oil}$ or vice versa.

As a mathematical expression, Eq. (1.6) does not look very complicated. Yet, this changes as soon as we write down an explicit expression for the radii of curvature in a specific coordinate system. Frequently, it is convenient to parameterize the shape of a liquid surface as a function $h(x, y)$ that specifies the local position of the interface above some reference plane. In this case, the explicit expression reads

$$\Delta P_L = 2\gamma\,\kappa = \gamma\,\frac{\partial_{xx}h(1 + (\partial_y h)^2) - 2\partial_x h\,\partial_y h\,\partial_{xy}h + \partial_{yy}h(1 + (\partial_x h)^2)}{(1 + (\partial_x h)^2 + (\partial_y h)^2)^{3/2}} \tag{1.7}$$

This equation reveals that the Young–Laplace law is in fact a nonlinear second-order partial differential equation for the surface $h(x, y)$. Such equations are difficult to solve and require in most cases a numerical approach. Given this complex structure of the governing equation, it is also not surprising that problems in capillary often lead to counterintuitive surface geometries.

Analytical solutions of the Young–Laplace equation usually only exist in case of specific symmetries or otherwise simplifying conditions. A common simplification that applies in many problems of thin film flows is the small slope approximation. In this case, the gradients of the surface are small, i.e. $|\,\partial_x h\,|, |\,\partial_y h\,| \ll 1$. As a consequence, we can neglect all the (quadratic) terms involving gradients of h. Equation (1.7) then reduces to

$$\Delta P_L = 2\gamma\,\kappa(x, y) = \gamma\,\nabla^2 h(x, y) \tag{1.8}$$

where $\nabla^2 = \partial_{xx} + \partial_{yy}$ is the Laplace operator in two dimensions. This equation is linear and hence much simpler to treat. Being a linear equation, any superposition of two solutions will also be a solution of the underlying equation, allowing for a convenient decomposition of the surface profile in Fourier modes.

Another common situation in applications is cylindrically symmetric problems, such as the one of the capillary bridges in Figure 1.4a. In this case, the Young–Laplace equation is best considered in cylindrical rather than Cartesian coordinates. Since cylindrically symmetric problems do not depend on the azimuthal angle ϕ by definition, the Young–Laplace equation reduces to an ordinary differential equation. If we parameterize the surface by a function $r(z)$, the Young–Laplace equation becomes

$$\Delta P_L = 2\gamma \left(\frac{1}{rS} - \frac{\partial_{zz} r}{S^3} \right) \tag{1.9}$$

where $S = \sqrt{1 + (\partial_z r)^2}$.

1.2.2 Applications of the Young–Laplace Equation: The Rayleigh–Plateau Instability

The instability of a cylindrical liquid jet against breakup into a series of drops is one of the most common examples of a process that is entirely driven by surface tension. The physical principles encountered here are very generic and govern many drop formation processes in microfluidics, including, among others, EW-driven lab-on-a-chip systems. Figure 1.5a shows a series of snapshots of a free jet of one liquid in a second immiscible ambient liquid at different times ranging from zero to the final moment when the jet is broken up completely. As the images show, the decomposition starts with a small perturbation of a well-defined wavelength λ of the initially perfectly cylindrical jet. This perturbation grows with time until the jet eventually breaks up into a combination of large drops located at the grown antinodes of the perturbation and – in this case – a number of smaller satellite drops located at the nodes of the original perturbation. Qualitatively, it is clear from the discussion above that the system reduces its interfacial energy by transforming the long and slender jet with a large surface-to-volume ratio into drops. In fact, one big drop would correspond to the absolute minimum of surface energy. But it is also intuitively plausible that forming one big drop is very unlikely because it would involve the transport of fluid over very long distances. Why is there a characteristic finite length wavelength of the perturbation? How

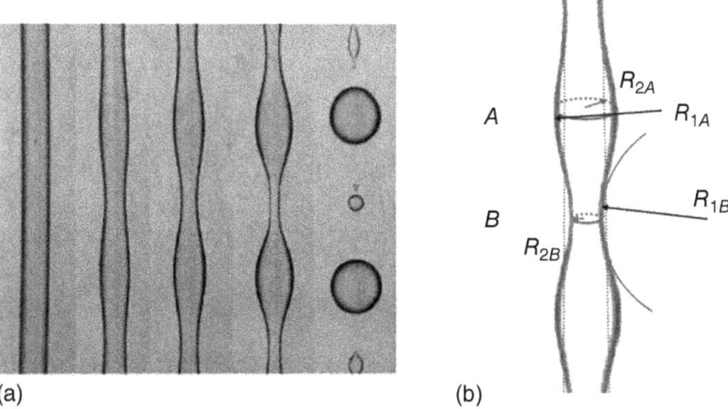

Figure 1.5 (a) Snapshots of a cylindrical jet breaking up into drops with increasing time from left to right. Source: Reprinted with permission from [7]. (b) Illustration of the local curvature as a function of the perturbation.

does this length scale depend on the characteristic parameters of the problem? Why are there satellite droplets?

The answers to some of these problems require a detailed full analysis of the fluid dynamic problem. A first set of answers, however, can already be obtained by considering the variation of either the surface energy or the pressure distribution for small perturbations of the originally cylindrical jet. For the energy approach, all we need to do is calculate the surface area as a function of the amplitude and wavelength for one period of the perturbation. If it decreases with increasing amplitude, the perturbation will spontaneously grow. If it increases, capillary forces will restore the original unperturbed configuration. This very common mathematical approach is known as linear stability analysis. We start with small a periodic perturbation $\delta r(z) = \varepsilon \cos qz$, where $q = 2\pi/\lambda$ is the wavenumber of the perturbation and $\varepsilon \ll r_{cyl}$ is the initially small amplitude. The resulting surface profile of the jet is then given by $r(z) = r_{cyl}(\varepsilon) + \delta r(z)$. Here the average radius of the cylinder r_{cyl} is a function of ε because the total liquid volume must not depend on the amplitude of the perturbation. Imposing volume conservation leads to $r_{cyl}(\varepsilon) = (r_0^2 - \varepsilon^2/2)^{1/2} \approx r_0(1 - (\varepsilon/r_0)^2/4)$, where r_0 is the unperturbed initial radius. To calculate the surface area, we note that the area element on the surface is $dA = r d\phi(1 + (\partial_z r)^2)^{1/2} dz$. Integrating over one full period of the perturbation, we find to lowest order in ε

$$A(\varepsilon) = 2\pi r_0 \lambda \left(1 + \frac{\varepsilon^2}{4r_0^2}(q^2 r_0^2 - 1) \right) \tag{1.10}$$

Whether A increases or decreases with increasing ε thus depends exclusively on the term $(q^2 r_0^2 - 1)$. It is negative for small wavenumbers and positive for large ones. If $\lambda > 2\pi r_0$, i.e. if the wavelength of the perturbation is longer than the circumference of the original cylinder, the perturbation reduces the energy of the system. Vice versa, the energy increases for $\lambda < 2\pi r_0$. Qualitatively, this result should not surprise us. If the wavelength is very short, that is, $\lambda \ll r_0$, any perturbation can obviously only increase the surface area. In this case, the surface is effectively locally flat on the scale of the perturbation. Hence, introducing ripples can only increase the surface area and energy. Conversely we already argued that a sphere has the smallest surface-to-volume ratio. Hence, on large scales, it must be energetically favorable for the jet to decompose into drops. From the perspective of a linearized analysis of the energy balance, we can thus understand that perturbations above a certain wavelength are favorable. The characteristic wavelength separating the stable from the unstable regime is $2\pi r_0$. This is reasonably close to but at the same time clearly less than the length scale, as seen in Figure 1.5a.

Let us analyze the same problem from the perspective of the Young–Laplace equation, i.e. from the perspective of local force and pressure balances. As the radius of the cylinder becomes perturbed, the local radii of curvature R_1 and R_2 both vary along the jet. At point A, the antinode of the perturbation where $r(z)$ is maximum, R_{1A} is positive, and at location B, the node, R_{1B} is negative. R_2 is positive both at A and at B. Yet, R_{2A} increases with increasing amplitude of the perturbation, whereas R_{2B} decreases. For small perturbations, it is easy to show that $1/R_{1A} = \partial_{zz} r(z)\,|_A = +\varepsilon q^2$. Similarly, $1/R_{1B} = -\varepsilon q^2$, i.e. the local radius of curvature at B in that plane is negative, as it should be (see Figure 1.5b). Since $R_2(z)$ is simply given by $r(z)$, we can write down after some algebra the local curvatures at A and B to lowest order in ε as $\kappa_A = 1/r_0 \cdot (1 + \varepsilon/r_0 \cdot (r_0^2 q^2 - 1))$ and $\kappa_B = 1/r_0 \cdot (1 + \varepsilon/r_0 \cdot (1 - r_0^2 q^2))$. Hence, the pressure difference between A and B is

$$\Delta P = \Delta P_{L,A} - \Delta P_{L,B} = \frac{2\gamma\varepsilon}{r_0^2}(r_0^2 q^2 - 1) \tag{1.11}$$

In other words, the pressure in A is larger than the pressure in B if $\lambda < 2\pi r_0$. In this case, the pressure gradient in the fluid will drive liquid back from the bulge of the perturbation to the neck, thus restoring the original shape. This is the stable configuration. Vice versa, for $\lambda > 2\pi\, r_0$ the pressure in B is larger than in A, and therefore even more liquid will flow from the neck toward the bulge, thus allowing the perturbation to grow, in agreement with our conclusions above.

Both linearized energy analysis and linearized pressure balance thus yield the same information regarding the critical wavelength separating stable from unstable perturbation modes. To obtain fully quantitative predictions of the final breakup, we also need to consider dynamic aspects. To illustrate this, let us go back to Eq. (1.11). This equation tells us that the driving pressure difference will be larger and larger if we increase λ. One might therefore expect that the entire jet should transform into one big drop. This, however, would require transport of fluid over very large distances, which is dynamically very unfavorable. In fact, the quantity that drives the fluid flow is not the pressure difference between node and antinode but the pressure *gradient*. The pressure gradient, however, scales as $q \times \Delta P$ and therefore vanishes for $\lambda \to \infty$. The full hydrodynamic solution of the problem obtained independently by Rayleigh and Plateau in the nineteenth century shows that the fastest-growing perturbation has a wavelength $\lambda_{max} = 9.02\, r_0$. This wavelength also sets the size of the drops in the final state. A detailed analysis of this problem can be found in standard textbooks of fluid dynamics [8].

1.3 Young–Dupré Equation: The Basic Law of Wetting

So far, we only considered interfaces between two fluid phases, liquid and vapor, or two immiscible liquids. Wetting is about contact between three different phases, in most cases a solid, a liquid, and a vapor phase. The latter can be replaced by a second immiscible liquid phase. Furthermore, we may also consider two immiscible liquids and vapor such as oil drops on a water surface, or even three immiscible liquids.

1.3.1 To Spread or Not to Spread: From Solid Surface Tension to Liquid Spreading

First of all, we note that the physical concept of a surface tension as we have introduced in Section 1.1 is not limited to liquids. Also for solids, molecules or atoms at the surface lack binding partners as compared with the bulk. Hence, solids equally

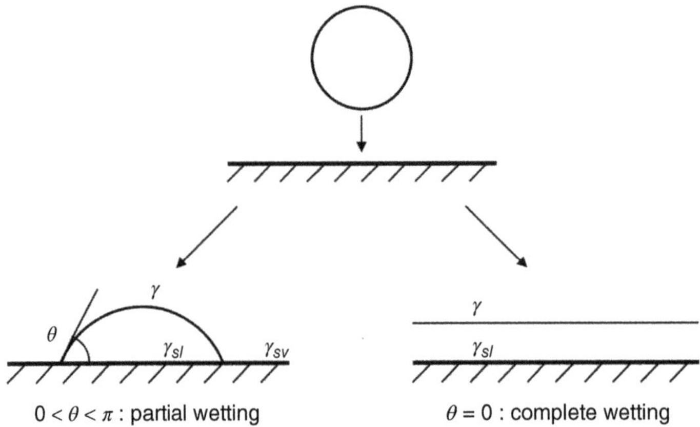

Figure 1.6 Deposition of a liquid drop on a solid surface leads to either partial wetting (left) or complete wetting and spreading of the liquid into a thin film (right).

display surface and interfacial energies with adjacent phases. Since the origin of the surface energies is the same, it is also not a surprise that the order of magnitude of solid surface tensions is roughly the same as the one for liquids – higher cohesive energies for solids with covalent bonds pending. However, in contrast to liquids, solids also possess a finite elasticity that usually prevents noticeable deformations of solid surfaces due to surface tension forces. This circumstance makes the absolute measurement of the surface tension of solids rather difficult.

A basic wetting experiment starts with the deposition of a drop of liquid on a solid surface, as sketched in Figure 1.6. Following drop deposition, the liquid either spreads completely to form a liquid film that covers the entire solid surface, or it spreads only partially and forms a drop with a finite contact angle θ. The former case is called complete wetting, and the latter partial wetting. If the surface is homogeneous, we can already anticipate from our considerations in the previous section that the drop will assume a spherical cap shape to guarantee a constant pressure everywhere inside the liquid.

At first glance, complete wetting may seem physically less interesting and practically less relevant. Yet, this is not the case. In fact, complete wetting is very desirable in many situations including in particular coating technology and lubrication. Complete wetting with zero contact angle is a prerequisite to achieve a thermodynamically stable films on top of a solid surface. In systems involving two immiscible fluid phases such as water and oil in many EW experiments, complete wetting of the continuous phase allows for keeping the dispersed drop phase separated from the solid surfaces by a thin oil film that can lubricate the motion of the drop and reduce the fouling of solutes onto a solid substrate. Partial wetting with a finite contact angle, on the other hand, is *the* generic situation scenario in standard EW experiments.

To understand the difference between complete and partial wetting, we consider the macroscopic energy of the system before and after the deposition of the drop. As previously, the equilibrium is determined by the minimum of the free energy of the system. If the substrate surface is very large and the drop volume is rather small, the energy in the initial state is essentially the solid–vapor interfacial energy, i.e. $G_i \approx A_{sv}\,\gamma_{sv}$. Here, the solid–vapor interfacial area A_{sv} is identical with total the surface area A of the solid. For partial wetting, total surface energy of the system is basically not affected by the deposition of the drop because the drop–substrate and the drop–vapor interfacial area are negligibly small compared with A_{sv}. In contrast, for complete wetting, the solid–vapor interface is replaced by two macroscopic interfaces, namely, the solid–liquid interface and the liquid–vapor interface. Hence, the surface energy in the final state is $G_f = \gamma\,A_{lv} + \gamma_{sl}A_{sl} = A(\gamma + \gamma_{sl})$. The difference between the two normalized by the surface area is known as the spreading parameter:

$$S = \frac{G_i - G_f}{A} = \gamma_{sv} - (\gamma_{sl} + \gamma) \tag{1.12}$$

If $S > 0$, the solid–vapor interface is energetically unfavorable compared with the sum of solid–liquid and liquid–vapor interface. Hence, the liquid spreads to a thin film covering the entire surface, i.e. the system displays complete wetting. Vice versa, for $S < 0$, localizing the liquid in a partially wetting drop with a finite contact angle is energetically more favorable.

A simple gedankenexperiment allows us to relate the occurrence of complete and partial wetting to the relative strength of the various molecular interaction forces. Let us consider a block of some material α, as sketched in Figure 1.7. We cleave the block into two halves. From a macroscopic perspective, the difference between the initial state and the final state is simply the existence of two additional surfaces of area A of material I, with a corresponding surface energy γ_I. Assuming that we can cleave the block without any energy losses due to dissipation, the work we performed in the process is $\delta W_{I-I} = 2\gamma_I A$. From a microscopic perspective, we broke the bonds between adjacent molecules everywhere along the cleavage plane. This means that we effectively transferred molecules from the minimum $U_{I-I}(a)$ of their molecular interaction potential in the bonded state to infinity, where the molecular interaction potential vanishes, $U_{I-I}(\infty) = 0$. Hence, we have $2\gamma_I = U_{I-I}$. If we repeat the same experiment with the same kind of experiment for a block of two different materials I and II that are initially in contact, cleavage eliminates an I–II interface in the initial state and creates instead an I-surface and a II-surface. Hence, we have macroscopically $\delta W_{I-II} = \gamma_I + \gamma_{II} - \gamma_{I-II}$. Microscopically, we separated the α- and β-molecules at the interface from their original minimum interaction energy U_{I-II} to infinity, i.e. $\delta W_{I-II} = U_{I-II}$.

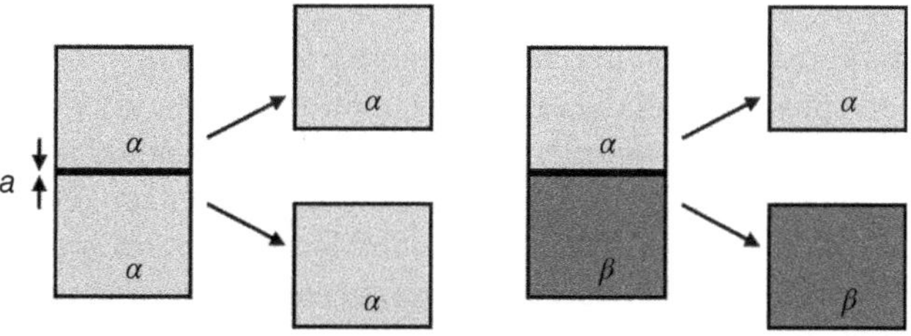

Figure 1.7 Gedankenexperiment of cleaving a homogeneous block of material α (left) and a heterogeneous block α/β (right) into two separate blocks.

If we now identify α with the solid s and β with the liquid l in a wetting experiment, we can express the spreading parameter in terms of the molecular interaction energies as

$$S = \gamma_{sv} - (\gamma + \gamma_{sl}) = U_{sl}(a) - U_{ll}(a) \tag{1.13}$$

We find thus a very natural result: Complete wetting, $S > 0$, occurs whenever the molecular interaction of the molecules of the solid with the molecules of the liquid is stronger than the interaction among the molecules of the liquid. In the opposite case, $S < 0$, partial wetting occurs. For the ubiquitous van der Waals forces, we can go even one step further: van der Waals interaction between two molecules is caused by the coupling of spontaneous fluctuations of the electron cloud of one molecule to the induced dipole moment of the other. Hence, van der Waals interaction is proportional to the polarizabilities α of both molecules: $U_{I-II\,vdW} \propto \alpha_I\,\alpha_{II}$. Inserting this into our expression Eq. (1.13), we find $S \propto \alpha_l$ $(\alpha_s - \alpha_l)$. Hence, in the case of pure van der Waals interaction, liquids completely wet any solid with a higher polarizability. This result leads to the conclusion that liquid helium, the chemically most inert material that interacts only via van der Waals interaction and displays the lowest polarizability of all materials, should wet any solid. This was found to be generally true. (Interestingly, the alkali metals Cs and Rb provide an exception to this rule because the specific electronic structure of these materials leads to an initially unexpected quantum mechanical contribution to the interaction forces that dominates over van der Waals interaction.)

1.3.2 Partial Wetting: The Young Equation

Complete wetting systems involve three different phases. Yet, one of the phases separates the two others completely such that there are only two types of interfaces. In partial wetting all three combinations of interfaces exist. Moreover, there is a one-dimensional region, the edge of the drop, where all three phases involved meet. For obvious reasons, this region is called the three-phase contact line, or shorter, the contact line. Since the three interfaces meet, we can define – within each phase – the angle between the two adjacent phases. For the generic case of a flat surface, the angle between the sv interface and the sl interface is 180° by definition. For a liquid drop in ambient vapor, the angle between the sl and the lv interface is known as the contact angle of the liquid. It is typically denoted as Young's angle θ_Y, honoring the English physicist Thomas Young (1773–1829). The complementary angle between the sv and the lv interface is obviously $\pi - \theta_Y$. Like in the case of the sign convention of the radius of curvature, we should note that the definition is not necessarily unique. In particular for liquid–liquid systems, it is important to define properly through which phase *the* contact angle is measured. For now, we will think of a liquid drop in ambient vapor and speak about the angle measured in the liquid phase. Our primary concern is to understand how the contact angle θ_Y is related to the three interfacial tensions in the system. Since θ_Y is an equilibrium property, it is clear that it must be determined by the minimum of the total (surface) energy for any given fixed drop volume.

The total surface energy of a sessile drop is given by the sum of all the interfacial energies:

$$E_{surf}[A] = \sum_i A_i \gamma_i \tag{1.14}$$

Here the sum i runs over all three types of interfaces, namely, solid–liquid (sl), solid–vapor (sv), and liquid–vapor (lv) – the index that we usually suppress. The area of the solid–vapor interface is given by $A_{sv} = A_{tot} - A_{sl}$, the total surface area of the substrate minus the solid–liquid interfacial area. The former can be neglected because it provides only a constant offset that does not depend on the drop shape. Neglecting this irrelevant contribution, we can rewrite Eq. (1.14) as

$$E_{surf}[A] = A_{lv}\gamma + A_{sl}(\gamma_{sl} - \gamma_{sv}) \tag{1.15}$$

E_{surf} is a functional of the shape A of the drop, which is determined by A_{lv} and A_{sl}. A is thus the unknown function that minimizes the functional E_{surf}. Finding functions that minimize functionals is the general subject of variational calculus. We will describe this somewhat elaborate formalism in detail in Section 1.5.

Here, we focus on the vicinity of the contact line. On this local scale, the curvature of the liquid–vapor interface is negligible, and hence drop surface can be considered as flat, as shown in Figure 1.8. As we already did before, we consider the problem first from the perspective of energy minimization and subsequently from the perspective of local force balance. If the drop is in an equilibrium configuration that minimizes the surface energy, the variation of the surface energy δE_{surf} upon displacing the contact line by a small amount must vanish. If we displace the contact line by dx, say, to the left, as sketched in Figure 1.8a, we convert sv interface into sl interface along dx. Moreover, we generate a little extra piece of liquid–vapor interface with the length $dx \cos\theta$. The resulting variation of the interfacial energy is given by

$$\delta E_{surf} = dx\,(\gamma_{sl} - \gamma_{sv} + \gamma\cos\theta) \tag{1.16}$$

Table 1.2 Geometric functions characterizing spherical caps and circles.

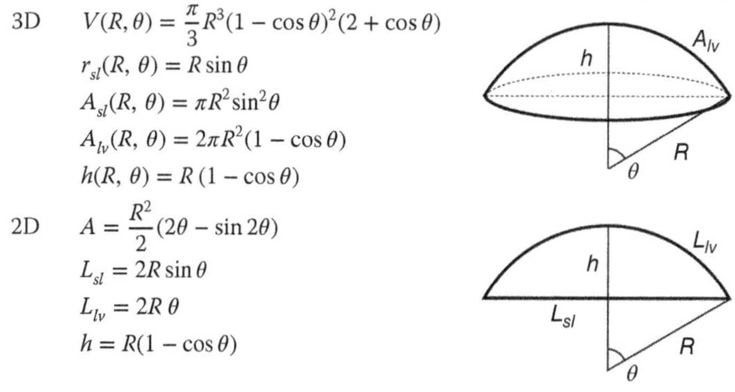

3D	$V(R,\theta) = \frac{\pi}{3}R^3(1-\cos\theta)^2(2+\cos\theta)$ $r_{sl}(R,\theta) = R\sin\theta$ $A_{sl}(R,\theta) = \pi R^2\sin^2\theta$ $A_{lv}(R,\theta) = 2\pi R^2(1-\cos\theta)$ $h(R,\theta) = R(1-\cos\theta)$	
2D	$A = \frac{R^2}{2}(2\theta - \sin 2\theta)$ $L_{sl} = 2R\sin\theta$ $L_{lv} = 2R\,\theta$ $h = R(1-\cos\theta)$	

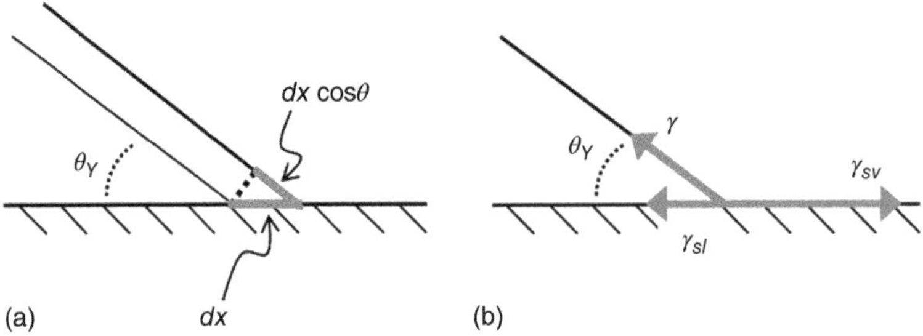

Figure 1.8 (a) Illustration of the variation in interfacial areas upon displacing the contact line by *dx*. (b) Force balance at the contact line.

Equating δE_{surf} to zero we find the desired relation between the equilibrium contact angle θ_Y and the three interfacial tensions in the system, the Young–Dupré equation:

$$\cos\theta_Y = \frac{\gamma_{sv} - \gamma_{sl}}{\gamma} \tag{1.17}$$

Upon inspecting the Young–Dupré equation, or simply *Young's* equation, we can immediately make a couple of simple observations: First of all, for $\gamma_{sl} = \gamma_{sv}$, we find $\theta_Y = 90^\circ$. This is natural because the creation of solid–liquid and solid–vapor interfacial area is equally costly under these conditions. That is, the solid has no preference for either of the two phases. If solid–vapor interface is more costly than solid–liquid interface, θ_Y is smaller than 90° and vice versa. Note also that the value of $\cos\theta_Y$ is bounded between −1 and +1, whereas the value of the RHS of Eq. (1.17) is not. This is not a problem. We derived Eq. (1.17) under the explicit assumption that the system *is* in the partial wetting regime. If θ_Y reaches a value of either 0° or 180°, the basic assumption underlying the derivation no longer holds. This simply means that the system undergoes a transition from partial wetting to complete wetting of one of the two phases.

As we discussed in the preceding sections, we can equivalently interpret interfacial tensions as tensile forces per unit length within the interface. When we discussed the force-based macroscopic definition of the interfacial tensions in the context of Figure 1.2, we already noticed what happens when a liquid surface with a surface tension ends on a solid surface: The tension is transmitted as a force per unit length of the contact line. Likewise, the liquid–vapor interface along the edge of our drop pulls on the three-phase contact line with its tension tangential to the liquid–vapor interface, as sketched in Figure 1.8b. In the same fashion, the solid–liquid and the solid–vapor interfacial tension pull on the contact line along their respective directions. In mechanical equilibrium, all forces must balance. Otherwise the contact line would start to move along the surface. Projecting γ onto the horizontal direction and adding the two other interfacial tensions, we recover Young's equation, Eq. (1.17), from the balance of the horizontal components of the interfacial tensions at the contact line.

In the force balance picture, we can easily understand the transition from partial wetting to complete wetting. Suppose we gradually increase γ_{sv} by some process or γ_{sl} decreases as it effectively does in EW upon applying a voltage. In this case the contact angle has to decrease to align the direction of γ better with

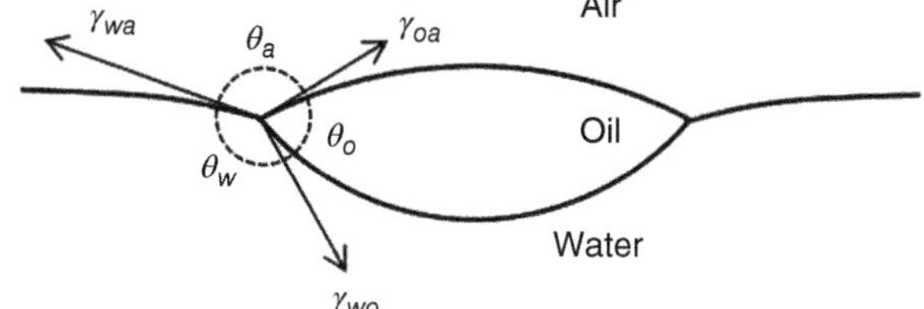

Figure 1.9 Schematic illustration of the force balance at contact line of an oil drop floating on water.

solid–liquid interface in order to preserve force balance. Eventually, θ_Y will have decreased to zero and even the fully aligned γ_{sl} and γ together are no longer able to balance the tension of the solid–vapor interface. This is the critical condition when the spreading parameter S (see Eq. (1.12)) vanishes and the liquid undergoes the transition from partial to complete wetting. Such a transition is known as wetting transition. Wetting transitions are thermodynamic phase transitions. They are typically driven by gradual variations of interfacial tensions with an external control parameter such as temperature of the chemical composition of one of the phases.

The reader may also wonder about the force balance in the direction normal to the solid surface. Obviously, the surface tension of the liquid also pulls in that direction with a force per unit length of $f_\perp = \gamma \sin\theta$. This component is balanced by elastic restoring forces of the solid substrate. Under normal circumstances, i.e. with typical *hard* substrate materials such as glass, this deformation is extremely small and for all practical purposes negligible. The characteristic length scale of these deformations is given the ratio of the surface tension divided by the elastic modulus G of the solid. For typical hard solids G is of the order of several gigapascal. Dividing a typical surface tension of $0.1\,\mathrm{J\,m^{-2}}$ by, say, 10 GPa, we obtain a characteristic length of 10^{-11} m. For softer substrates, such as weakly cross-linked polymers (e.g. polydimethylsiloxane (PDMS)) with $G \approx 10^5$ Pa, recent experiments showed that the surface tension-induced deformation of the substrate can indeed be finite [9]. The most extreme example of a *soft* substrate is another liquid. Indeed, everyday kitchen experience tells us that small oil drops floating on water pull the water surface upward around the contact line. In this case, the oil–water (ow) interface, the water–air (wa) interface, and the oil–air (oa) interface are all deformable. Balancing all three surface tension forces at the contact line leads to the so-called Neumann triangle relating the three characteristic angles to the three interfacial tensions involved:

$$\frac{\gamma_{oa}}{\sin\theta_w} = \frac{\gamma_{wa}}{\sin\theta_o} = \frac{\gamma_{ow}}{\sin\theta_a} \tag{1.18}$$

Here, the subscripts of the angles indicate the phase in which the angle is measured. See Figure 1.9 for an illustration of the definitions. Useful geometric relations of spherical caps are given in Table 1.2.

1.4 Wetting in the Presence of Gravity

So far, we discussed wetting systems in the presence of surface tension forces only. In practice, other forces such as gravity and in EW electrostatic forces are

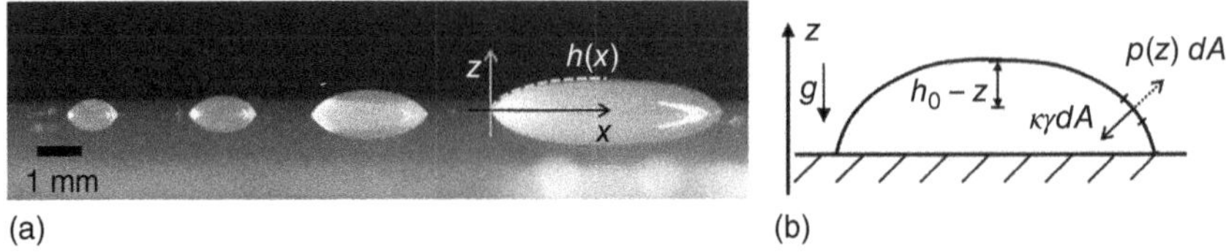

Figure 1.10 (a) Sessile water drops of variable size on a polymer substrate. (b) Schematic of sessile drop illustrating inward- and outward-oriented forces on a surface element *dA*.

omnipresent. From daily experience, we are familiar with the effect of gravity. While small sessile drops assume spherical cap shapes on homogeneous surfaces as discussed above, larger drops assume the shape of puddles that are flat on the top; see Figure 1.10. Daily experience tells us that the transition from the capillarity-dominated regime of spherical caps to the gravity-dominated regime of flattened puddles takes place on a characteristic length scale of a few millimeters, which is known as capillary length λ_c. The exact value of λ_c depends on the properties of the liquid. Yet, it turns out that the order of magnitude of a few millimeters actually holds for the majority of common liquids. In this section we demonstrate how external forces can be incorporated into the laws of wetting and capillarity as we have discussed them so far. Specifically, we want to understand the origin of the capillary length, and we want to derive a quantitative formalism that allows us to calculate equilibrium drop surface profiles $h(x)$ in the presence of gravity, as shown in Figure 1.10a.

To understand heuristically the effect of gravity on the equilibrium shape of drops, we return to our discussion of the derivation of the Young–Laplace equation based on force balance. A key element in our derivation was the point that the pressure p within the drop is constant and equal to the pressure drop $\Delta P_L = \gamma\kappa$ across the surface. In the presence of gravity, however, the pressure within the liquid is no longer constant but increases with increasing depth below the surface due to the hydrostatic pressure. Every volume element dV of liquid experiences a gravitational body force $d\vec{f}_g = -\rho g dV \vec{e}_z$. At a depth $\Delta h = h_0 - z$ below surface, this *directional* gravitational force gives rise to an *isotropic* hydrostatic pressure $p(z) = p_0 - \rho g z$. (h_0 and p_0 denote the highest position of the liquid surface and the reference pressure at that height.) (In Appendix 1.A to this chapter, we provide a somewhat more formal discussion of these considerations introducing a description in terms of the stress tensor that is also relevant for electric fields in EW.) Qualitatively, any surface element dA experiences an outward-oriented force $p(z)dA$ that is balanced by the inward-oriented surface tension force $\kappa\gamma\, dA$, as illustrated in Figure 1.10b. In the presence of gravity, the original Young–Laplace equation in three-dimensional space is thus replaced by

$$2\gamma\kappa = p_0 - \rho g z \tag{1.19}$$

(The factor 2 applies for three-dimensional systems.) If we write the equation in this manner, we assume that the density in the ambient phase is negligible. If this is not the case, the hydrostatic pressure in the ambient phase also varies with z according to the density ρ_a of the ambient fluid. In this case, ρ in Eq. (1.19) has to be replaced by $\Delta\rho = \rho - \rho_a$. Note that $\Delta\rho$ can be either positive or negative, depending on the relative densities of the drop phase and the ambient liquid.

The gravitational body force thus appears as an additional term in the Young–Laplace equation. As we will demonstrate explicitly in a more formal derivation in Section 1.5, gravity does not appear in the balance of forces at the contact line. Gravity thus distorts the surface of the drop but does not affect the contact angle. In Chapter 5, we will see that the same applies for electric fields in the context of EW if we look sufficiently closely in the vicinity of the contact line.

1.4.1 Bond Number and Capillary Length

Before analyzing the specific problem of the shape of the puddle in some detail, we estimate the relative importance of the three terms in Eq. (1.19). To do so, we need to identify the characteristic parameters of the problem and rewrite the equation in nondimensional units. The radius R of the drop is obviously a characteristic length scale. If necessary, we can define R based on the cube root of the liquid volume. Hence, we can use R to define a nondimensional height $\tilde{z} = z/R$. Since the curvature has the dimension of an inverse length, we have $\tilde{\kappa} = R\kappa$. The characteristic value of the capillary pressure is γ/R. This leads to a nondimensional pressure $\tilde{p} = p\,R/\gamma$. With these definitions and after rearrangement of the terms, the nondimensional version of Eq. (1.19) becomes

$$\tilde{p} = 2\tilde{\kappa} + \frac{\rho g R^2}{\gamma}\tilde{z} = 2\tilde{\kappa} + Bo\,\tilde{z} \tag{1.20}$$

Since $\tilde{z}$ and $\tilde{\kappa}$ have numerical values of order unity, we see immediately that the gravity term dominates if $Bo = \rho g R^2/\gamma$, the *Bond number*, is large compared with unity and vice versa capillary forces dominate for $Bo \ll 1$. For $Bo = 1$, gravity and capillarity are equally important. Since Bo scales as R^2, we find what we expected: Gravity is negligible on small scales and dominates on large scales. Using the definition of Bo, we can identify a characteristic length scale, the capillary length

$$\lambda_c = \sqrt{\frac{\gamma}{\rho g}} \tag{1.21}$$

λ_c is *the* characteristic length in any problem involving capillary forces and gravity. It defines the transition from the capillarity-dominated small scales to the gravity-dominated large scales. Inserting numbers for water, we find $\lambda_c = 2.3$ mm. Since λ_c depends only on the square root of the ratio of surface tension and density, it turns out that λ_c almost universally adopts a value of order 1 mm for all common liquids, as can be verified by inserting the surface tensions and densities tabulated in Table 1.1. Much larger values are realized in space. For this reason many fundamental experiments to test the laws of capillary have actually been carried out in space. On Earth, the easiest manner to achieve large values of the capillary length is to use liquid–liquid systems with matched densities of the two phases. (In this case, the density ρ in the expression for λ_c is again replaced by $\Delta\rho$, as noted above.) Taking specific care, e.g. by tuning using mixtures of solvent or by adding solutes densities of oil and water, can be matched up to a precision of $\Delta\rho \approx 10^{-3}\rho$. This leads to values of λ_c of several centimeters. This is, for instance, crucial for optimizing the design of (e.g. EW-driven) optofluidic devices such as tunable lenses and beam deflectors that we will discuss in Chapter 8. Note that matching the densities of liquids not only reduces the effect of gravity, but it also reduces the effect of inertial forces. As a consequence density-matched optofluidic devices are particularly insensitive to mechanical shocks and vibrations.

1.4.2 Case Studies

1.4.2.1 The Shape of a Liquid Puddle

What is the equilibrium shape of the liquid puddles shown in Figure 1.10, and what is their height h_∞ far away from the contact line? Let us assume that the puddle is very large (i.e. $R \gg \lambda_c$) such that it becomes completely flat far away from the contact line. In this limit it is reasonable to consider the one-dimensional version of Eq. (1.19), i.e.

$$-\gamma \frac{h''}{(1+h'^2)^{\frac{3}{2}}} = p_0 - \rho g\, h \tag{1.22}$$

where the prime indicates a derivative with respect to x (cf. Eq. (1.7)). Note the minus sign on the left-hand side (LHS) that ensures a positive pressure inside the drop. Since we assume the puddle to become flat far away from the contact line, we know that $h''(x), h'(x) \to 0$ for $x \to \infty$, while $h(x) \to h_\infty$. From this boundary condition, we can infer immediately that $p_0 = \rho g h_\infty$. Furthermore, we know that the slope of the surface at the contact line must correspond to Young's angle, i.e. $h'(0) = \tan\theta_Y$. Equation (1.22) is an ordinary nonlinear second-order differential equation for the surface profile $h(x)$. Noting that $h''/(1+h'^2)^{3/2} = d/dx(h'/(1+h'^2)^{1/2})$, we can integrate Eq. (1.22) once to obtain

$$(\cos\phi(x) - \cos\theta_Y) = \frac{h(x)\left(h_\infty - \frac{h(x)}{2}\right)}{\lambda_c^2} \tag{1.23}$$

Here, $\phi(x)$ with $\cos\phi = 1/\sqrt{1+h'^2}$ is the local slope angle of $h(x)$, as indicated in Figure 1.11. Equation (1.23) is now a first-order differential equation for $h(x)$. It can be integrated numerically to obtain the full profile.

While there is no analytical solution for $h(x)$, we can nevertheless deduce some useful information from Eq. (1.23). In particular, we can calculate the height h_∞ of a puddle without calculating the full profile by considering the limit $x \to \infty$. In this limit $\phi \to 0$ and $h \to h_\infty$. Inserting these boundary conditions in Eq. (1.23), we obtain after a little algebra

$$h_\infty = 2\lambda_c \sin\frac{\theta_Y}{2} \tag{1.24}$$

The thickness of a liquid puddle is thus independent of its size, in agreement with daily experience. Moreover, we find that h_∞ is given by a combination of the wettability of the surface and λ_c. Thus, λ_c appears prominently in this capillarity-/gravity-dominated problem, as anticipated above.

Next to the formal manner of solving the Young–Laplace equation discussed above, we can also arrive at Eq. (1.23) directly by balancing the horizontal component of all external forces acting on the various control volumes shaded in gray

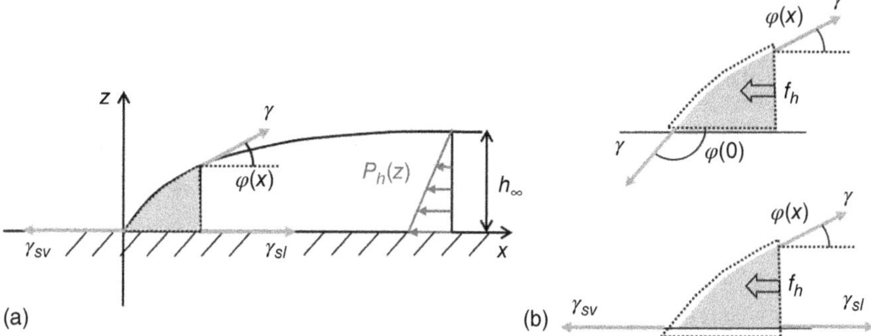

Figure 1.11 Surface profile of a puddle of liquid. (a) Global view indicating the height-dependent pressure. (b) Two alternative selections of the control volume with the bottom just above (top) and just below (bottom) the solid–liquid interface.

in Figure 1.11. In mechanical equilibrium, the sum of all forces acting on a control volume must obviously add up to zero. The details of the calculation depend on our specific choice of the control volume. Let us first consider the gray-shaded control volume in the bottom of Figure 1.11b, which is a zoomed view of the one shown in Figure 1.11a. The lower boundary is chosen inside the solid substrate just below the solid–liquid interface, and the curved boundary follows the surface profile just outside the liquid. The right vertical boundary connects the ends of the two other sections. With this choice, the boundary of the control volume crosses the solid–liquid, the liquid–vapor, and the solid–vapor interface. At every intersection with an interface, the corresponding interfacial tension pulls tangentially along the direction of the interface, as indicated in the figure. In addition to the surface tension forces, the liquid experiences the hydrostatic pressure force from the liquid column along the right boundary of the control volume. This force is given by

$$f_h = \int_0^{h(x)} (\Delta p - \rho g\, z)\, dz = \rho g\, h(x)\, (h_\infty - h(x)/2)$$

Balancing horizontal components of all forces, we arrive at the equation

$$\gamma_{sl} + \gamma \cos \phi(x) = \gamma_{sv} + \rho g h(x)(h_\infty - h/2) \tag{1.25}$$

Using Young's equation to eliminate γ_{sl} and γ_{sv}, we recover Eq. (1.23). Alternatively, we could have chosen the bottom edge of the control volume in the liquid phase, just above the solid–liquid interface, as shown in the top of Figure 1.11b. In this case, the surface of the control volume intersects neither with the solid–liquid nor with the solid–vapor interface. Instead, there is a second intersection with the liquid–vapor interface close to the contact line, where the slope of the surface profile corresponds to Young's angle. Inserting this boundary condition, we find that the net surface tension force acting on the control volume is $f_\gamma(x) = \gamma(\cos\phi(h(x)) - \cos\theta_Y)$. Balancing this expression with the pressure force f_h, we recover again Eq. (1.23).

1.4.2.2 The Pendant Drop Method: Measuring Surface Tension by Balancing Capillary and Gravity Forces

The competition between gravity and capillary forces provides us with a very convenient and accurate method to measure the surface tension of a liquid or the interfacial tensions between two liquids. In principle, γ can be obtained by fitting the numerical solution of Eq. (1.23) to the shape of a puddle. If the density of the liquid is known, γ and θ_Y are the only free parameters. In practice, this approach is not very convenient because puddles need to be rather large in diameter, and they tend to display rather small curvatures, in particular if the contact angle is small. Additional uncertainties arise if the solid surface is not perfectly homogeneous. These problems can be avoided if we simply hang a drop of liquid from the needle of a syringe, as shown in Figure 1.12. If the drop radius R is small compared with λ_c, the drop is perfectly spherical. Upon gradually increasing the volume, the drop gets more and more deformed under the influence of gravity as R becomes comparable to λ_c. At some point, the drop becomes too heavy and falls off the needle. This simple experiment offers two possibilities to measure surface tension. First of all, the falling drop has a well-defined volume that is determined by the maximum capillary force at the interface between the needle and the drop. The capillary force holding the drop on the needle is given by $f_\gamma = 2\pi r\gamma \cos\phi$, as illustrated in Figure 1.12. Obviously, this force has a maximum value of $f_\gamma = 2\pi r\gamma$ for $\phi = 0$. Since the weight of the drop is $f_\gamma = 4/3\pi R^3 \rho g$, balancing the two forces yields a drop radius of $R^* = \lambda_c(3r/2\lambda_c)^{1/3}$, where r is the radius of the needle. By measuring the weight of the falling drops, we can thus determine the surface tension. In principle, this sounds very easy. Yet, the problem is that only a part of the originally unstable drop ends up falling. A finite fraction – up to 40% – remains stuck on the needle. The exact value depends on the details of the detachment dynamics and on the viscous properties of the liquid. This limits the accuracy of the measurements.

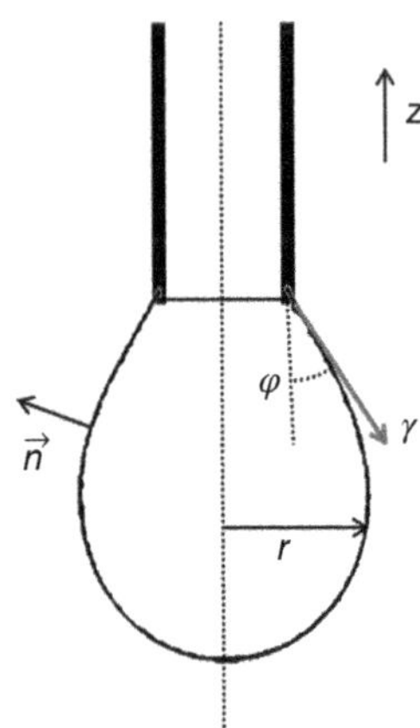

Figure 1.12 Pendant drop hanging from a needle deformed under the influence of gravity. $\vec{n}$ is the surface normal unit vector.

The second much more reliable method is to determine the equilibrium shape of the drop under static conditions before it falls off the needle. In this case, we can fit the numerical solution of the capillary equation in cylindrical coordinates (Eq. (1.19)) to the measured shape of the drop. Knowing the needle radius r and the liquid density, γ is the only fit parameter in this procedure. Since both the acquisition of digital images and numerical fitting are nowadays performed on computers very easily and accurately, this so-called pendant drop method is one of the most popular methods for interfacial tension measurements. A practical added value is that the measurement can be done conveniently in the same instrument as a contact angle measurement.

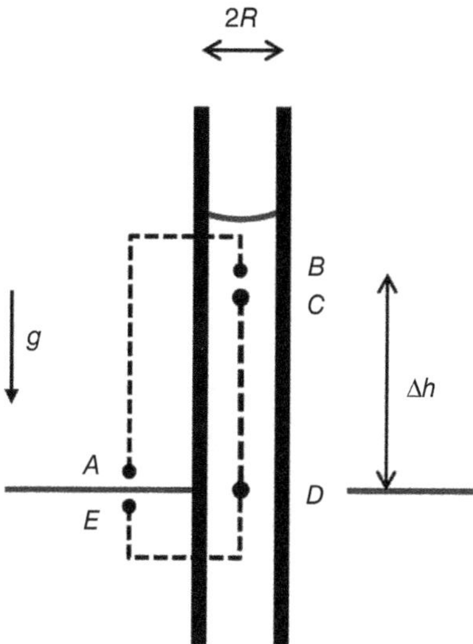

Figure 1.13 Capillary rise in thin tube.

1.4.2.3 Capillary Rise

The final example to be discussed is capillary rise. Capillary rise describes the phenomenon of liquid that is pulled upward into a thin capillary against the direction of gravity as sketched in Figure 1.13. The thinner the capillary, the higher the liquid rises. This is only true if the contact angle of the liquid on the surface is less than 90°. For $\theta_Y > 90^\circ$, the opposite phenomenon, capillary depression, is observed. A substantial height of rise (or depression) is only important for capillary diameters $2R \ll \lambda_c$. The problem of calculating the equilibrium height can be analyzed in many ways.

First, we analyze the variation of the pressure along the path indicated by the dashed lines and the points A–E in Figure 1.13. The pressure in the ambient air at point A is $p(A) = p_0$. If we neglect the density of the air, the pressure at point B is the same, i.e. $p(B) = p(A)$. As we cross the air–water interface between B and C, the pressure jumps due to the curvature of the interface. For a circular tube of radius R and contact angle $\theta_Y < 90^\circ$, we have $p(C) = p(B) - 2\gamma \cos\theta/R$. Here, we implicitly assumed that the liquid meniscus is spherical. This is justified as long as $2R \ll \lambda_c$. Moving from point C to D, the pressure increases due to the hydrostatic pressure: $p(D) = p(C) + \rho gh$. D and E are at the same level. Hence, $p(D) = p(E)$. Since the surface of the liquid bath is flat, there is no pressure jump at that interface: i.e. $p(E) = p(A)$. Combining all these expressions, we find the height of capillary rise

$$h = \frac{2\gamma \cos\theta_Y}{\rho g R} \tag{1.26}$$

This equation is known as Jurin's law.

This derivation is very easy. Yet, it tends to hide the true physical driving forces of the process. The pressure jump across the meniscus is merely a consequence of the boundary condition θ_Y that we imposed on the liquid along the edge of the meniscus. The actual microscopic force driving the process is the difference between γ_{sv} and γ_{sl}. Since $\theta_Y < 90^\circ$, the capillary prefers to be wetted by water rather than by the ambient air. The role of this actual driving force appears more explicitly if we derive Jurin's law by considering the sum of all the forces acting on the liquid column inside the capillary from the level of the bath up to the meniscus. Along the three-phase contact line at the meniscus, the solid–vapor interfacial tension pulls upward with a total force $f_{sv} = 2\pi R\gamma_{sv}$. Along the edge of the lower end of the control volume at the level of the bath, the solid–liquid interfacial tension pulls downward with a force $f_{sl} = -2\pi R\gamma_{sl}$. Finally, there is the body force $-\rho g dV$ that pulls on each volume element of liquid. Integrating over the entire column, this leads to $f_g = -\pi R^2 h\rho g$. Adding up all the terms yields

$$2\pi R\gamma_{sv} = 2\pi R\gamma_{sl} + \pi R^2 h\rho g \tag{1.27}$$

Using Young's equation to replace $\gamma_{sv} - \gamma_{sl}$ by $\gamma \cos\theta_Y$, we recover Jurin's law. From this derivation it is very clear that the actual driving force is the difference of the interfacial tensions γ_{sv} and γ_{sl}. This observation has an important consequence. It implies that the actual force pulling in the liquid does not depend on the curvature of the meniscus. Even if the liquid were frozen such that the meniscus would remain flat, it would still be pulled upward by the same force. If there were no friction, we would expect the same height of rise. In this sense, the fact that the liquid meniscus becomes curved is *merely* a consequence of the characteristic response of liquids to interfacial tension forces and to the specific geometry. The same argument has played an important role in early discussions of the origin of the ponderomotive forces responsible for the actuation of drops in EW and its relation to the variations of the contact angle. We will come back to this discussion in Chapter 5.

It is worth noting that wetting as a driving force for the penetration of confined spaces is important in a much broader context than capillary rise. Capillary forces also suck liquids into porous media, such as rock, paper, and textiles if the liquid wets the walls. The narrower the pores, the stronger the suction pressures. This phenomenon is known as *wicking*. The same process also gives rise to liquid fingers spreading in the corners of rectangular microfluidic channels. Conversely, very large pressures are needed to force liquid into narrow pores with contact angles on the walls exceeding 90°.

1.5 Variational Derivation of the Young–Laplace and the Young–Dupré Equation

As mentioned above, the two basic laws of wetting, Laplace's law and Youngs law, can both be derived in a consistent manner from the principle of energy minimization. The conceptual approach is similar to the minimization of some ordinary function of a set of variables, say, $f(x, y, z)$. A necessary condition to have a minimum (or maximum) some point (x_0, y_0, z_0) is the requirement that the gradient of f vanishes, i.e. $\nabla f(x_0, y_0, z_0) = 0$. To find extrema of functions, we therefore generally solve this condition for (x_0, y_0, z_0). The minimization problem of the surface energy is similar, except that E_{surf} is not simply a function of a set of variables, but it depends on the shape A of the liquid surface. For instance, we can parameterize A by a function $h(x, y)$. E_{surf} a function of the function $h(x, y)$, is called a functional. The analogue of our criterion $\nabla f(x_0, y_0, z_0) = 0$ is a differential equation that the equilibrium shape, the function $h_0(x, y)$, has to fulfill for $E_{surf}[h_0(x, y)]$ to be minimum. This equation is given by equating the first *functional* or *variational* derivative to zero, i.e. $\delta E_{surf} = 0$. As we will see in the end, there are in fact two equations that $h(x, y)$ has to fulfill, namely, the Young–Laplace and the Young–Dupré equations, Eqs. (1.6) and (1.17). Both basic equations thus arise directly from the same energy minimization procedure. Like in the case of an ordinary function, solving a specific problem then implies solving Laplace's equation and Young's equation for the equilibrium function $h_0(x, y)$ subject to specific boundary conditions. This is the general concept. The remainder of this section is somewhat technical and may be skipped by readers who are satisfied with understanding the general background.

In Eq. (1.15) we already noted that the surface energy of a sessile drop with a finite contact angle can be written as $E_{surf}[A] = A_{lv}\gamma + A_{sl}(\gamma_{sl} - \gamma_{sv})$. We can rewrite this expression in an integral form as

$$E_{surf}[A] = \int \gamma \, dA_{lv} + \int (\gamma_{sl} - \gamma_{sv}) dA_{sl} \tag{1.28}$$

We seek the minimum of this functional for a fixed volume V of the drop. This constraint is included by minimizing the functional

$$G[A] = E_{surf}[A] - \lambda V \tag{1.29}$$

Here, λ is a so-called Lagrange parameter, which turns out to be the pressure Δp in the liquid. $G[A]$ is the Gibbs free energy functional of the system. In addition to surface energies, we want to include gravitational body forces as well. The energy density of the gravitational energy is $\rho g z$. Hence, the total functional that we aim to minimize is

$$G[A] = E_{surf}[A] + E_g[A] - \Delta p V = \int \gamma dA_{lv} + \int (\gamma_{sl} - \gamma_{sv}) dA_{sl} + \int (\rho g z - \Delta p) dV \tag{1.30}$$

We assume that the solid–liquid interface is flat and choose to make it coincide with the (x, y)-plane of our coordinate system. To proceed, we rewrite the first and the last integral in Eq. (1.30) as integrals over the drop–substrate interfacial area A_{sl} like the second one. Furthermore, we rewrite the integrand as a function of $h(x, y)$ and $h'(x, y)$, our explicit parameterization of the drop shape. To do so, we note that $dA_{lv} = (1 + (\partial_x h)^2 + (\partial_y h)^2)^{1/2} dx\, dy$ and $dA_{sl} = dx\, dy$. Furthermore, we can write $dV = dx\, dy\, dz$ in the third integral and carry out the z-integration from 0 to $h(x, y)$. Using these expressions, we find

$$G[h(xy), \partial_x h, \partial_y h] = \iint_{A_{sl}} dx\, dy \left\{ \gamma \sqrt{1 + (\partial_x h)^2 + (\partial_y h)^2} + \gamma_{sl} - \gamma_{sv} + \frac{1}{2}\rho g\, h^2 - \Delta p\, h \right\} \tag{1.31}$$

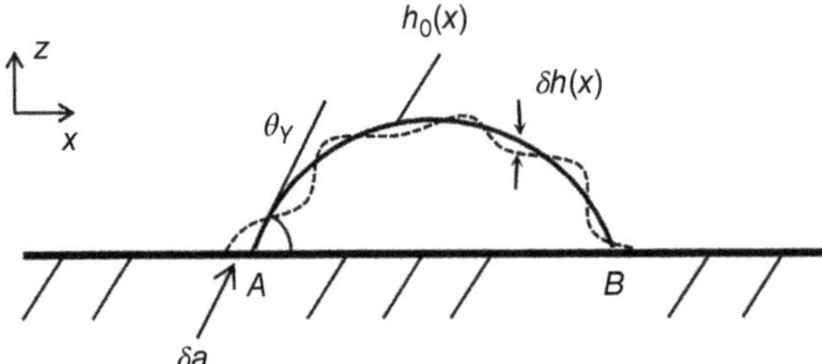

Figure 1.14 Illustration of a sessile drop with equilibrium surface profile $h_0(x)$ and arbitrary variation $\delta h(x)$.

To minimize notational effort, we will restrict ourselves in the following to a two-dimensional problem such that the interface position is a function of one spatial coordinate x only as sketched in Figure 1.14.

In this case, the integration domain on the substrate extends from the left contact line position at $x = A$ to the right one at $x = B$. The surface profile is given by a function $h(x)$. Dropping all y-dependencies and writing $h' = \partial_x h$, the variation of Eq. (1.31) becomes

$$\delta G = \delta \int_A^B \left[\gamma\sqrt{1 + h'^2} + \gamma_{sl} - \gamma_{sv} + \frac{1}{2}\rho g h^2 - \Delta p\, h\right] dx = 0 \tag{1.32}$$

To carry out the variational minimization, we write $h(x) = h_0(x) + \delta h(x) = h_0(x) + \alpha\eta(x)$, where $h_0(x)$ is the equilibrium shape that we seek and $\eta(x)$ is an arbitrary variation around that equilibrium. α is a small parameter. We denote the integrand of Eq. (1.32) as $\tilde{g}(h, h')$. The variational derivative then reads

$$\delta G = \delta b\, \tilde{g}(B) - \delta a\, \tilde{g}(A) + \int_A^B \left[\frac{\partial \tilde{g}}{\partial h}\frac{\partial h}{\partial \alpha} d\alpha + \frac{\partial \tilde{g}}{\partial h'}\frac{\partial h'}{\partial \alpha} d\alpha\right] dx \tag{1.33}$$

where the first two terms arise from the variations δa and δb of the contact line positions. The integral can be further simplified by partial integration of the second term

$$\int_A^B \frac{\partial \tilde{g}}{\partial h'}\frac{\partial h'}{\partial \alpha} d\alpha dx = \left.\frac{\partial \tilde{g}}{\partial h'}\frac{\partial h}{\partial \alpha} d\alpha\right|_A^B - \int_A^B \frac{d}{dx}\left(\frac{\partial \tilde{g}}{\partial h'}\right)\frac{\partial h}{\partial \alpha} d\alpha dx \tag{1.34}$$

Inserting into Eq. (1.33) we find

$$\delta G = \delta b\, \tilde{g}(B) - \delta a\, \tilde{g}(A) + \left.\frac{\partial \tilde{g}}{\partial h'}\delta h\right|_A^B + \int_A^B \left(\frac{\partial \tilde{g}}{\partial h} - \frac{d}{dx}\left(\frac{\partial \tilde{g}}{\partial h'}\right)\right) \delta h dx \tag{1.35}$$

Here, the variations of the contact line positions δa and δb are not independent of the variation of $\delta h(A)$ and $\delta h(B)$. Rather, we have $\delta h(A) = -h'(A)\,\delta a$ and $\delta h(B) = -h'(B)\delta b$ by geometry. Inserting this yields

$$\delta G = \delta b\left(\tilde{g}(B) - \frac{\partial \tilde{g}}{\partial h'}h'\right) - \delta a\left(\tilde{g}(A) - \frac{\partial \tilde{g}}{\partial h'}h'\right) + \int_A^B \left(\frac{\partial \tilde{g}}{\partial h} - \frac{d}{dx}\left(\frac{\partial \tilde{g}}{\partial h'}\right)\right) \delta h(x) dx = 0 \tag{1.36}$$

Since δG has to vanish for arbitrary variations δh, δa, and δb, all the expressions in the parentheses in Eq. (1.36) must vanish independently. The expression under the integral in the third term is the Euler–Lagrange equation for the $\tilde{g}(h, h')$. Inserting the definition for $\tilde{g}$ from Eq. (1.32) yields

$$\gamma\,\kappa(x) = \Delta p - \rho\, g\, h(x)$$

where we used $\kappa = -h''/(1 + h'^2)^{3/2}$ to take into account the proper sign convention for the radius of curvature. Hence, we recover the Young–Laplace equation including the gravitational term as the Euler–Lagrange equation of our minimization procedure. Inserting the definition of $\tilde{g}$ into the conditions that arise from the variations δa and δb, and taking into account $\tilde{g}(x_A) = \gamma(1 + h'(x_A)^2)^{1/2} = \gamma / \cos\theta$, we recover Young's equation, Eq. (1.17). As we already mentioned in our heuristic considerations in Section 1.4, we thus find from the general formalism of variational energy minimization that gravity does indeed *not* affect the contact angle – only the shape of the drop away from the contact line.

In summary, we have now seen how both basic equations of capillarity arise from one calculation based on the general principle of energy minimization. Note that the procedure to include the gravitational body force as an energy density in Eq. (1.30) is rather general. The same strategy can be applied to include molecular interaction forces in thin films (see Section 1.6) as well as the electrostatic body forces that are responsible for EW and dielectrophoresis.

1.6 Wetting at the Nanoscale

So far, we considered wetting problems from a macroscopic perspective. However, we know from our brief discussion in Section 1.1 that interfacial tensions arise from molecular interaction forces and that these forces typically have a range of a few molecular diameters. Therefore, the question arises, how these molecular forces affect the wetting behavior on scales that are comparable to the range of the interaction forces. From an applied perspective, this question may seem somewhat academic at first glance since drop dimensions of interest are typically much larger. Yet, nanoscale wetting properties are important for the macroscopic behavior of drops for several reasons: First of all, liquids with a finite vapor pressure tend to adsorb from the vapor phase onto solid surfaces. In daily life, this leads to the formation of molecularly thin films of water on almost any surface. Depending on the ambient humidity and the nature of the molecular interaction forces (hydrophilicity/hydrophobicity of the surface), these films can become several nanometers thick. Since sessile drops of liquid generate their own atmosphere of saturated vapor around them, the solid surface next to a partially wetting drop is in general also covered by such a thin film, as sketched in the inset of Figure 1.15. Conceptually, this is quite a dramatic observation by itself: An equilibrated solid–vapor interface is not the same as a perfectly dry solid–vacuum interface. Similarly, the equilibrium solid–vapor interfacial tension γ_{sv} is *not* the same as the solid–vacuum interfacial tension: The spontaneous adsorption of the thin film always reduces the interfacial tension. In practice, the presence or absence of such films can also have important macroscopic consequences, e.g. for the friction and lubrication of solid surfaces and for electrostatic

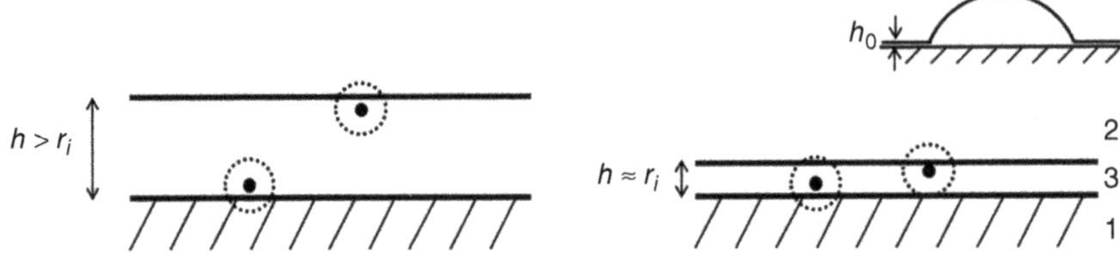

Figure 1.15 Relevance of molecular interaction forces for the surface energy of thin liquid films. Illustration of necessity of a correction term for $h \approx r_i$. The numbers on the far right indicate the media 1 (solid), 2 (ambient), and 3 (film). Inset: illustration of a partially wetting drop accompanied by molecularly thin wetting films.

charging of surfaces, which is a key concern when handling sensitive electronic devices in dry environments. In fact, it turns out that the existence of such thin films is not limited to volatile liquids. Even drops of partially wetting nonvolatile fluids are usually surrounded by such films if given enough time to equilibrate. Their thickness h_0 is particularly large if one of two liquid phases wets a solid much better than the other. A typical examplc is oils used as ambient medium in EW devices that wet hydrophobic polymer surfaces much better than water.

1.6.1 The Effective Interface Potential

To develop a quantitative description of the thin wetting films, we consider the free energy of a solid surface covered by a liquid film with a homogeneous thickness h. If h is large compared with the range r_i of the molecular interaction forces, the solid–liquid interface and the liquid–vapor interface are both well defined and separated from each other, as illustrated on the left of Figure 1.15. The free energy per unit area is then simply given by the sum of the interfacial energies of the solid–liquid and the liquid–vapor interface, i.e. $G \equiv \gamma_{sl} + \gamma$ independent of the exact value of h. If $h \approx r_i$, however, both interfaces are within the range of interactions of the molecules in the film. Hence, there are no longer two separate interfaces. There is only one composite interface that includes the thin liquid film. A very convenient manner to describe this situation is to introduce a new quantity, the so-called effective interface potential $\Phi(h)$. Φ is an excess quantity that takes into account all necessary corrections due to the molecular forces with respect to the reference state of two separate interfaces. Before discussing an explicit expression for Φ in terms of some specific molecular interaction force, let us use this general definition to re-express the thickness-dependent free energy of the interface as

$$G(h) = \gamma_{sl} + \gamma + \Phi(h) \tag{1.37}$$

We can readily infer a few necessary properties of $\Phi(h)$ from the limiting cases without knowing any details about the specific type of molecular interactions. First, the absolute value of Φ should be of the same order of magnitude as the interfacial tensions, because both have the same physical origin. Since the effect of molecular interactions vanishes for $h > r_i$, we know that $\Phi(h) \to 0$ for $h \gg r_i$. Furthermore, we know that $G(h) \to \gamma_{sv}$ for $h \to h_0$. (Note that we include the thin

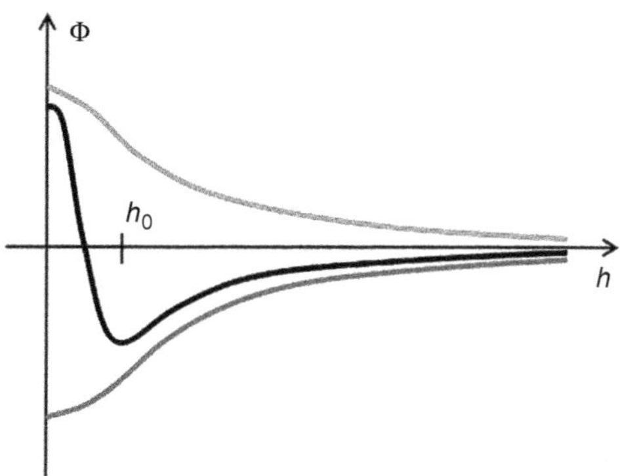

Figure 1.16 Effective interface potential versus film thickness for generic cases of complete wetting (top), pseudopartial wetting (middle), and partial wetting (bottom).

liquid film explicitly in the definition of γ_{sv}.) Comparing Eqs. (1.12) and (1.37), we find $\Phi(h_0) = S$. For complete wetting systems, where the spreading parameter S is positive, Φ is typically a function that decays monotonically from γ_{sv} to zero within the range $h = 0, \ldots, r_i$; see top curve in Figure 1.16. This shape of $\Phi(h)$ implies that reducing h to values of the order of r_i or less increases the surface energy of the system and is thus unfavorable. The minimum of the interfacial energy corresponds to an *infinitely* thick liquid film – exactly as expected for a completely wetting system. (*Infinite* in this context means $\gg r_i$.) This explains the name *effective interface potential*: $\Phi(h)$ is the interaction potential per unit area between the solid–liquid and the liquid–vapor interface. Correspondingly, we can define a pressure, the so-called disjoining pressure $\Pi(h)$, as the negative gradient of the potential

$$\Pi(h) = -\frac{d\Phi}{dh} \tag{1.38}$$

For partial wetting systems we know that $S = \Phi(h_0) < 0$. This means that the system gains energy by reducing the film thickness once it comes within the range of the interface potential. In partial wetting conditions, the interfacial tensions are related to the contact angle by Young's equation. Eliminating γ_{sl} and γ_{sv}, we find

$$\Phi(h_0) = S = \gamma(\cos\theta_Y - 1) \quad \text{(for partial wetting)} \tag{1.39}$$

This is a remarkable equation. It states that Young's angle is completely determined by the value of the minimum of $\Phi(h)$. The actual functional form of Φ is irrelevant for the macroscopic angle that is observed in conventional contact angle goniometry measurements. The middle and the bottom curves in Figure 1.16 both show examples of interface potentials for partially wetting systems. The bottom curve shows a pure partial wetting case, for which the thickness of the equilibrium film strictly vanishes. The middle curve illustrates the more common case with a finite value of h_0. Sometimes, this regime is denoted as *pseudopartial* wetting. It is characterized by a combination of an attractive molecular interaction force at long distances and a sharply increasing repulsive contribution at short distance. This short-range part of the effective interface potential is typically related to the same short-range chemical forces as the repulsive part of the molecular interaction potential shown in Figure 1.1c.

1.6.2 Case Studies

1.6.2.1 The Effective Interface Potential for van der Waals Interaction

We already inferred quite a few general properties of the effective interface potential without specifying the microscopic molecular interactions. Nevertheless it is very instructive to derive a quantitative expression for $\Phi(h)$ for a specific type of molecular interaction. We will choose van der Waals interactions as an example because of their ubiquity and because they give rise to a simple functional form of Φ. (The case of screened Coulomb interactions, which is important for the electrolytes typically used in EW, will be addressed separately in Chapter 4.) Fundamentally, van der Waals interaction arises from quantum mechanical fluctuations of the electron cloud of molecules that give rise to fluctuating dipole moments. These fluctuating dipole moments generate electric fields that polarize other molecules. In the end, this leads to an interaction that is always attractive between two isolated atoms or molecules and that decays as function of the distance r with r^{-6}. This gives rise to the attractive part of the molecular interaction potential in Figure 1.1. To first approximation, the van der Waals interaction energy of a system of many interaction molecules can be considered as pairwise additive. Hence, if we want to know the interaction energy of a liquid film of a certain thickness h covering a solid substrate, we need to add the van der Waals interaction energy of each molecule in the liquid with each molecule in the substrate and with the rest of the adsorbed liquid film. Details of this calculation as well as a discussion of the more advanced quantum mechanical derivations of the van der Waals interaction can be found in References [2, 3]. For a system consisting of a substrate (medium 1) interacting through a thin film (medium 3) with a second immiscible phase (ambient air, oil, etc.; medium 2) with parallel interfaces, as shown on the RHS of Figure 1.15, the resulting effective interface potential is

$$\Phi(h) = \frac{A_H}{12\pi h^2} \tag{1.40}$$

The corresponding disjoining pressure is given by

$$\Pi(h) = -\frac{d\Phi}{dh} = \frac{A_H}{6\pi h^3} \tag{1.41}$$

Here, A_H is the so-called Hamaker constant. It is characteristic for the specific three-phase system. Note that Φ decays rather slowly as h^{-2} with increasing film thickness, much slower than the r^{-6}-dependence of the underlying molecular interaction potential suggests. This weaker exponent arises from the summation over all the interacting pairs of molecules in the substrate and in the thin film. This makes van der Waals interaction a *long-range* force in the terminology of wetting science. Still, *long range* typically means no more than several nanometers.

The complete theory of van der Waals interaction has been worked out by Lifshitz using a quantum mechanical analysis that takes into account the response of the materials to electromagnetic fields integrated over all possible frequencies. The result is that the Hamaker constant $A_H = A_{132}$ for three materials interacting with each other can be written as

$$A_{132} = -C_1 \frac{\varepsilon_1 - \varepsilon_3}{\varepsilon_1 + \varepsilon_3} \frac{\varepsilon_2 - \varepsilon_3}{\varepsilon_2 + \varepsilon_3} - C_2 \frac{(n_1^2 - n_3^2)(n_2^2 - n_3^2)}{\sqrt{n_1^2 + n_3^2}\sqrt{n_2^2 + n_3^2}(\sqrt{n_1^2 + n_3^2} + \sqrt{n_2^2 + n_3^2})} \tag{1.42}$$

Here, $C_1 = \frac{3k_BT}{4} \approx 3.5 \times 10^{-21}$ J and $C_2 = 3\hbar\omega/16\sqrt{2} \approx 1.40 \times 10^{-20}$ J, where $\hbar$ is Planck's constant and $\omega \approx 10^{15}\ \text{s}^{-1}$ is a typical optical frequency. (We choose the sign of A_{132} here consistent with the wetting literature; see, e.g. [1], which is opposite to [2, 3].) The order of the subscripts in A_{123} indicates that medium 1 interacts with medium 2 through the ambient medium 3 according to the conventional notation. The first term in Eq. (1.42) involves the static polarizability of the materials quantified in terms of the dielectric permeabilities ε_i ($i = 1, 2, 3$). The second term describes the response at optical frequencies and is expressed in terms of the refractive indices n_i. Upon inserting numbers, one finds that the high frequency optical term involving the refractive indices usually dominates. Exceptions occur in case of index matching, i.e. when the refractive index of the ambient medium 3 equals the refractive index of one of the two other materials. In that case, the high frequency part obviously vanishes. Furthermore, Eq. (1.42) also shows that nature *likes* to arrange phases in the order of their polarizabilities. That is, complete wetting with $A_{132} > 0$ is found for $n_1 > n_2 > n_3$ or $r\ n_1 < n_2 < n_3$ (and analogue for the low frequency permittivities). Vice versa, if medium 3 has either the highest or the lowest polarizability, the system displays partial wetting with $A_{132} < 0$. In this case, a film of medium 3 on the substrate is unstable and transforms into a drop with a finite contact angle. That is, medium 3 is expelled from the space in between mediums 1 and 2 (except for a possible thin film that is stabilized by a short-range chemical force).

In EW experiments, the substrates are frequently made of fluoropolymers such as Teflon AF with a very low refractive index $n_1 = 1.31$ in the visible range and a dielectric constant of $\varepsilon_1 = 1.93$. If we perform an experiment with aqueous drops in ambient oil, the water is medium 2 with a refractive index of $n_2 = 1.33$ and a very large dielectric constant of $\varepsilon_2 = 81$. Typical alkane or silicone oils have a refractive index of $n_3 = 1.4$–1.45 and a dielectric constant of $\varepsilon_3 = 2$–3. These values imply that the optical contribution favors partial wetting, whereas the static contribution favors complete wetting. Due to the large value of the dielectric constant of water, the balance is actually very subtle. Using $\varepsilon_3 = 2$ and $n_3 = 1.43$ (hexadecane) yields a value of $A_{132} = -2.7 \times 10^{-23}$ J. That is, the system is expected to display partial oil wetting, albeit with a rather small value of the Hamaker constant. Using $\varepsilon_3 = 2.5$ and $n_3 = 1.4$ (for some specific silicone oils) yields a value of $A_{132} = +3.8 \times 10^{-22}$ J. That is, the system is expected to display complete oil wetting and thus a stable oil film in between the water drop and the underlying fluoropolymer substrate. While additional molecular interactions may modify the details of the picture, the calculation shows that a proper choice of materials has important consequences for the microscopic wetting behavior that can affect the outcome of experiments – and the reliability of devices.

1.6.2.2 Equilibrium Surface Profile Near the Three-Phase Contact Line

The disjoining pressure appears as an additional contribution in the Young–Laplace equation if the thickness of a liquid film is of the order r_i. For a two-dimensional system, the corrected form of the equation reads

$$\gamma \frac{h''}{(1+h'^2)^{3/2}} = \frac{d\Phi}{dh} - \Delta p \tag{1.43}$$

Equation (1.43) can be derived formally by variational minimization of the Gibbs free energy in analogy with our calculation for gravity as an external force in Section 1.5. Replacing the latter by the contribution due to the interface potential, we can rewrite Eq. (1.31) in two dimensions as

$$G[h, h'] = \int dx \, \{\gamma\sqrt{1+h'^2} + \gamma_{sl} - \gamma_{sv} + \Phi(h)\} \tag{1.44}$$

Functional minimization of this expression along the lines of Section 1.5 leads to Eq. (1.43). (Note that we can set $\Delta p = 0$ because the surface of the macroscopic drop is local flat on the very small scale that is affected by the molecular forces.) Hence, the thickness profile of molecularly thin films is governed by the balance of capillary pressure and disjoining pressure alone.

This has major consequences for the local surface profile: Far away from the contact line, the disjoining pressure is negligible. Hence, the liquid surface is flat, and the balance of all interfacial tensions acting on the gray-shaded control volume in Figure 1.17 yields Young's angle as a macroscopic contact angle. Note that we included explicitly the molecularly thin wetting film with a finite thickness h_0 corresponding to a *pseudopartial wetting* condition. Within the range of the molecular interactions, Φ is non-negligible, and the disjoining pressure distorts the surface profile $h(x)$, leading a position-dependent local slope angle $\phi(x)$. Very much like the case of a drop in the presence of gravity, we can obtain this profile by integrating Eq. (1.43). We find

$$\cos\phi(x) = \cos\theta_Y - \frac{\Phi(h(x))}{\gamma} \tag{1.45}$$

Since $\cos\phi(x) = 1/\sqrt{1+h'^2}$, Eq. (1.45) is still a first-order differential equation for $h(x)$. Yet, we can already read from the equation that positive values of $\Phi(h)$ give rise to local slope angles $\phi > \theta_Y$ and vice versa. Because of the short range of the molecular interaction forces, such disjoining pressure-induced deformations

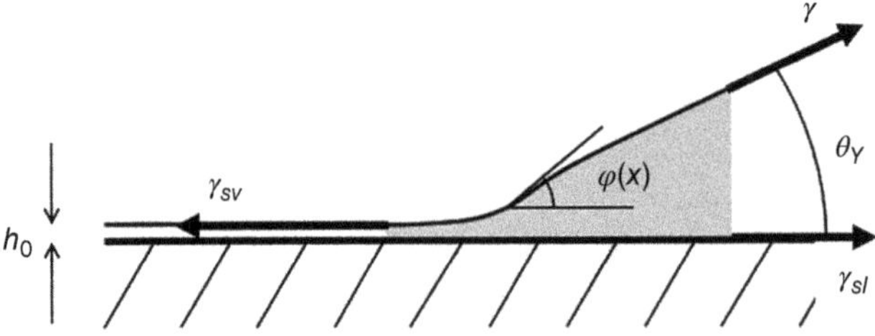

Figure 1.17 Profile of a liquid surface in the vicinity of a three-phase contact line. Note that γ_{sv} includes the contribution from the thin film, i.e. $\gamma_{sv} = \gamma + \gamma_{sl} + \Phi(h_0)$.

of liquid surfaces are extremely difficult to observe. As mentioned above, optical imaging of drops does not allow to access these scales and therefore always shows Young's angle. In Chapter 5, we will use a very similar approach as discussed here to calculate the surface profile in the vicinity of the contact line in EW. In that case, electrical stresses will take over the role of the disjoining pressure. In contrast to the molecular interaction forces, the electrostatic forces have a much larger range. Therefore, deformations of the surface due to local electric fields close to the contact line can be visualized optically for suitable conditions.

1.7 Wetting of Heterogeneous Surfaces

All our considerations so far dealt with perfectly homogeneous and flat surfaces. The contact angle and the drop shape were completely determined by the interfacial tensions of the three phases and the desire of the system to minimize its free energy in equilibrium. In contrast to this idealized situation, real surfaces are never perfectly flat and homogeneous. They display heterogeneity, usually of both topographic and chemical nature. Such heterogeneities occur on various length scales with defect sizes ranging from atomic to macroscopic dimensions. Wetting on real surfaces is always wetting of heterogeneous surfaces. If we are interested in the wetting behavior of drops of some finite size, it is useful to distinguish between heterogeneity on scales that are either much smaller than the drop size or comparable to it. Small-scale heterogeneities are unavoidable. They give rise to usually invisible microscopic deformations of the drop along the three-phase contact line. Their most important macroscopic consequence is contact angle hysteresis. Large-scale heterogeneity gives rise to macroscopically visible deformations of the drops. In a technological context, macroscopic heterogeneities are frequently imprinted onto the surface to achieve certain properties. Examples include superhydrophobicity caused by small-scale heterogeneity as well as EW-based lab-on-a-chip devices, in which actively switchable large-scale heterogeneity is used to deform and transport drops. In this section, we will analyze the physical principles governing the wetting of heterogeneous surfaces in static conditions. Most basic aspects will be established for two-dimensional systems. Toward the end of the chapter, we address contributions arising from deformations of the liquid surface in three dimensions.

1.7.1 Young–Laplace and Young–Dupré Equation for Heterogeneous Surfaces

The first question to wonder is whether the basic laws of capillarity still apply for heterogeneous surfaces. Let us consider a sessile drop on a heterogeneous surface in two dimensions, as sketched in Figure 1.18. The free energy of the drop is formally given by a similar integral as before:

$$G = \int ds_x[\gamma\sqrt{1 + (\partial_x h)^2} + \gamma_{sl}(x) - \gamma_{sv}(x) - \Delta p\, h] \tag{1.46}$$

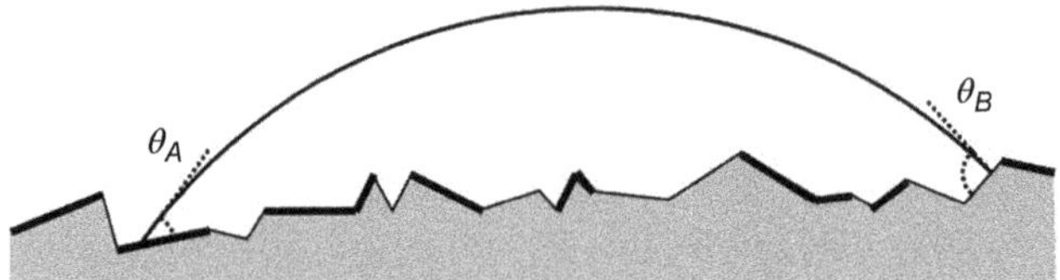

Figure 1.18 Sessile drop on topographically and chemically heterogeneous surface (bold lines on surface: material A – small contact angle θ_A; thin lines: material B with $\theta_B > \theta_A$).

Compared with Eq. (1.31), there are two differences caused by the topographic roughness and the chemical heterogeneity, respectively: First, the integral extends along the contour of the actual rough solid–liquid interface (denoted as ds_x) rather than simply along x. Following all the ups and downs on the rough surfaces gives rise to an enhancement of the true solid–liquid interfacial area as compared with the projected area. Second, the chemical heterogeneity is reflected in position-dependent values $\gamma_{sl}(x)$ and $\gamma_{sv}(x)$.

Variational minimization of Eq. (1.46) along the formalism described in Section 1.5 shows that the Laplace equation remains unaffected by the surface heterogeneity, i.e. the pressure drop across the liquid surface is still given by $\Delta P_L = \gamma\kappa$. This should not surprise us. The drop surface itself is not directly affected by the presence of the heterogeneity on the substrate surface. The influence is only indirect, via the boundary condition at the solid–liquid interface. Hence, it is Young's equation that is affected by the presence of heterogeneity. And even for Young's equation, the modification is – in some sense – minor: If the contact line is located somewhere within a locally homogeneous region on the surface, i.e. away from boundaries between patches of different chemical composition or surface orientation, the principle of the local variation of the contact line position still holds. Hence, Young's equation simply applies locally with the position-dependent local Young angle $\theta_Y(x)$ according to the local composition and orientation of the surface, i.e. $\cos\theta_Y(x) = (\gamma_{sv}(x) - \gamma_{sl}(x))/\gamma$. As illustrated in Figure 1.18, the local contact angle $\theta_Y(x)$ is to be measured relative to the local orientation of the surface.

The range of the molecular interactions provides a good estimate for the minimum size of patches of well-defined local properties. If two types of atoms A and B with different interaction potentials are arranged on a surface in patches with a size of a few times the molecular interaction range, well-defined A and B regions arise (Figure 1.19a). On these patches, local contact angles θ_A and θ_B assume the same values as on macroscopic homogeneous surfaces. In contrast, placing A and B atoms next to each other in an alternating fashion produces an energetically more or less homogeneous surface with mixed properties. While the observation of surface heterogeneity is in practice often limited by the resolution of the

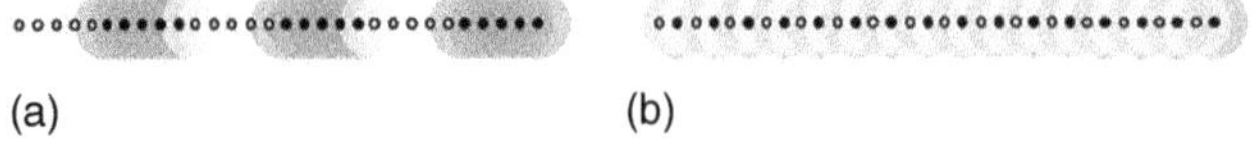

Figure 1.19 Schematic of a surface consisting of two different types of molecules A and B patched in groups of 5 (a) and alternating (b).

instruments that are used to characterize the surfaces, the range of the molecular interactions provides a fundamental lower limit.

The existence of chemical domains with surface areas of a few nanometers raises another important problem: Since the balance of surface energies on a lyophobic (*liquid-repelling*) patch is less favorable for wetting, such a patch represents an energy barrier for the spreading liquid. The height of such an energy barrier is of the order of $\Delta E \approx \gamma\, \Delta \cos\theta_Y \cdot l^2$, where l is the size of the patch. For $l = 5$ nm and a typical wettability contrast $\gamma\Delta\cos\theta = 0.01\ \mathrm{J/m^{-2}}$, we find $\Delta E \approx 2.5 \times 10^{-19}$ J. This value is much larger than the thermal energy at room temperature, which is $k_B T \approx 4 \times 10^{-21}$ J. Hence, the contact line is unable to overcome such a wetting defect by thermal activation. As a consequence, it gets trapped in some metastable nonequilibrium position: The drop is unable to explore the surface to reach the energetically most favorable configuration. The existence of a large number of such metastable states is characteristic for the wetting on heterogeneous surfaces. As in many other fields of physics, the existence of metastable states implies hysteresis and history dependence of the actual configuration of the system. This is arguably the most important physical consequence of surface heterogeneity.

1.7.2 Gibbs Criterion for Contact Line Pinning at Domain Boundaries

On surfaces with sufficiently large patches (i.e. $l \geq$ several nm) of well-defined chemical composition and orientation, the contact angle on each patch is well defined, as illustrated in Figure 1.20. If the contact line is located at the boundary between two adjacent domains, however, this is no longer true. In this case, it is not clear whether the more lyophilic or the more lyophobic angle should be chosen for a surface with chemical patches A and B. Similarly, for a topographically patterned surface, it is not clear with respect to which *facet* of the surface the contact angle should be measured.

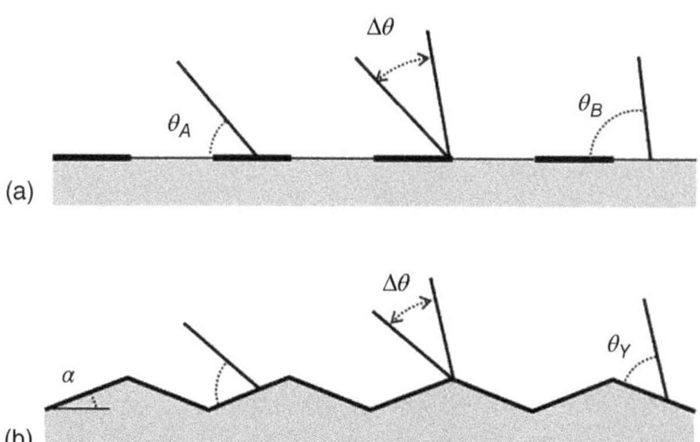

Figure 1.20 Local contact angle on heterogeneous surfaces and Gibbs pinning at domain boundaries for chemically heterogeneous flat surfaces ((a); thick lines, philic regions; thin lines, phobic regions) and chemically homogeneous rough surfaces (b).

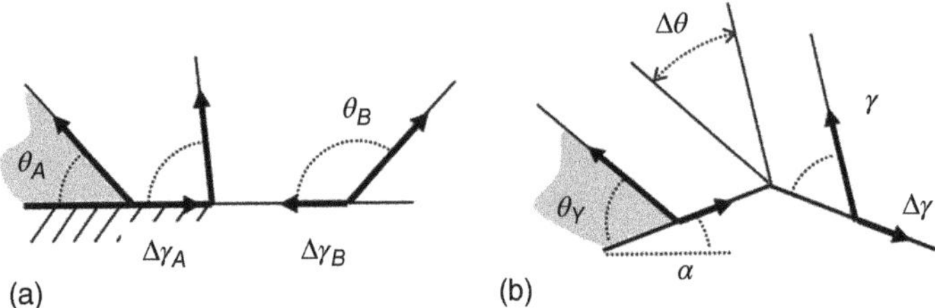

Figure 1.21 Force balance for a contact line upon approaching a domain boundaries for chemical heterogeneity (a) and a topographic corner (b).

We gain some more physical insight if we consider the local force balance as a contact line is approaching an A–B domain boundary as illustrated in Figure 1.21a. Let us assume $\theta_A < 90^\circ$ and $\theta_B > 90^\circ$. When the contact line is on the more lyophilic side of the boundary, the contact angle is θ_A such that the projection of the surface tension $\gamma \cos \theta_A$ balances the net force $\Delta\gamma_A = \gamma_{sv}^A - \gamma_{sl}^A$ exerted on the liquid by the solid substrate. $\Delta\gamma_A$ is thus pointing toward the right, i.e. toward the domain boundary. Conversely, if the contact line is on the more lyophobic side of the boundary, the corresponding force $\Delta\gamma_B$ exerted by the substrate points toward left – i.e. again toward the domain boundary. In equilibrium this force is of course balanced by the liquid assuming a contact angle θ_B. If the contact line is exactly at domain boundary, it will experience a force toward pushing it toward the domain boundary for any contact angle θ satisfying

$$\theta_A < \theta < \theta_B \tag{1.47}$$

Any slight perturbation that tries to move the contact line away from the domain boundary is restored by the surface tension force. Hence, the contact line is stably *pinned* to the domain boundary for any value of θ within this interval. Note that this reasoning only holds for domain boundaries where the liquid occupies the more lyophilic side of the domain boundary. The same reasoning for the opposite situation where the liquid occupies the more lyophobic side of the boundary shows that this configuration is unstable and the contact line is always driven away from the boundary. The analogue reasoning for a topographically patterned substrate with well-defined inclination angles leads to pinned configurations at the tops of the pattern for contact angles with the respect to the horizontal within the range

$$\theta_Y - \alpha < \theta < \theta_Y + \alpha \tag{1.48}$$

as illustrated in Figure 1.21b. Equations (1.47) and (1.48) are generally known as Gibbs criterion for contact line pinning at domain boundaries.

1.7.3 From Discrete Morphology Transitions to Contact Angle Hysteresis

The consequences of Gibbs pinning for the morphologies of drops on heterogeneous surfaces and for the occurrence of contact angle hysteresis were investigated in great detail in the 1960s and 1970s for macroscopically patterned surface,

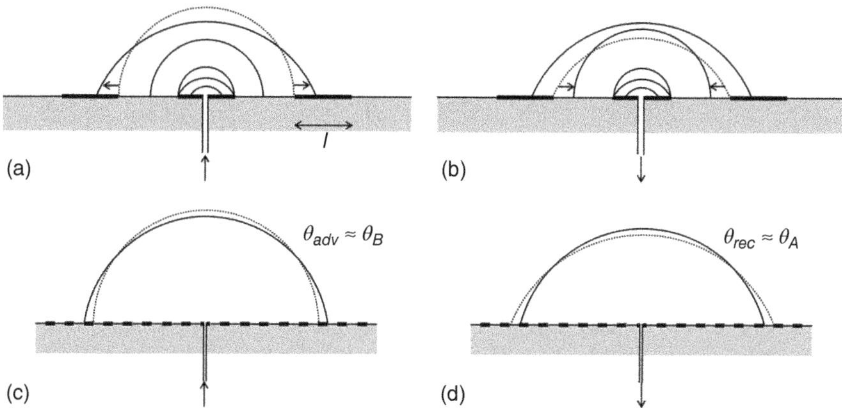

Figure 1.22 Evolution of the drop shape upon increasing and decreasing the drop volume on a surface with alternating lyophilic (θ_A) and lyophobic (θ_B) domains of width l. (a) and (b) for drop size $R \approx l$, (c) and (d) for $R \gg l$.

e.g. in seminar contributions by Johnson and Dettre [10] and Huh and coworkers [11].

Let us consider a drop on a surface with alternating lyophilic and lyophobic stripes. The drop is gradually inflated and deflated by pumping liquid in and out through a hole in the substrate in the center of a lyophilic domain, as sketched in Figure 1.22. For very small volumes V the drop only wets the central lyophilic domain (smallest drop in Figure 1.22a). The contact angle $\theta = \theta_A$ is constant as the volume and the drop radius increase as shown in Figure 1.23. The drop spreads until the contact line reaches the first domain boundary with the lyophobic domains at $V = V_1$. At that moment, the contact line gets pinned. Subsequently, the drop accommodates the increasing volume by increasing its contact angle until θ reaches the contact angle θ_B of the lyophobic domain for $V = V_2$. At that moment, the contact line gets depinned, and the drop spreads further at constant $\theta = \theta_B$ until it reaches the next domain boundary at $V = V_3$. Since the contact line reaches that boundary from the lyophobic side, it is unstable and experiences at force pulling it onto the next lyophilic domain. This gives rise to an abrupt transition from the shape indicated by the dotted line in Figure 1.22a with $\theta = \theta_B$ to the outermost one drawn as solid arc with $\theta = \theta_A$. Simultaneously, r_{sl} increases by some geometry-dependent amount that is dictated by volume conservation.

If we reverse the pumping direction, the contact line recedes again, and the drop shrinks at constant $\theta = \theta_A$. As the contact line reaches the lyophilic–lyophobic domain (dotted circle in Figure 1.22b), it becomes unstable again. Another abrupt transition in the morphology of the drop occurs in this case toward a small drop radius with $\theta = \theta_B$, as indicated by the arrows. From there on, the shrinking drop volume can be accommodated again by reducing the drop diameter on the lyophobic domain at constant $\theta = \theta_B$ until the next boundary with the innermost lyophilic domain is reached. There, the contact line gets pinned in a stable configuration until θ has decreased down to θ_A. From that moment on, the drop only wets the inner lyophilic domain and shrinks again

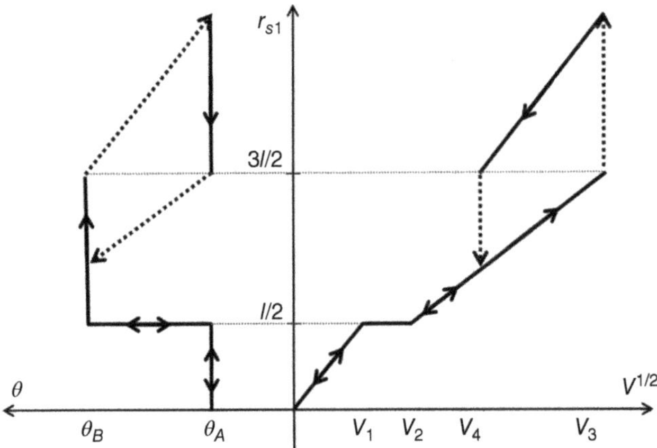

Figure 1.23 Contact angle and drop base radius versus square root of the drop volume (in two dimensions).

with constant $\theta = \theta_A$. For volumes $V_3 > V > V_4$, there are thus two mechanically stable drop morphologies with different drop radius and contact angle. The actually realized morphology depends on the history of the drop. Upon cycling the volume up and down, the drop follows a hysteresis loop. We denote the abrupt transition between two distinct drop shapes wetting a different number of stripes as a morphology transition. Morphology transitions take place whenever a variation of a control parameter, i.e. the drop volume in the present example, renders the initial morphology unstable. In the present case this happens due to a violation of the Gibbs pinning criterion.

So far, we discussed a situation in which the stripe width, i.e. the characteristic length scale of the heterogeneity, is a sizeable fraction of the drop radius. In this case, a morphology transition affects the drop shape on a global scale. If we repeat our reasoning for $l \ll r_{sl}$, the scenario sketched above still holds. Yet, there is one important difference: The increase in r_{sl} required to decrease the contact angle from θ_B to θ_A is now larger than l. Since θ is a monotonously decreasing function of r_{sl}, however, the jumping contact line becomes trapped again at the next lyophobic domain boundary, while θ has only decreased by a small amount, as illustrated in Figure 1.22c. In the limit of microscopic roughness with $l/r_{sl} \to 0$, the amount of contact angle relaxation per transition becomes negligible. As a consequence, we observed a constant *advancing* contact angle θ_a while the drop volume is increased. For the two-dimensional system of alternating stripes, this advancing contact angle is given by the contact angle of the lyophobic domain, i.e. $\theta_a = \theta_B$. The analogue reasoning for decreasing drop volumes shows that the contact angle of the then *receding* contact line also assumes a constant value, the receding contact angle θ_r, which is given by the contact angle θ_A of the lyophilic domain. Depending on whether the contact line advances or recedes, we thus observe θ_a or θ_r. A sessile drop, for which the contact line is neither advancing nor receding θ, can assume any value between θ_a and θ_r, depending on the history. The difference between advancing and receding contact angle is the contact angle hysteresis $\Delta\theta = \theta_a - \theta_r = \theta_B - \theta_A$. This equation also implies that. $\Delta\theta$ is a direct measure of the degree of heterogeneity of the surface.

The simple example of a model heterogeneity of alternating lyophilic and lyophobic stripes thus illustrates two important general aspects of wetting of heterogeneous substrates: First, it shows how surface heterogeneity gives rise to competing metastable drop configurations. Second, it illustrates that the existence of these metastable drop configurations causes both macroscopic morphology transitions between different drop shapes and contact angle hysteresis. The difference between global morphology transitions and local relaxations of the contact line is not a fundamental one, but rather of practical nature related to the resolution at which we investigate the system. The principle of destabilization due to a local violation of Gibbs pinning criterion (see Figure 1.21) is the same in both cases.

The specific type of surface heterogeneity is obviously idealized. For instance, we assumed infinitely sharp transitions between adjacent domains of low and high contact angle. In practice, the boundaries between adjacent domains will always be smeared out to some degree. As discussed in the context of Figure 1.19, the finite range of the molecular interaction forces always gives rise to some degree of smoothing even for atomically sharp domain boundaries. The continuous character of domain boundaries (or topographic corners) introduces two new aspects. On the atomic scale, we notice that the contact line position is usually not perfectly *pinned* at one position. It is actually free to adjust its microscopic position depending on external constraints. Contact line *pinning* at a continuous domain boundary should be considered as localization of the contact line within the transition zone between adjacent domains. This finite rigidity of contact line pinning has a second consequence that is arguably even more fundamental: It allows for a competition between pinning and deformation of the drop shape. In our discussion above, we focused completely on the contact line and neglected the fact that competing drop configurations, say, in Figure 1.22, also have different liquid–vapor interfacial areas. This is indeed justified for perfectly sharp domain boundaries. For domain boundaries with a gradual variation of the wettability, however, the stiffness of the boundary competes with the stiffness of the liquid–vapor interface. If the latter is stronger, the metastability disappears and the uniqueness of the equilibrium drop morphology is reestablished. This gives rise to the notion of *strong defects* versus *weak defects* that was introduced in the seminal paper on wetting of heterogeneous surfaces by Joanny and deGennes [12].

The difference between strong and weak defects is easily illustrated by a two-dimensional *drop* of liquid filling the space between a solid block of finite length and a flat solid surface that contains a wetting defect, as sketched in Figure 1.24. The free liquid surface is pinned to the bottom right corner of the block at point X. The pressure in the liquid is kept at zero such that the liquid–vapor interface is flat. The contact line position x_{eq} adjusts itself according to the local Young angle on the substrate. At the contact line, the liquid surface tension pulls toward the left (for the situation sketched with $\theta_Y < 90^\circ$) with a force $f_\gamma = \gamma \cos\theta(x) = \tilde{x}/\sqrt{1+\tilde{x}^2}$, where $\tilde{x} = x/H$ is the normalized

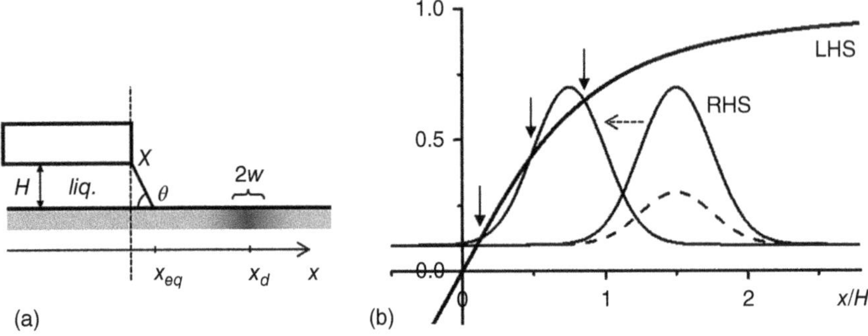

Figure 1.24 Strong defect versus weak defect. (a) Illustration of geometry with liquid filling the gap between a solid block and the solid surface with a wetting defect at $x = x_d$ with a width $2w$. (b) Force balance at the contact line versus normalized contact line position. Thick solid line: surface tension force according to left-hand side (LHS) of Eq. (1.49). RHS: defect force according to right-hand side of Eq. (1.49) for Gaussian wetting defects of variable strength and position ($\Delta\cos\theta_{1,2} = 0.6$; $\Delta\cos\theta_3 = 0.2$; $\tilde{x}_{d;1,3} = 1.5$; $\tilde{x}_{d;2} = 0.75$; $\tilde{w}_{1,2,3} = 0.25$).

position of the contact line. The solid surface pulls toward the right with a force $f_{surf} = \gamma_{sv}(x) - \gamma_{sl}(x) = \gamma \cos\theta_Y(x)$. The forces are balanced if

$$\frac{\tilde{x}}{\sqrt{1+\tilde{x}^2}} = \cos\theta_Y(\tilde{x}) \tag{1.49}$$

This is an implicit equation for the contact line position $\tilde{x}$ with the obvious solution $\cos\theta(x) = \cos\theta_Y(x)$. Yet, this solution is not always unique. To be specific, we consider a specific wetting defect at $x = x_d$ where the local contact angle is lower than on the rest of the surface. To be specific, we choose a Gaussian variation of the surface energy such that

$$\cos\theta_Y(x) = \cos\theta_Y^\infty + \Delta\cos\theta_Y\ \exp(-(x-x_d)^2/2w^2) \tag{1.50}$$

The equilibrium position of the contact line can be obtained most easily by a graphical solution of Eq. (1.49). Figure 1.24b shows the LHS (thick line) and the RHS (thin line) of Eq. (1.49) for a strong and for a weak defect with large and small amplitude, respectively. If the defect is sufficiently far away from the origin, there is only one unique contact line position given by the intersection point of the two curves. If we shift the center of the defect further toward the left, however, there appear at some point three intersection points, as indicated by the arrows in Figure 1.24b. While the middle one is unstable, the two others represent two competing mechanically stable configurations of the contact line. As usual, the existence of such metastable states implies pinning the appearance of a hysteresis loop of the contact line position upon moving the defect from right to left and back. Abrupt depinning events occur whenever the physically realized contact line position becomes unstable, i.e. whenever the corresponding intersection point in Figure 1.24b disappears. Assuming that the contact line position is initially given by the leftmost arrow, this happens if we shift the Gaussian defect a little further toward the left. At some point, the leftmost and the middle intersection points merge into one. Subsequently, this solution disappears. This is characteristic for a strong defect. For a weak defect such as the one characterized by the dashed defect force in Figure 1.24b, there is always only one single stable solution for the contact line position, for all values of x_d. This leads a very clear criterion to distinguish between weak and strong defects: A defect is strong, i.e. stiffer than the liquid–vapor interface, if the maximum slope of the defect force (RHS in Figure 1.24b) exceeds the maximum slope of the surface tension force (LHS). Vice versa, if the slope is lower, the defect is weak. The reader is invited to analyze this problem in more detail (see Problem 1.5). Wetting defects with tunable strength are easy to fabricate using EW. Experiments showed the appearance and disappearance of contact line pinning and trapping of drops [13, 14].

1.7.4 Optimum Contact Angle on Heterogeneous Surfaces: The Laws of Wenzel and Cassie

Even if a drop can in general not reach its true equilibrium contact angle on a heterogeneous surface by means of thermal excitation, there is nevertheless a specific contact angle that minimizes the total surface energy, both for rough and for chemically heterogeneous surfaces.

For flat surfaces with variable heterogeneity as shown in Figure 1.22c,d, we can rewrite Eq. (1.46) as

$$G = \gamma \left[L_{lv} - \int_{x_A}^{x_B} dx \cos\theta_Y[x] \right] \tag{1.51}$$

where the width of the drop $L = x_B - x_A$ and the drop surface arc length L_{lv} are connected by the constraint of constant volume. Figure 1.25 shows a typical result plotted as a function of the macroscopic apparent contact angle θ for $L \gg l$. The thick vertical arrow indicates the position of the absolute energy minimum. Similarly looking curves can be generated for topographically structured surfaces. While any local minimum in Figure 1.25 is mechanically stable, drops can nevertheless be brought close to the absolute minimum by a suitable, e.g. mechanical or electrical excitation that allows them to cross energetic barriers between adjacent minima. The value of the optimum apparent contact angle can be derived in the same way as in our energy-based derivation of Young's law. Let us consider a surface consisting of a random distribution of lyophilic patches with area fraction

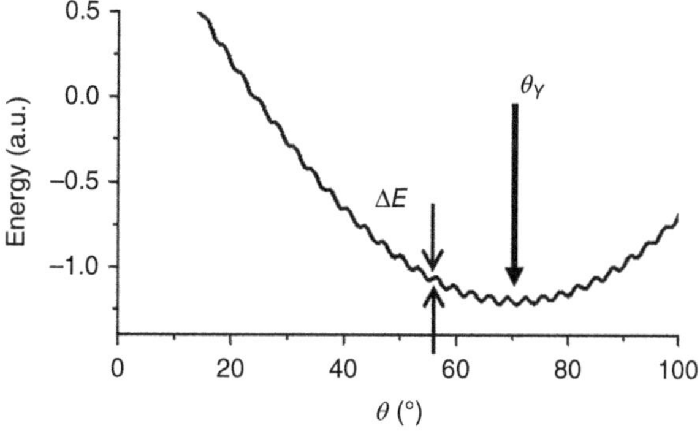

Figure 1.25 Free energy of a sessile drop on a surface with alternating stripes of high and low contact angle leading to a variety of locally stable drop configurations in the vicinity of the absolute energy minimum.

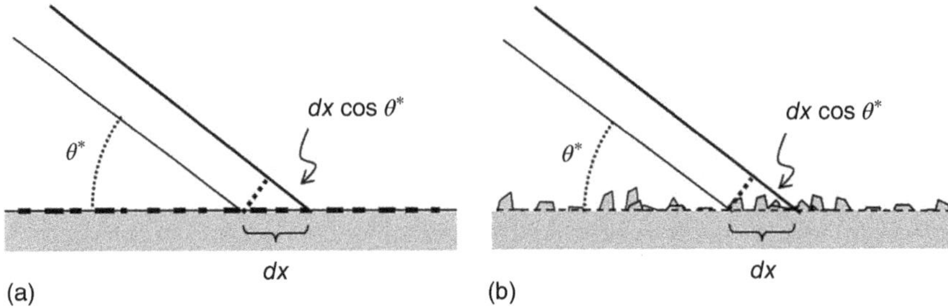

Figure 1.26 Virtual contact line displacement *dx* and derivation of optimum apparent contact angle θ^* on chemically heterogeneous (a) and rough (b) surface.

f_A and of lyophobic patches B with an area fraction $f_B = 1 - f_A$ (Figure 1.26a). If we assume that the drop has enough energy to explore all possible configurations and if we assume further that any virtual displacement dx comprises a surface area sufficiently large to represent the average composition of the surface, the variation of the surface energy upon a contact line displacement dx reads

$$\delta E_{surf} = dx\ [f_A(\gamma_{sl}^A - \gamma_{sv}^A) + f_B(\gamma_{sl}^B - \gamma_{sv}^B) + \gamma \cos\theta] \tag{1.52}$$

Equating this expression to zero yields the optimum apparent contact angle θ^* on the heterogeneous surface with

$$\cos\theta^* = f_A \cos\theta_Y^A + f_B \cos\theta_Y^B \tag{1.53}$$

This expression is known as Cassie's law for the contact angle on heterogeneous surfaces. If the surface becomes homogeneous, i.e. if either $f_A \to 1$ and $f_B \to 0$ or vice versa, we recover Young's contact angle of the corresponding homogeneous surface, as it should be.

An analogue expression for topographically rough surfaces was derived by Wenzel along the same spirit. Wenzel noticed that omnipresent microscopic roughness enhances the actual surface area of the solid as compared with the apparent projected surface area A (Figure 1.26b). To characterize the topographic roughness, he introduced the parameter $r = A_m/A$, which specifies the ratio between the true microscopic surface area A_m and the apparent projected one. As a consequence of this enhancement of the interfacial area, the weight of the contributions of γ_{sl} and γ_{sv} in the variation of the surface energy increases:

$$\delta E_{surf} = dx(r\cdot(\gamma_{sl} - \gamma_{sv}) + \gamma \cos\theta) \tag{1.54}$$

Equating this expression to zero yields Wenzel's equation for the optimum apparent contact angle on a rough surface:

$$\cos\theta^* = r\ \cos\theta_Y \tag{1.55}$$

Since $r > 1$ by definition, Wenzel's law implies that roughness *drives* the contact angle away from 90°: For $\theta_Y > 90^\circ$, θ^* becomes even larger than θ_Y with increasing r, and conversely, for $\theta_Y < 90^\circ$, we have $\theta^\star < \theta_Y$. This is exactly what we should expect given the enhanced weight of the drop–substrate interface due to the larger absolute interfacial area.

Note, however, that both derivations are based on the rather strong assumptions that the contact line is able to explore the representative fraction of the surface. This is in contrast to our earlier observation that energy barriers due to surface heterogeneities larger than a few nanometer are much higher than thermal energy. Equations (1.53) and (1.55) thus describe equilibrium states that often cannot be reached in practice. It is therefore not surprising that measurements of contact angles on heterogeneous surfaces frequently deviate from these predictions.

1.7.5 Superhydrophobic Surfaces

Wetting of rough hydrophobic surfaces leads to the well-known phenomenon of superhydrophobicity, which has attracted increasing attention since novel micro- and nanofabrication technologies have enabled the production of increasingly complex surface patterns. Superhydrophobicity arises for very rough hydrophobic surfaces. Figure 1.27 shows a generic geometry of simple superhydrophobic surfaces consisting of an array of regularly spaced pillars. According to our discussion of Wenzel's law in the preceding section, roughness increases the contact angle of hydrophobic surfaces because it increases the area of the energetically unfavorable solid–liquid interface. The optimum contact angle θ^*_W in the so-called Wenzel state, in which liquid fills the cavities of the surface roughness as sketched in Figure 1.27a, is readily calculated by inserting the roughness factor r into Wenzel's equation. For a simple array of pillars with height h and periodicity λ as shown in Figure 1.27, we find $r = 1 + 2h/\lambda$.

It turns out, however, that the actual contact angle observed on superhydrophobic surfaces is often substantially higher than predicted by Wenzel's law, Eq. (1.55). The reason is that the drop does not necessarily fill the cavities of the rough surface topography. Provided that the surface is sufficiently rough,

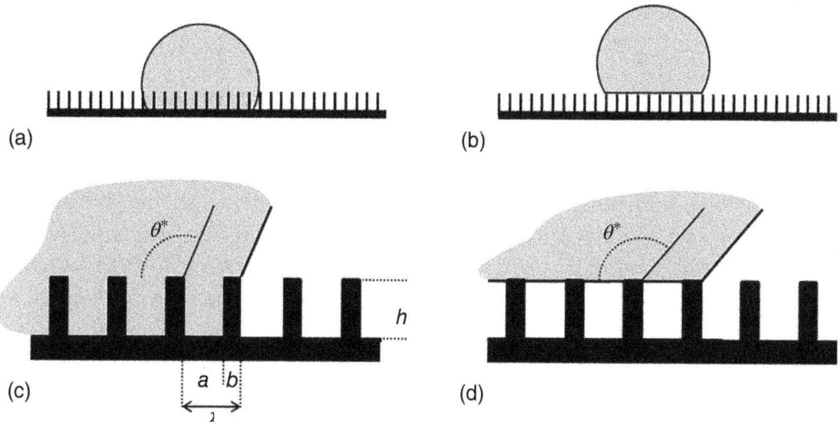

Figure 1.27 Wetting of rough hydrophobic surface and derivation of optimum contact angle θ^*. (a) Wenzel state with liquid filling the cavities between the pillars. (b) Superhydrophobic Cassie–Baxter state with air entrapped in the cavities. (c,d) Microscopic views near the contact line for Wenzel and Cassie–Baxter state, respectively.

the drop can instead assume a configuration, in which it rests on the tops of the pillars while the space in between remains filled with air, as sketched in Figure 1.27b. This is the superhydrophobic or *Cassie–Baxter* state. It is characterized by a massively reduced solid–liquid interfacial area as compared with a flat surface. This reduction of the solid–liquid interfacial area is the basis of many advantageous properties of the superhydrophobic state, including the very high contact angle θ^*, the very low contact angle hysteresis, the self-cleaning properties, hydrodynamic slip, etc. The drop–substrate interface in the Cassie–Baxter state consists of a series of alternating domains of solid–liquid interface and liquid–vapor interface (see Figure 1.27d). Hence, we can calculate the optimum value of θ^* using Cassie's equation, Eq. (1.52), using air with a contact angle of 180° as one of the two materials of this *chemically heterogeneous* substrate. If we denote the area fraction of the pillars as $f = b/\lambda$, the corresponding one of air is $1 - f = a/\lambda$. Inserting these ingredients in Eq. (1.53), we find that the optimum contact angle θ^*_{CB} in the Cassie–Baxter state is determined by

$$\cos\theta^*_{CB} = f(\cos\theta_Y + 1) - 1 \tag{1.56}$$

Here, θ_Y is Youngs angle on a flat surface made of the material of the pillars. The Cassie–Baxter state is stabilized by the fact that the contact line is pinned at the edges of the pillars, as illustrated in Figure 1.27c,d. θ_Y obviously needs to be larger than 90° to support a Cassie–Baxter state. Otherwise, the contact line would not remain pinned, and the liquid would be sucked into the cavities by the then negative Laplace pressure. θ^*_{CB} approaches 180° upon minimizing the area fraction f of the pillars. This is qualitatively plausible: For $f \to 0$, the drop essentially rests on a layer of air – with few pillars in between to support it. Yet, it is also plausible that such a state is not very stable and prone to collapse into the Wenzel state.

It is worthwhile to reiterate an important aspect form our discussion in the preceding section: Drops can assume many locally mechanically stable configurations on superhydrophobic surfaces. The energy barriers between these states are large compared with thermal energies. There is a family of Wenzel-like morphologies covering a variable number of pillars, and likewise, there is a family of Cassie–Baxter-like morphologies covering a variable number of pillars. Depending on the surface geometry, even mixed configurations with partially liquid-filled and partially air-filled cavities can be stable. The configurations with the optimum contact angles θ^*_W and θ^*_{CB}, respectively, are thus the states that minimize to total surface energy in the spirit of Figure 1.25. Because the solid–liquid interfacial area is very small in the superhydrophobic Cassie–Baxter state, it turns out that the energy barriers are also rather low, and hence the practically realized contact angles are close to θ^*_{CB}. This is also reflected in the low contact angle hysteresis in the superhydrophobic state. The barrier between the Cassie–Baxter-like and the Wenzel-like morphologies is often very high. The morphology of the drop thus always depends on the history, e.g. the manner of depositing the drop on the surface. In that sense, the expression *superhydrophobic surface* is misleading. Superhydrophobicity characterizes a specific state of wetting of a drop on a rough hydrophobic surface rather than an intrinsic property of the surface. Almost all surfaces denoted as *superhydrophobic* support drops both in the superhydrophobic Cassie–Baxter state and in the Wenzel state.

Figure 1.28a illustrates the transition of a drop from the Cassie–Baxter state to the Wenzel state upon evaporation. While the drop volume gradually decreases without noticeable variation of the contact angle in the top row of images, a sudden dramatic reduction of the contact angle takes place between the first two images of the second row (see arrow). This reduction of the contact angle is caused by sudden transition from the Cassie–Baxter state to the Wenzel state: The Cassie–Baxter morphology became unstable due to the reduction of the drop size. To understand the origin of this transition, we note first that the pressure inside the drop has to be constant in mechanical equilibrium. (We assume that the drop is small compared with the capillary length.) Hence, Laplace's law tells us that the curvature $\kappa = 1/R$ of all the liquid micromenisci spanning the gaps between adjacent pillars at the drop–substrate interface has to be the same as the macroscopic curvature of the drop. The latter increases as the radius of the drop decreases during the evaporation. As a consequence, the micromenisci bend and deflect more and more as the drop shrinks. From our discussion of Gibbs pinning criterion, we can immediately identify a limit of this process: If the deflection Δh of the menisci becomes too large, the angle α in Figure 1.28b exceeds the contact angle on the vertical sidewall of the pillar. As a consequence, the contact line is no longer pinned, and the liquid invades the cavity: The drop undergoes a transition to the Wenzel state. Evaluating elementary geometric relations, we find that the critical drop radius R_c at the transition is given by $R_c = -a/2\cos\theta_Y$. Since $\cos\theta_Y \approx -0.5$ for typical hydrophobic materials, contact line depinning only takes place when R_c has decreased to a value of the order of the pillar spacing. Note, however, that this scenario only takes place if the pillars are not too sparsely spaced on the surface. If the aspect ratio h/λ is very low, as it is desirable to maximize θ^*_{CB}, contact line depinning is preceded by the micromenisci touching the bottom of the cavity, i.e. when the deflection Δh reaches the height of the pillars,

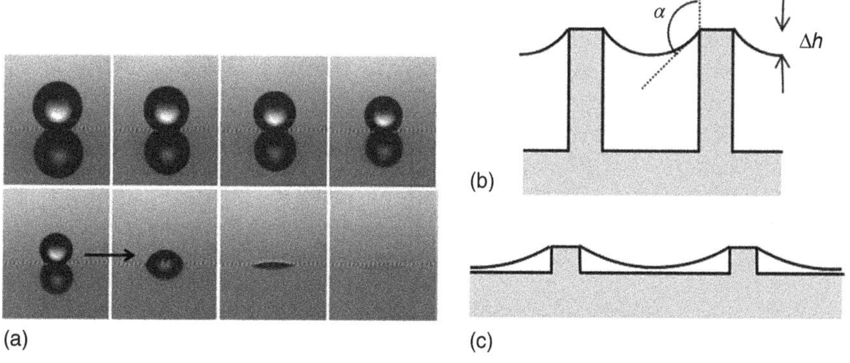

Figure 1.28 Cassie-to-Wenzel transition on superhydrophobic surfaces. (a) Video snapshots. Source: Reproduced with permission from [15]. (b) Stability limit for surface pattern with large aspect ratio. (c) For low aspect ratio.

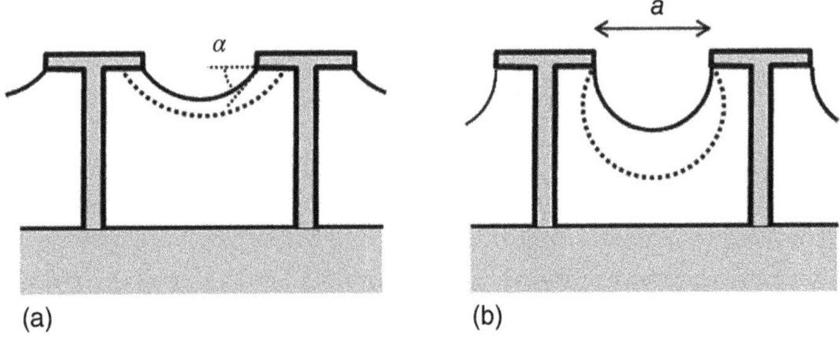

Figure 1.29 Stability limits of structured surfaces with overhangs leading to superoleophobic behavior. (a) Depinning according to Gibbs criterion. (b) Destabilization due to negative pressure gradient criterion.

as sketched in Figure 1.28c. Geometric considerations show that the critical drop radius in this case is given by $R_c = a^2/8h$ for $a \ll R$.

In the examples given above, the Cassie–Baxter configuration is destabilized because the decreasing drop volume increases the pressure. In other situations, the pressure in the liquid phase may be controlled by some parameters such as the hydrostatic pressure for superhydrophobic surfaces completely submerged under water or the impact speed for drops impinging onto a superhydrophobic surface. In all cases, the stability of superhydrophobic surfaces can be enhanced by decreasing the spacing between the pillars. Since the width of the pillars has to be scaled down simultaneously to maintain a high contact angle, superhydrophobic surfaces that combine high contact angles and good stability against pressure typically require surface patterns on the nanometer scale. Particularly stable surfaces can be achieved by creating surface patterns with overhangs, as shown in Figure 1.29. A surface with overhangs can sustain a Cassie–Baxter state – meaning a state with entrapped air underneath the drop – even if the liquid displays a contact angle $\theta_Y < 90^\circ$. As long as the pressure is not too high, the angle α in Figure 1.29a remains smaller than θ_Y, and the contact line remains pinned at the bottom edge of the overhang. This principle allows to design not only superhydrophobic but also superoleophobic surfaces, i.e. surfaces that display a very large contact angle and very low hysteresis for low surface tension liquids. Upon exposure to liquid at enhanced pressure, these surfaces display another instability mechanism in addition to the ones explained above: By geometry the radius of curvature of the meniscus in Figure 1.29 is $r = a/2 \sin \alpha$, and the corresponding pressure drop across the interface is $\Delta p = \gamma/r$. Δp has a maximum for $\alpha = 90^\circ$, where $r = a/2$, as sketched in Figure 1.29b. For larger deflections of the meniscus, $d\Delta p/dV$ becomes negative. In this case, the Cassie–Baxter state becomes unstable, and the meniscus spontaneously expands and displaces the air in the cavities of the superhydrophobic surface.

1.7.6 Wetting of Heterogeneous Surfaces in Three Dimensions

Our considerations so far were limited to two-dimensional systems. While illustrating many important aspects, wetting of heterogeneous surfaces in three dimensions involves one important additional aspect: Any local displacement of the contact line entails a distortion of the adjacent free liquid surface. Such deformations give rise to an additional interfacial area and thus an excess energy. This results in a wavevector-dependent stiffness of the interface that tends to restore the average shape of the surface away from the contact line.

We will limit our discussion to the case of small deformations, for which we can calculate the response of the liquid surface and hence its stiffness analytically. We consider the specific problem a liquid surface parameterized by some function $h(x, y)$. On average, we assume that the surface is flat with $h \equiv h_0$. Let us now introduce a periodic distortion of the surface in one direction, i.e. we impose a constraint $h(x = 0, y) = h_0 + \Delta h \cos qy$ with a wavevector $q = 2\pi/\lambda$. Such as situation can arise, for instance, if a plate with a suitable wettability pattern is immersed into a liquid bath, as sketched in Figure 1.30. (Note that we do not specify the actual wettability pattern; we simply assume that there is a wettability pattern that ensures a sinusoidal variation of the contact line. From the previous discussions, it should be plausible that this is enough to calculate the response of the liquid surface.)

The question is how the amplitude of the corrugation decays as a function of the distance from the (y, z) plain. With these specifications, the problem is actually rather simple: All we need to do is solve the Young–Laplace equation with the specific boundary conditions that the surface follows the imposed perturbation at $x = 0$ and that it is flat for $x \to \infty$. The latter implies $\Delta P_L = \gamma\kappa = 0$. If we also assume that the amplitude of the imposed perturbation is small, i.e. $\Delta h\, q \ll 1$, we simply need to solve the homogeneous Laplace equation $\nabla^2 h(x, y) = 0$. Using separation of variables, i.e. writing $h(x, y) = f(x)g(y)$, the reader can quickly find the solution for the surface profile:

$$h(x,y) = \Delta h \cos qy\, e^{-qx} \tag{1.57}$$

Hence, we find that the amplitude of the corrugation decays exponentially with increasing distance from the imposed perturbation with a decay constant that is given by the wavevector q of the perturbation. Perturbations with a high spatial

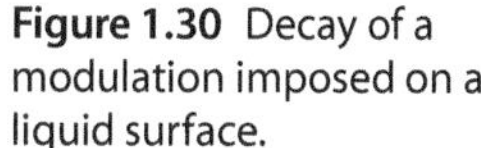

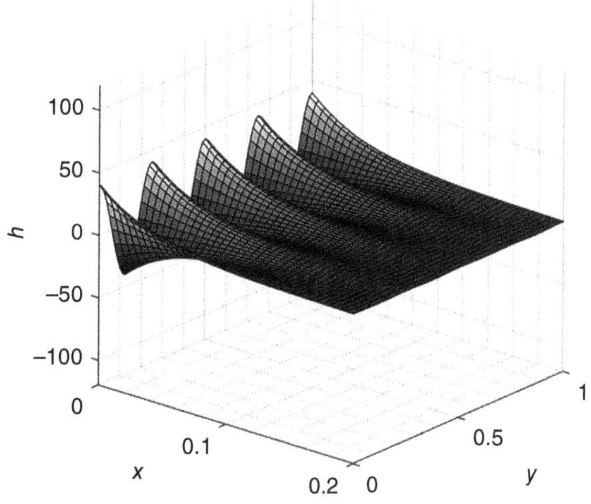

Figure 1.30 Decay of a modulation imposed on a liquid surface.

frequency thus decay quickly, and vice versa, long wavelength perturbations affect the surface profile over longer distances. We obtained this result here in a linearized model. The linearization implies that the superposition principle applies. To the extent that the amplitude is sufficiently small, we can thus calculate the deformation of the surface for arbitrary imposed perturbations. As usual, many qualitative aspects of the linearized model survive for larger perturbations.

Knowing the full solution of the surface profile, it is easy to show that the resulting excess surface energy scales as $\gamma q\ \Delta h^2$, corresponding to a surface with an effective stiffness $k(q) = \gamma q/2$; see Problem 1.6.

1.7.7 Wetting of Complex Surfaces in Three Dimensions: Morphology Transitions, Instabilities, and Symmetry Breaking

The discrete transitions between competing wetting morphologies discussed in the context of Figures 1.22 and 1.23 and for the superhydrophobic surfaces in Figures 1.28 and 1.29 were somewhat simplified because we limited ourselves to two-dimensional problems. Neglecting gravity, this implied that any section of a free liquid surface had the shape of a circular arc in order to guarantee a constant pressure within the fluid. In three dimensions, the constant mean curvature requirement expressed in the Young–Laplace equation allows for a much broader class of liquid morphologies. As already discussed the resulting surface shapes can be quite complex and at times counterintuitive to predict, in particular in the presence of complex boundary conditions imposed by structured solid surfaces. For instance, one of the goals of the experiments shown in Figure 1.4b was to explore the use of surfaces with wettability patterns to control the flow of fluids for microfluidic applications in an open configuration without restrictive solid channel walls. However, as the fluid volume on the hydrophilic ducts was increased, the initial fluid configuration consisting of sections of cylinders with a translationally invariant cross section became unstable, and thick localized *bulges* of liquid appeared. While perhaps disappointing from an applied perspective, it turns out that this kind of spontaneous transition between different types of liquid morphologies with a different symmetry is an important intrinsic characteristic that needs to be taken into account whenever designing fluidic devices involving free liquid surfaces. Such transitions can occur upon variations of many different control parameters, including the liquid volume, pressure, the geometry, and – in particular for EW – the contact angle. In view of the freedom of the liquid surface to adjust its shape in three dimensions, a quantitative analysis of the possible equilibrium configurations and the transitions among them generally requires a detailed numerical analysis.

Figure 1.31a shows the result of a numerical minimization of the surface energy of a drop of liquid of fixed volume deposited onto a lyophilic patch (see top figure) with $\theta_Y = 10^\circ$. Most of the liquid accumulates in an elongated drop on the wide section of the patch with a rather small curvature. Only a small fraction of the fluid wets the narrower horizontal lyophilic stripe. For the same liquid volume but a slightly larger contact angle of 30° on the patch, the narrow section remains

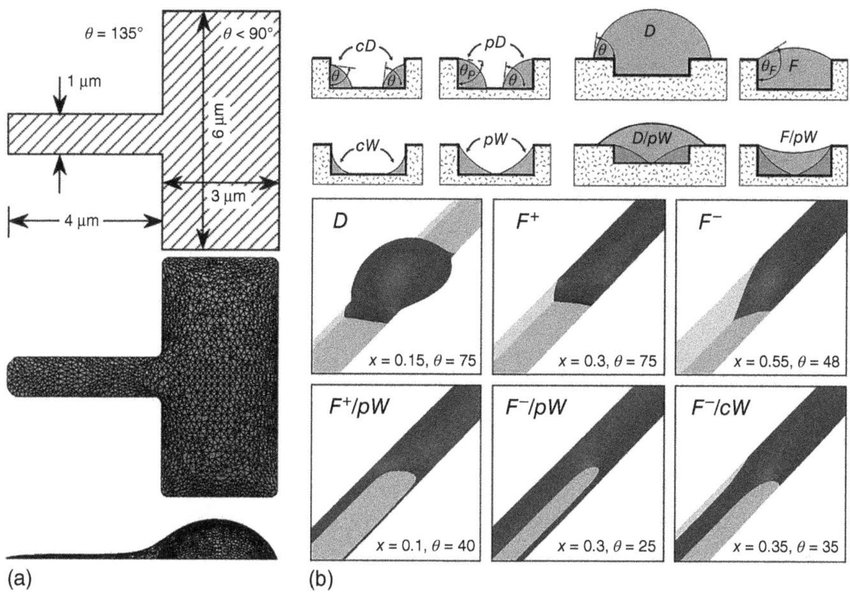

Figure 1.31 Examples of wetting morphologies on structured surfaces. (a) Calculated wetting morphology of a drop on a rotated-T-shaped lyophilic patch (top panel) on a hydrophobic surface. Middle and lower panels: top view and side view of the equilibrated fluid shape. Source: Darhuber et al. 2000 [16]. Adapted with permission of AIP Publishing. (b) Competing wetting morphologies on a surface with a rectangular groove of variable aspect ratio and contact angle. Top panels: cross-sectional views; bottom panels: 3D views of a series of morphologies. Source: Seemann et al. 2005 [17]. Adapted with permission of PNAS.

completely bare of fluid. Figure 1.31b shows a series of liquid configurations for a drop deposited onto a solid surface with a rectangular groove. Depending on the contact angle and on the aspect ratio X of the groove, an amazing variety of liquid morphologies appears, including localized drops, drops in coexistence with liquid filaments that can have positive or negative curvature with their edges being pinned or not being pinned to the upper corner of the groove. These examples illustrate the richness of possible morphologies encountered upon wetting structured surfaces.

Despite the richness there are a couple of generic phenomena that are frequently observed in the context of morphology transitions in complex geometries. One of them is spontaneous symmetry breaking at a morphology transition. Figure 1.32a,b shows two examples, a pair of two communicating liquid menisci and a drop confined between a flat surface and a sphere. The first example can either be considered as a two-dimensional system with liquid confined between two parallel solid surfaces or as a cross section through a short cylindrically symmetric tube with a spherical cap-shaped meniscus on each side. In both cases, the liquid assumes a mirror-symmetric configuration with two menisci of equal radius if the liquid volume is below a certain critical volume V_c. For larger volumes, the liquid assumes an asymmetric configuration as shown in the right panel of Figure 1.32a. In this case, the two menisci are complementary spherical caps

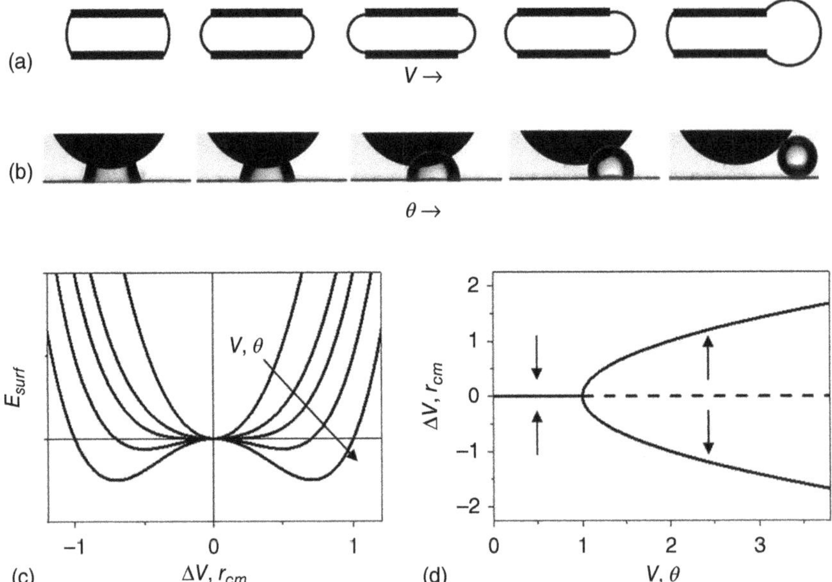

Figure 1.32 (a) Schematic illustration of spontaneous symmetry breaking of communicating drops upon increasing the liquid volume V along the arrow. (b) Snapshots of symmetry-breaking expulsion of a drop confined between a sphere and a flat surface upon increasing the contact angle from $\theta < 90^\circ$ to $\theta > 90^\circ$. Source: Experimental data taken from de Ruiter et al. [18]. (c) Schematic of the energy landscape for the morphology transitions in (a) and (b). Control parameters V or θ increase along the arrow. (d) Pitchfork diagram of the equilibrium value of the asymmetry parameter versus the control parameter. The critical value of V_c and θ_c was set to 1. Solid lines, stable solutions; dashed line, unstable solution branch; arrows indicate direction of evolution toward equilibrium upon perturbation.

of one sphere (or circle) to ensure that the Laplace pressure is the same on both sides. The stability of this configuration can be assessed in the same manner as our analysis of the Rayleigh–Plateau instability of a liquid jet; see Figure 1.5. The symmetric configuration is stable as long as the transfer of a small volume of liquid δV from the left meniscus to the right generates a pressure difference between the two menisci that counteracts the perturbation. This analysis, which the reader is invited to carry out in Problem 1.7, shows that V_c corresponds to the volume for which the radius of the two protruding menisci equals the radius of the tube (or half the distance between the plates). The nature of the symmetry-breaking instability is closely related to the maximum drop pressure criterion controlling the collapse of reentrant superhydrophobic surfaces discussed in Figure 1.29b.

Mathematically, symmetry-breaking morphology transitions can be described in the language of bifurcations of dynamic systems theory. In the vicinity of the critical volume, we can expand the surface energy of the system in even powers of an asymmetry parameter $\Delta V = V_{left} - V_{right}$:

$$E_{surf}\,(\Delta V) = \alpha\,(V_c - V)\,\Delta V^2 + \beta\,\Delta V^4 \tag{1.58}$$

where α and β are positive coefficients. For $V < V_c$ the prefactor in front of the quadratic term is positive. As a consequence, E_{surf} has a unique minimum at

$\Delta V = 0$ as shown in Figure 1.32c,d. As V is increased beyond V_c, the prefactor in front of the quadratic term changes sign, and hence the energy landscape assumes a double-well character with two equivalent minima at symmetric values:

$$\Delta V_{eq} = \pm\sqrt{\alpha(V - V_c)/2\beta} \tag{1.59}$$

(Physicists will notice the analogy of this description with Landau's mean field theory of second-order phase transitions in thermodynamics.) If the system is driven into the symmetry-broken state by increasing the control parameter beyond its critical value, it spontaneously chooses one of the two in principal equivalent local minima. Upon increasing the control parameter further, the absolute value of the asymmetry parameter increases further in a continuous manner as specified in Eq. (1.59).

The same type of analysis applies to a large variety of morphology transitions involving symmetries. The drop confined between a sphere and a plate shown in Figure 1.32b is such an example, where the contact angle rather than the drop volume is the control parameter and the radial coordinate r_{cm} of the center of mass of the drop is the asymmetry parameter. The drop is in a cylindrically symmetric state at $r_{cm} = 0$ if the contact angle is below a certain critical angle θ_c. Upon increasing θ beyond θ_c, e.g. by means of EW [18], the symmetry is spontaneously broken, and the drop moves away from the symmetry axis in a continuous manner analogue to Eq. (1.59). In this case, breaking the symmetry involves the spontaneous selection of an arbitrary azimuthal angle out of a continuum of possible states.

For the cases discussed above, the asymmetry parameter evolves continuously and reversibly as a function of the control parameter. This is different from the morphology transitions discussed earlier that involved an abrupt depinning of contact lines from pinning sites (Figures 1.22, 1.23, and 1.29) and thus abrupt and irreversible variations of the drop shape beyond a critical of some control parameter. The latter transitions are discontinuous. General criteria for the occurrence of continuous versus discontinuous transitions are described on an abstract level in mathematical textbooks on bifurcation theory [19].

An important example that introduces additional aspects to the phenomenology of morphology transitions is the wetting of curved solid surfaces such as fibers. The finite curvature of the fibers introduces the peculiar phenomenon that any coating film of constant thickness is intrinsically unstable even if the liquid completely wets the fiber. This aspect causes important practical challenges in coating technology, e.g. for textile fibers and for insulating coatings on electrical wires. Physically, the origin of this instability is perfectly plausible. If we review our discussion of the Rayleigh–Plateau instability of liquid jets (see Figure 1.5), we can see that the exact same analysis of the underlying linear instability applies to thin liquid layers on a fiber as well as long as the amplitude of the perturbation does not exceed the film thickness. As a consequence, even for $\theta_Y = 0$, the liquid is localized in a drop on the fiber. Qualitatively, there are two different morphologies for wetting a fiber, a cylindrically symmetric *barrel* morphology and a symmetry-broken *clamshell* morphology, in which the drop is attached to the side of the fiber, as can be observed, e.g. for dew drops on spider webs (see Figure 1.33). There are two control parameters governing the wetting of such structures, the

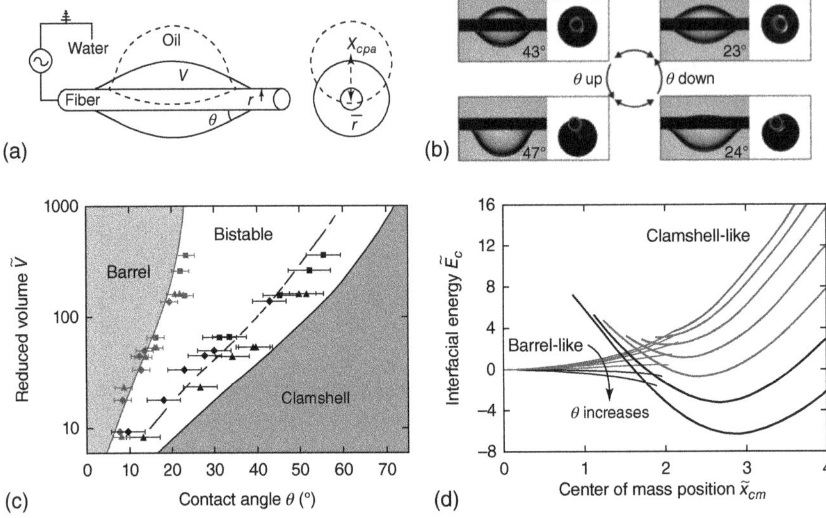

Figure 1.33 Fiber wetting. (a) Cylindrically symmetric, topologically multiply connected barrel morphology (solid) versus symmetry-broken singly connected clamshell morphology (dashed). (b) Experimental snapshots of competing morphologies including hysteresis upon varying θ. (c) Morphology diagram indicating the ranges of stability of the competing morphologies. Solid lines, numerical stability limits; dashed, equal absolute energy of both barrel and clamshell morphology. Symbols, experimental data. (d) Numerically calculated energy landscape of barrel-like and clamshell-like morphologies versus radial position of center of mass of the fluid for variable contact angle. Source: Eral et al. 2011 [20]. Reproduced with permission of Royal Society of Chemistry.

contact angle θ_Y on the fiber and the ratio between the drop size and the fiber radius r. For the latter, it is convenient to think in terms of the dimensionless drop volume V/r^3. For certain values of the parameters, the limiting wetting behavior is easy to determine: For $\theta \to 0$, the system will try to maximize the fiber–drop interfacial area, thus favoring the barrel state. Similarly, the clamshell state will be preferred for high contact angles. Furthermore, it is clear that the clamshell state is preferred for $\tilde{V} = V/r^3 \to 0$ because the curvature of the fiber becomes negligible in this limit on the scale of the drop (see Figure 1.33c). Away from these limits, the competition between the two morphologies is more complex. Clamshells are essentially somewhat distorted sessile drops with a single three-phase contact line. The barrel morphology is qualitatively different. The solid fiber effectively pierces a hole through the liquid, giving rise to two separate contact lines at each end. Topologically speaking, a clamshell drop is simply connected, whereas a barrel drop is not. As a consequence of this fundamental topological distinction between the two competing morphologies, a simple continuous evolution from the symmetrical to the symmetry-broken morphology is not possible. Instead, there is a wide region in the morphology diagram for which both morphologies are mechanically stable. This gives rise to a pronounced hysteresis if, for instance, the contact angle of a drop of fixed volume is decreased from a large to a small value and subsequently increased again, as illustrated in Figure 1.33b. Correspondingly, the energy diagram as a function of the asymmetry parameter, illustrated in Figure 1.33d, looks qualitatively different from Figure 1.32c. While the barrel morphology undergoes a linear instability as the contact angle exceeds a certain critical value similar to the symmetric state in the examples given above, its energy curve is intercepted by the energy curve of the clamshell state, which has a minimum at some finite contact angle-dependent radial coordinate. As a consequence, the center of mass of the drop undergoes an abrupt and discontinuous transition to the clamshell state. In the opposite direction, the clamshell state becomes unstable when the drop starts to engulf the fiber at decreasing contact angle. At some point, two opposing sections of the contact line merge on the opposite side from the drop (see bottom right snapshot in Figure 1.33b) to induce the abrupt transition to the barrel state.

From a practical perspective, the discontinuity of this type of morphology transitions involves energy dissipation upon switching, which may be undesirable in certain applications. On the other hand, bistability may be attractive in applications that require digital switching between two discrete states of the system with very distinct properties. Depending on the requirements of a specific application, clever engineering will allow the reader to identify suitable combinations of surface geometries and wettabilities to develop systems with more analogue or more digital response characteristics.

1.A Mechanical Equilibrium and Stress Tensor

To analyze force balance in continuum mechanics, it is frequently useful to consider certain control volumes and all the forces acting on them, as we discussed in Section 1.4 in the context of drops under the influence of gravity. In this appendix, we take a step back and consider first a small control volume $dV = dx\, dy\, dz$ around a point (x_0, y_0, z_0) in the absence of gravity. dV is bounded by six surface elements $d\vec{S}_j = dS\, \vec{n}_j$ with normal vectors $\vec{n}_j$. Each surface element experiences forces due to the stresses T_{ij} within the fluid (Figure 1.A.1). The stress tensor T_{ij} indicates the stress in the coordinate direction $\vec{e}_i$ acting on a surface element with the surface normal along $\vec{e}_j$. The total force acting on dV is therefore

$$dF_i = \sum_k T_{ik} dS_k \tag{A.1}$$

In mechanical equilibrium the stress tensor is isotropic, because finite shear stresses would generate fluid motion. Hence, we know that $T_{ij} = \delta_{ij} T_{ij}$, where δ_{ij} is the Kronecker delta symbol. Equation (A.1) then reduces to

$$\begin{aligned} dF_i = & \left[T_{ix}\left(x_0 - \frac{dx}{2}\right) - T_{ix}\left(x_0 + \frac{dx}{2}\right)\right] dS\, \vec{e}_x \\ & + \left[T_{iy}\left(y_0 - \frac{dy}{2}\right) - T_{iy}\left(y_0 + \frac{dy}{2}\right)\right] dS\, \vec{e}_y \\ & + \left[T_{iz}\left(z_0 - \frac{dz}{2}\right) - T_{iz}\left(z_0 + \frac{dz}{2}\right)\right] dS\, \vec{e}_z \\ = & -\partial_x T_{ix} dx\, dydz\, \vec{e}_x - \partial_y T_{iy} dy\, dxdz\, \vec{e}_y - \partial_z T_{iz} dz\, dxdy\, \vec{e}_z = 0 \end{aligned} \tag{A.2}$$

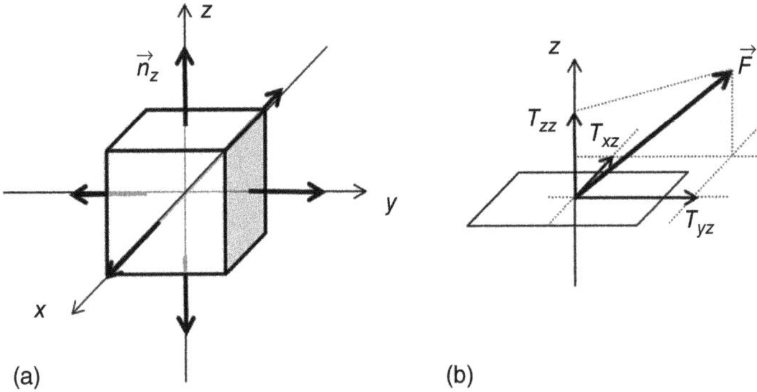

Figure 1.A.1 (a) Illustration of a control volume within a fluid. (b) Decomposition of a force acting on the top facet of the control volume into components of the stress tensor.

This implies that $\partial_x T_{ix} = \partial_y T_{iy} = \partial_z T_{iz} = 0$. That is, T_{ij} is constant in space. Because all spatial directions are equal, symmetry reasons allow us to conclude that all three diagonal components are equal. (If the stress in one direction would be larger than in another one, the volume element dV would deform and stretch along the direction(s) of lower pressure, which would contradict our assumption of mechanical equilibrium.) The common value is denoted as the (isotropic) hydrostatic pressure $p = T_{xx} = T_{yy} = T_{zz}$. That is, we can write the stress tensor as $T_{ij} = p\,\delta_{ij}$. In this manner, the Laplace pressure caused by the *directional* surface tension forces at a liquid surface translates into an *isotropic* hydrostatic pressure $p_0 = \Delta P_L$.

In the presence of gravity, the liquid inside the control volume experience the body force $d\vec{f}_g = -\rho g dV\,\vec{e}_z$. Making use of Eq. (A.2), the force balance of all forces acting on dV is then given by

$$\begin{aligned} d\vec{F} &= -\nabla p + d\vec{f}_g = 0 \\ &= -\partial_x p dx\, dydz\,\vec{e}_x - \partial_y p dy\, dxdz\,\vec{e}_y - (\partial_z p dz\, dxdy - \rho g dV)\,\vec{e}_z \end{aligned} \tag{A.3}$$

This implies $\partial_x p = \partial_y p = 0$ and $\partial_z p = \rho g$. Because of the isotropy of the pressure, we find a hydrostatic pressure

$$p = p_0 - \rho g z \tag{A.4}$$

where p_0 is again a suitably chosen reference pressure. Again, the directional gravitational body force translates into an *isotropic* pressure acting equally in all directions within the fluid.

Problems

1.1 Young's angle minimizes the surface energy of a spherical cap. For spherical caps, the values of A_{sl} and A_{lv} as well as the drop volume V are known geometrical functions of R and θ_Y (see Table 1.2). Hence we can rewrite Eq. (1.14) as $E_{surf}(R, \theta) = A_{lv}(R, \theta)\gamma + A_{sl}(R, \theta)(\gamma_{sl} - \gamma_{sv})$. (a) Rewrite this equation as a function of θ alone using volume conservation and plot $E_{surf}(\theta)$. (b) Show that the angle θ^* that minimizes $E_{surf}(\theta)$ is given by $\cos\theta^* = (\gamma_{sv} - \gamma_{sl})/\gamma$, i.e. $\theta^* = \theta_Y$.

1.2 Coffee mug. Consider the shape of a meniscus wetting a vertically immersed solid surface. An example from daily life would be the profile of the liquid surface in our coffee mugs. The coffee has a contact angle θ_Y on the wall of the coffee mug, which will be close to zero. In the center of the mug, the surface of the coffee is flat. Describe qualitatively the shape of the coffee surface upon approaching the wall. At which distance from the surface do you expect noticeable deformations of the surface? How high will the contact line rise above the flat reference level in the middle of the mug?

1.3 Force balance on a sessile drop. Consider a sessile drop in two dimensions, i.e. a segment of a circle, with contact angle θ_Y and height h in the absence of gravity. Show that the sum of all external forces on the right half of the drop vanishes.

1.4 Capillary rise. Derive Jurin's law, Eq. (1.26), by minimizing the total energy of the system, i.e. surface plus gravitational energy.

1.5 Strong versus weak wetting defects. Consider a Gaussian wetting defect of variable strength similar to the one sketched in Figure 1.24 with a wettability profile as given in Eq. (1.50) with $\theta_Y^\infty = 120^\circ$, $\Delta\cos\theta_Y = 1$, and $\tilde{w} = 0.025$. (a) Draw a graphical representation of the force balance equation analogue to Eq. (1.50). Is this defect strong or weak? (b) If applicable, draw the force balance curves for all the defect positions corresponding to depinning transitions. (c) Write a computer program to track numerically the position of the contact line upon varying the defect position from $\tilde{x}_d = 2$ to -2 and back.

1.6 Restoration of a perturbed liquid surface. (a) Consider a liquid surface with an imposed surface perturbation as shown in Figure 1.30. Show that the restoring force of the free surface can be interpreted as a harmonic spring with a wavevector-dependent spring constant $k_{eff} = \gamma q/2$. (b) Assume that the imposed contact line position is given by a superposition of sine functions approximating a square wave. Use the first five Fourier coefficients of the harmonic expansion with a global amplitude factor 0.1. Calculate and plot $h(x, y)$ for $y = 0,\ 1/2q, 1/q, 2/q$. Discuss the relation between the surface shape and $k_{eff}(q)$.

1.7 Instability of two communicating liquid menisci. (a) Consider the geometric configuration shown in Figure 1.32a with two menisci connected by a slit pore. Assume that the system is two-dimensional. Show that the equilibrium configuration of the liquid is symmetric liquid volumes V below a critical liquid volume V_c and that it is unstable for $V > V_c$. If L is the length of the slit pore and $2a$ the spacing between the solid walls, show that $V_c = 2aL + \pi a^2$. (b) Discuss possible analogies between this symmetry-breaking morphology transition and the bulging transition shown in Figure 1.4b. Source: The reader may want to consult Lenz and Lipowsky [21] for inspiration.

1.8 Wetting of wedges in two dimensions. (a) Determine the equilibrium shape and position of a drop in a wedge with an opening angle α and a contact angle θ_Y on the solid walls. Calculate in particular the forces between the drop and the solid surfaces in the tangential and normal direction.

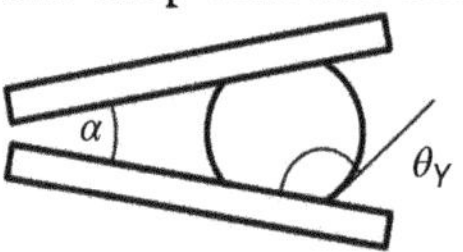

Source: See also Baratian et al. [22] for inspiration.

1.9 Wetting in a sphere–plate geometry. (a) Argue that the equilibrium shape and radial position r_c of a drop confined between a sphere of radius R and a plate at distance s are given by a truncated sphere at a position r_c displaying Young's angle θ_Y on both surfaces. (b) Show that $\tilde{r}_c = \sqrt{\tilde{r}^2 \sin^2\theta_Y - (2+\tilde{s})(2\tilde{r}\cos\theta_Y + \tilde{s})}$, where all length scales are given in units of R. $\tilde{r}$ is the position-dependent radius of the drop. (c) Show particular the critical contact angle θ_c below which $\tilde{r}_c(\theta) \equiv 0$ coincides with the onset of an attractive capillary force between the plate and the sphere.

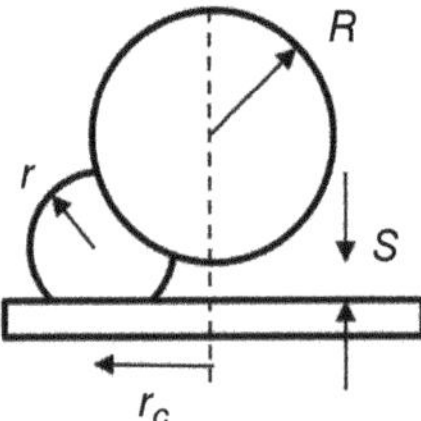

Source: See also de Ruiter et al. [18].

References

1 deGennes, P. G., Brochard-Wyart, F., and Quéré, D. (2004). *Capillarity and Wetting Phenomena*. New York: Springer.

2 Butt, H. J. and Kappl, M. (2010). *Surface and Interfacial Forces*. Wiley VCH Verlag GmbH & Co. KGaA.

3 Israelachvili, J. N. (1992). *Intermolecular and Surface Forces*, 2e. London: Academic Press.

4 Rowlinson, J. S. and Widom, B. (1982). *Molecular Theory of Capillarity*. Oxford: Clarendon.

5 Gau, H. (1998). Stabilität und Dynamik fluider Mikrostrukturen. PhD thesis. University of Potsdam, Germany.

6 Gau, H., Herminghaus, S., Lenz, P., and Lipowsky, R. (1999). Liquid morphologies on structured surfaces: from microchannels to microchips. *Science* 283: 46–49.
7 Hagedorn, J. G., Martys, N. S., and Douglas, J. F. (2004). Breakup of a fluid thread in a confined geometry: droplet-plug transition, perturbation sensitivity, and kinetic stabilization with confinement. *Phys. Rev. E* 69 (5): 18.
8 Lamb, H. (1932). *Hydrodynamics*. New York: Dover Publications.
9 Pericet-Camara, R., Best, A., Butt, H. J., and Bonaccurso, E. (2008). Effect of capillary pressure and surface tension on the deformation of elastic surfaces by sessile liquid microdrops: an experimental investigation. *Langmuir* 24 (19): 10565–10568.
10 Johnson, R. E. and Dettre, R. H. (1964). Contact angle hysteresis: III: study of an idealized heterogeneous surface. *J. Phys. Chem.* 68: 1744–1750.
11 Oliver, J. F., Huh, C., and Mason, S. G. (1977). Resistance to spreading liquids by sharp edges. *J. Colloid Interface Sci.* 59: 568–581.
12 Joanny, J. F. and de Gennes, P. G. (1984). A model for contact angle hysteresis. *J. Chem. Phys.* 81: 552–562.
13 't Mannetje, D. J. C. M., Banpurkar, A. G., Koppelman, H. et al. (2013). Electrically tunable wetting defects characterized by a simple capillary force sensor. *Langmuir* 29 (31): 9944–9949.
14 't Mannetje, D. J. C. M., Ghosh, S., Lagraauw, R. et al. (2014). Trapping of drops by wetting defects. *Nat. Commun.* 5: 3559.
15 Reyssat, M., Yeomans, J. M., and Quere, D. (2008). Impalement of fakir drops. *Europhys. Lett.* 81 (2): 26006.
16 Darhuber, A. A., Troian, S. M., Miller, S. M., and Wagner, S. (2000). Morphology of liquid microstructures on chemically patterned surfaces. *J. Appl. Phys.* 87 (11): 7768–7775.
17 Seemann, R., Brinkmann, M., Kramer, E. J. et al. (2005). Wetting morphologies at microstructured surfaces. *Proc. Natl. Acad. Sci. U.S.A.* 102 (6): 1848–1852.
18 de Ruiter, R., Semprebon, C., van Gorcum, M. et al. (2015). Stability limits of capillary bridges: how contact angle hysteresis affects morphology transitions of liquid microstructures. *Phys. Rev. Lett.* 114 (23): 234501.
19 Strogatz, S. H. (1994). *Nonlinear Dynamics and Chaos: with Applications to Physics, Biology, Chemistry and Engineering*. Boulder, CO: Westview Press.
20 Eral, H. B., de Ruiter, J., de Ruiter, R. et al. (2011). Drops on functional fibers: from barrels to clamshells and back. *Soft Matter* 7 (11): 5138–5143.
21 Lenz, P. and Lipowsky, R. (1998). Morphological transitions of wetting layers on structured surfaces. *Phys. Rev. Lett.* 80: 1920.
22 Baratian, D., Cavalli, A., van den Ende, D., and Mugele, F. (2015). On the shape of a droplet in a wedge: new insight from electrowetting. *Soft Matter* 11: 7717.

2
Electrostatics

Electrowetting (EW) is driven by electrostatic forces. While various basic aspects of electrostatics are taught in high school, EW involves a number of subtleties and somewhat more advanced concepts, e.g. regarding the details of the electric field distribution and forces in complex geometries as well as the response of materials to electric fields. This chapter therefore starts by recapitulating basic principles of electrostatics, focusing on the specific needs of electrowetting. Next to the fundamental laws, boundary conditions, and materials response to electric fields (conductor, dielectrics, leaky dielectrics), we discuss in particular complementary manners of calculating the resulting electrostatic forces based on general principles of energy gradients, global momentum conservation arguments, and local force balance. We will also discuss the solutions of a few specific electrostatic problems of particular interest for EW.

2.1 Fundamental Laws of Electrostatics

2.1.1 Electric Fields and the Electrostatic Potential

The physics of electrowetting is governed by the laws of electrostatics, i.e. by a simplified subset of the general Maxwell equations of electrodynamics. The time dependence of electric fields in EW is sufficiently slow that the dynamic coupling between electric and magnetic fields in electrodynamics can safely be ignored, even at the highest AC frequencies that are typically applied. Static magnetic fields may occasionally be applied. Yet, they do not couple in a relevant manner to the dominant electric fields.

Electric fields $\vec{E}$ emerge from electric charges. This is expressed in a differential manner in the first of Maxwell's equations:

$$\nabla \cdot \vec{E} = \frac{\rho(\vec{r})}{\epsilon_0} \tag{2.1}$$

where $\rho(\vec{r})$ is the electric charge density at the location $\vec{r}$ and ϵ_0 is the permeability of vacuum. Integrating this equation over a finite volume V and making use of Gauss' integral law to convert the volume integral over the divergence into

Electrowetting: Fundamental Principles and Practical Applications,
First Edition. Frieder Mugele and Jason Heikenfeld.

a surface integral over the boundary of the integration volume, we can rewrite (Eq. (2.1)) in an integral form:

$$\oint \vec{E}\, d\vec{A} = \int \nabla\vec{E}\, dV = \int \frac{\rho(\vec{r})}{\epsilon_0}\, dV = \frac{Q}{\epsilon_0} \tag{2.2}$$

This expression shows that the integral over the electric field over the closed surface $\vec{A}$ is completely determined by the total charge inside the enclosed volume. We can use Eq. (2.2), for instance, to calculate the electric field around the positive point charge Q in Figure 2.1a. For symmetry reasons, we know that the electric field is oriented along the radial direction and only depends on the distance. If we choose a sphere centered around the charge as integration volume, the integral reduces to a multiplication of $E(r)$ with the surface area $A = 4\pi r^2$, and we obtain $E(r) = Q/4\pi\epsilon_0 r^2$. This radial dependence of the electric field is true not only for a point charge but also for any spherically symmetric charge distribution. The statement of Eq. (2.2), however, is even more general. It holds for any charge distribution $\rho(\vec{r})$ and for any choice of the arbitrarily complex choice of the integration volume. As long as the total charge enclosed by the surface is Q, the integral will have the same value Q/ϵ_0 specified in Eq. (2.2). The contributions from several charges inside the integration volume simply add up to the total charge. This results from the linearity of Eq. (2.1), which is a characteristic of all Maxwell equations. Figure 2.1 illustrates this for a few particular cases.

The independence of Eq. (2.2) on the specific integration volume also implies that we can make convenient choices in order to solve a given problem in an easy manner, e.g. by making use of symmetries. Figure 2.2 illustrates this for the specific case of plates with fixed surface charge density $\pm\sigma$. Placing a box around one of the two plates shows that the electric field caused by the charge has the absolute value

$$\oint \vec{E}\, d\vec{A} = EA + EA = \frac{Q}{\epsilon_0} \rightarrow E_n = \frac{\sigma}{2\epsilon_0} \tag{2.3}$$

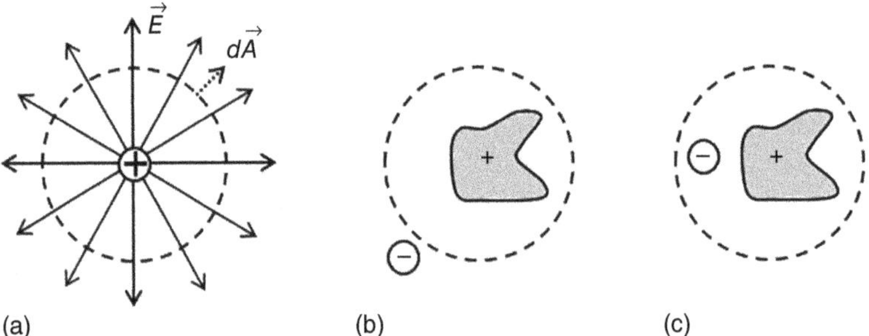

Figure 2.1 Illustration of Gauss' law for a variety of distributions of discrete and continuous electric charges. (a) Point charge with a radially symmetric electric field around it. Integration of the differential flux $\vec{E}d\vec{A}$ over the closed surface (dashed circle) yields the total flux of the electric field caused by the total charge enclosed by the control surface. (b) The same flux is achieved for a more complex charge distribution if the total charge is the same. Charges placed outside the control volume do not contribute to the total flux. (c) Opposite charges compensate each other yielding zero net flux for zero charge.

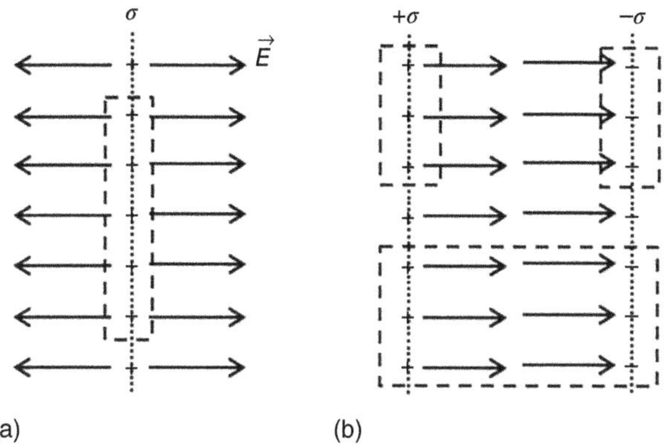

Figure 2.2 Electric field distribution around charge plates. (a) A single plate with charge density σ and a square control box. For symmetry reasons only the vertical sections of the control box contribute to the integral. (b) Two plates of opposite surface charge density $\pm\sigma$. Between the plates, the electric fields due to the two plates add up; outside they compensate each other.

The subscript n indicates that the field is oriented normal to the surface. The factor 2 arises from the left and from the right surface of our control volume. Assuming that the two plates are the only charges in the system, the total electric field is given by the superposition of these two fields. Taking into account the orientations of the electric fields, we see that the total field between the plates is $E = \sigma/\epsilon_0$ and strictly zero outside the plates. The latter result could have been obtained directly by choosing a control volume that contains both plates, resulting in a vanishing net enclosed charge; see bottom part of Figure 2.2b. (Note that the plates are assumed to be infinitely wide. For finite plates, there are fringe fields emerging from the edges of the plates, as in Figure 2.3.)

The second basic law of electrostatics states the absence of the rotation of the electric field:

$$\nabla \times \vec{E} = 0 \tag{2.4}$$

It implies that we can write the electric field as a gradient of a scalar function, the electric potential $\phi(\vec{r})$:

$$\vec{E} = -\nabla\phi(\vec{r}) \tag{2.5}$$

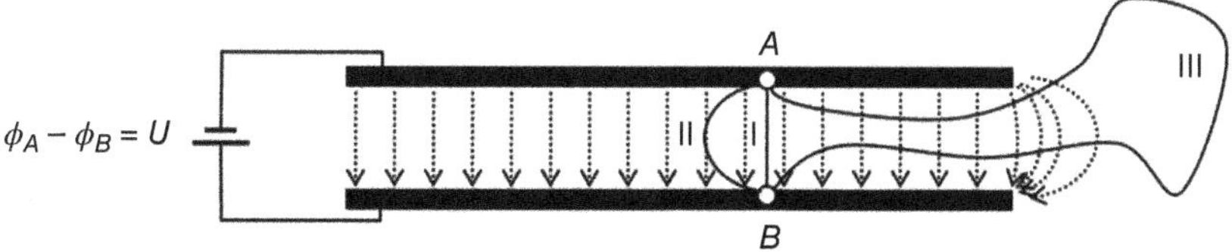

Figure 2.3 Schematic of two plates at distance d kept at potentials ϕ_A and ϕ_A. The value of the line integral over $\vec{E}d\vec{s}$ from A to B is equal to $\phi_A - \phi_B$ for each of the different integration paths I–III. Path independence also holds for integrals between arbitrary points, not necessarily located on the plates.

The potential is only defined up to an arbitrary constant. Frequently, it is convenient to choose this constant such that the potential vanishes at infinity. The fact that ϕ is a well-defined unique function of the position implies that line integrals over the electric field along an arbitrary paths from some point A to another point B do not depend on the specific choice of that path. For each path shown in Figure 2.3, the line integral has the same value that is given by the difference of the potential between A and B, which is commonly denoted as the voltage U, i.e.

$$\int_A^B \vec{E}\, d\vec{s} = -\int_A^B \nabla\phi\, d\vec{s} = \phi(A) - \phi(B) = U \tag{2.6}$$

For the specific geometry in Figure 2.3, we already know that the electric field between the plates is $E = \sigma/\epsilon_0$. Hence, we know that the potential difference, i.e. the voltage, between the plates of this charge is $U = d\sigma/\epsilon_0$.

If we insert the definition of ϕ (Eq. (2.5)) in (Eq. (2.1)), we obtain the Poisson equation for the distribution of the electrostatic potential:

$$\nabla^2\phi(\vec{r}) = -\frac{\rho}{\epsilon_0} \tag{2.7}$$

This equation allows us to calculate directly the potential distribution for any given charge distribution $\rho(\vec{r})$ and boundary conditions by solving simply a scalar partial differential equation rather than a set of vector equations.

2.1.2 Specific Examples

To illustrate the use of the electrostatic potential, we calculate the electric field distribution for the three specific configurations shown in Figure 2.4 that are commonly encountered in EW.

To calculate the electric field around the wedge (Figure 2.4a), it is useful to write Eq. (2.7) in cylindrical coordinates. Because the charge density vanishes everywhere in the empty space around the wedge, we can write

$$\frac{1}{r}\frac{\partial}{\partial r}\left(r\frac{\partial}{\partial r}\phi\right) + \frac{1}{r^2}\frac{\partial^2}{\partial\varphi^2}\phi = 0 \tag{2.8}$$

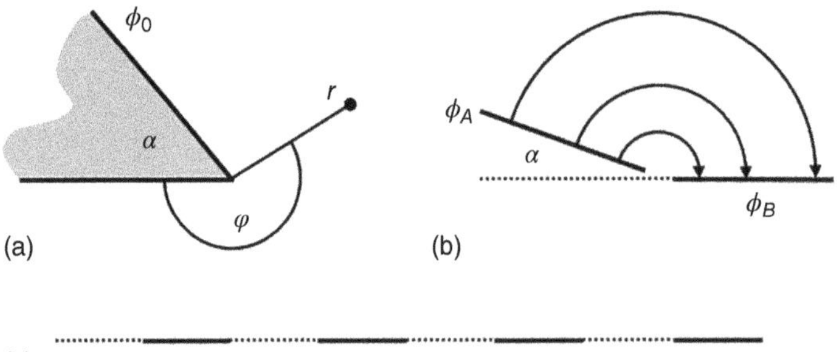

Figure 2.4 Electric field distribution for specific geometries. (a) A conductive wedge with opening angle α at constant potential ϕ_0. φ and r indicate the cylindrical coordinates with respect to the apex of the wedge. (b) Between two plates at an angle α kept at fixed potentials ϕ_A and ϕ_B. (c) A surface with alternating stripes of potential $\pm\phi_0$.

We can separate the variables by writing $\phi(r, \varphi) = f(r) \cdot J(\varphi)$. Inserting this ansatz yields

$$\frac{1}{f(r)} r \frac{\partial}{\partial_r} \left(r \frac{\partial}{\partial r} f(r) \right) = -\frac{1}{J(\varphi)} \frac{\partial^2}{\partial \varphi^2} J(\varphi) = \text{const}. \tag{2.9}$$

The radial part of the equation is fulfilled by any algebraic function $f(r) \propto r^{\mu}$. The azimuthal part is solved by linear combinations of sine and cosine functions $J(\varphi) = A \sin \mu\varphi + B \cos \mu\varphi$. If we define the potential on the wedge as $\phi_0 = C_0$, we must guarantee that $\phi(r, 0) = \phi(r, 2\pi - \alpha) = C_0$ for all values of r. This is fulfilled for the solution with the structure

$$\phi(r, \varphi) = \sum_n c_n r^{\mu_n} \sin \mu_n \varphi + C_0 \tag{2.10}$$

with the additional requirement

$$\mu_n = \frac{n\pi}{2\pi - \alpha} \tag{2.11}$$

For $n = 1$ and $\alpha < \pi$, this implies that $\mu_1 < 1$. Hence the electric field $E \propto \frac{\partial \phi}{\partial r} \propto r^{\mu_1 - 1}$ diverges upon approaching the apex of the wedge at $r = 0$. This is the origin of the well-known lightning rod effect: The electric field diverges in the vicinity of (conductive) wedges at fixed potential provided that the opening angle is <180°. As we will see later, this singularity appears also close to the three-phase contact line of drops in electrowetting. It turns out to play a crucial role for the local shape of the liquid surface in this region and for the occurrence of diverging electric fields that limit the performance of EW devices at high voltage.

The problem in Figure 2.4b is very similar from a mathematical perspective. We can use again the same cylindrical coordinate system. Upon imposing the boundary conditions $\phi(\varphi = 0) = \phi_B$ and $\phi(\varphi = \pi - \alpha) = \phi_A$, we find the solution $\phi = \frac{\phi_A - \phi_B}{\pi - \alpha} \cdot \varphi + \phi_B$, i.e. the potential does not depend on the distance from the origin, but only on the azimuthal angle. The solution implies that the electric field is oriented in the azimuthal direction, too, and decays as $E_\varphi(r) = \frac{1}{r} \frac{\partial \phi}{\partial \varphi} = \frac{1}{r} \frac{\phi_A - \phi_B}{\pi - \alpha}$. Like Figure 2.4a, this electrostatic problem is a characteristic for the field distribution close to the edge of a drop in EW experiments. We will return to these two solutions in Chapter 5.

The calculation of the potential for the configuration shown in Figure 2.4c is best addressed in Cartesian coordinates. We are interested in the electrostatic potential in the half space above the horizontal line that consists of segments of length L with alternating electrical potential $+\phi_0$ and $-\phi_0$. (Technically, the segments should be separated by small gaps of width $\delta L \ll L$.) In addition to the potential values on each segment, we assume that the potential vanishes at infinity. Before calculating the formal solution, we can already develop a general intuition of what the solution is expected to look like. Very close to the surface, the potential should follow the externally imposed potential. From a distance much larger than L, the line of segments of alternating potential looks electrically neutral. Hence, we expect the modulation of the potential to decay and eventually vanish as we move away from the surface. Denoting the coordinate along the surface as x and the one perpendicular as z and noting once again that there are no free charges outside the surface, the electrostatic problem can be written as

$$\left(\frac{\partial^2}{\partial x^2} + \frac{\partial^2}{\partial z^2}\right) \phi(x, z) = 0 \tag{2.12}$$

The boundary condition can be written as a Fourier series:

$$\phi(x, 0) = \phi_0 \frac{4}{\pi} \sum_{n=1,3,5,\ldots}^{\infty} \frac{1}{n} \sin q_n x \tag{2.13}$$

where $q_n = \pi/L$. (Equation (2.13) is the representation of a square wave in terms of sine functions.) Similar to Eq. (2.10), we expect that we can write the full solution again as a superposition of base functions ϕ_n:

$$\phi(x, z) = \sum_n c_n \, \phi_n(x, z) \tag{2.14}$$

In this case, separation of variables shows that the base functions can be written as

$$\phi_n(x, z) = e^{-q_n z} \sin q_n x \tag{2.15}$$

The solution thus consists of a superposition of base functions that decay exponentially along z with a decay constant corresponding to the wavevector of the periodicity in the x-direction. (Formally, exponentially growing functions are also solutions of Eq. (2.13). Yet, these solutions violate our boundary condition of vanishing potential at infinity.) Hence, each Fourier mode contributes to the electric field only in a range of order $1/q_n$ from the surface. Because the longest wavelength is given by the periodicity of the surface, the electrical field is essentially localized within a distance of order L from the surface. This very basic property of the Laplace equation (Eq. (2.12)) is of crucial importance for the design of electrowetting and dielectrowetting systems involving interdigitated electrodes covered by insulating layers. To find the complete solution, we make use of the boundary condition Eq. (2.13) for each mode individually and identify the coefficients $c_n = 4\phi_0/\pi n$.

2.2 Materials in Electric Fields

So far, we considered only free charges and electric fields in vacuum. Materials become polarized when they are exposed to electric fields. In the macroscopic theory of electrodynamics, one typically considers two types of materials, conductors and nonconductors (i.e. dielectrics or insulators). Conductors are characterized by the presence of freely mobile charge carriers, such as electrons in metals and dissolved ions in electrolyte solutions. In perfect dielectric materials, all electrons are tightly bound to *their* atoms and hence cannot move freely.

2.2.1 Conductors

Perfect conductors are characterized by the absence of macroscopic electric fields in their interior in equilibrium. The absence of electric fields results

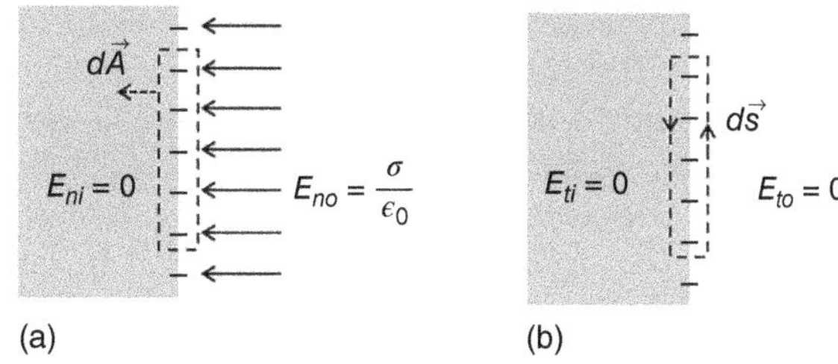

Figure 2.5 Electric field at the surface of a conductor (gray). (a) Normal component. (b) Tangential component.

immediately from the presence of free charge carriers. If there is a finite electric field at some moment in time, the freely mobile charge carriers respond to this field and redistribute until the internal field $\vec{E}_i$ vanishes. In many, but not all, electrowetting situations, even water can be considered as a perfect conductor – with the dissolved ions rather than electrons acting as mobile charge carriers. The vanishing of $\vec{E}_i$ in combination with the fundamental laws (Eqs. (2.2) and (2.4)) gives rise to specific boundary conditions for the behavior of normal and tangential electric fields at conductor–insulator interfaces. If we integrate again Eq. (2.2) over a square box containing such an interface as in Figure 2.5a and make use of Gauss' law and the fact that the normal component of the internal field vanishes, i.e. $\vec{E}_{ni} = 0$, we find for the normal component of the electric field on the outside the conductor surface

$$E_{no} = -\frac{\partial}{\partial n}\phi|_s = \frac{\sigma}{\epsilon_0} \tag{2.16}$$

here $\partial/\partial n$ denotes the derivative normal to the surface, and the subscript s indicates that the value is to be taken at the surface. Note that this expression does not contain the factor 2 in the denominator that we saw earlier for charged plate in vacuum with the same surface charge; see Eq. (2.3).

Similarly, we can calculate the line integral over Eq. (2.4) along the rectangle that encloses the interface in Figure 2.5b. Because the tangential field E_{ti} inside the conductor vanishes, it follows that the tangential field also just outside the conductor surface vanishes, too, i.e.

$$\oint \nabla \times \vec{E}\, d\vec{s} = 0 \rightarrow E_{to} = 0 \tag{2.17}$$

Together, Eqs. (2.16) and (2.17) form the general macroscopic boundary conditions for electric fields at the interface between conductors and nonconductors. Next to electrode surfaces, they generally apply at the surface of conductive drops in EW experiments.

Two additional remarks are useful before turning to dielectrics: First, the interface between a conductor and a nonconductor is obviously not infinitely sharp on a microscopic scale. While the atoms, say, at a metal–air interface, are located at well-defined lattice positions and thereby provide a rather sharp distribution of the positive background charge density, the density of the conduction electrons is slightly smeared out, leading to a thin space charge layer with a finite dipole moment at the interface. For metals, the degree of smearing is rather small. For the corresponding characteristic length scale, the Thomas–Fermi screening length λ_{TF} is typically a fraction of an angstrom, which is negligible in the vast majority of practical applications. Upon applying an external electric field, the entire *sea of free electrons* slightly shifts toward the interface or away from it, depending on the orientation of the field. This results in the net charge density at the surface that screens the electric field from the inside of the metal. In case of electrolyte solutions, the corresponding screening length is known as Debye screening length λ_D. It is much thicker, with dimensions ranging from approximately 1 nm to a few hundreds of nanometer, depending on the salt concentration. The calculation of the distribution of ions within the screening layer of electrolytes, the *electric double layer*, will be discussed in detail in Chapter 4.

Second, the fact that the macroscopic electric field vanishes inside conductors holds only in equilibrium. If we impose nonequilibrium conditions, e.g. by forcing a direct current through a conductor or by switching the potential repeatedly at high frequencies, E_i will be nonzero. This will lead to a current density j due to the free charge carriers proportional to the electric field: $\vec{j}_{el} = \lambda\,\vec{E}$. Here λ is the electrical conductivity of the material. (Only for superconductors, λ is infinite, and hence current can flow without internal electric field.) If we integrate $\vec{j}$ over the cross section of a wire, we obtain the total electrical current I:

$$I = \int \vec{j}_{el}\, d\vec{A} \int \lambda\vec{E}\, d\vec{A} = \lambda\, A\frac{U}{L} \tag{2.18}$$

In the last step, we assumed that the wire has the length L and made use of the fact that the electric field is constant inside a homogeneous conductor with a homogeneous current density. Obviously, Eq. (2.18) leads to the well-known expression $R = L/\lambda A$ for the resistance of a wire of length L and cross section A. This reasoning is the basis for representing a piece of a conductor as a resistor in an electric circuit model. Conceptually, this representation is equally appropriate for electrolytes as for metals.

Next to this steady-state nonequilibrium situation of imposing a constant current through a conductor, one may also consider the situation of time-dependent external electric fields and imposed currents. The question is then how quickly the charge carriers in the conductor can respond to such variations of the external conditions. Thanks to the small mass and high mobility of electrons, it turns out that metals respond as perfect conductors up to optical frequencies, the so-called plasma frequency. That is, for metals all typical AC frequencies encountered in electrowetting experiments can be considered as quasi-static. For electrolyte solutions this is not always the case, as we will discuss below in the section on leaky dielectrics.

2.2.2 Dielectrics

Dielectrics do not possess a *sea of free electrons* that can collectively shift to screen the electric fields. Nevertheless, net charge density of the electrons in the material responds to the applied electric field. In each volume element within the material, the center of the negative and the positive charge density becomes slightly shifted, as indicated in Figure 2.6a. Microscopically, this effect arises from the polarizability of each individual atom, as we will address in more detail below. For the time being, the *volume element of material* that we discuss should be considered as

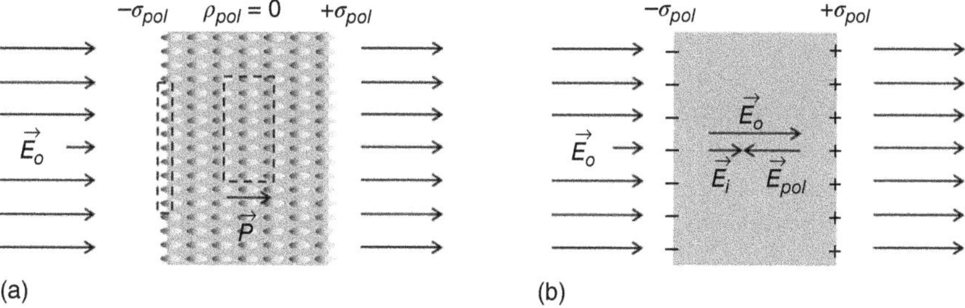

Figure 2.6 Distribution of polarization charges and electric fields in a dielectric medium (gray) placed in an external electric field. (a) Illustration of the polarization of volume elements that leads to a finite surface charge. (b) Resulting internal fields, including polarization charge field.

a macroscopic ensemble of many atoms or molecules. Provided that the electric field is not too strong, the polarization will be linear in the electric field, and we can define a polarization vector:

$$\vec{P} = \epsilon_0 \chi \vec{E}_i \tag{2.19}$$

pointing along the same direction as the electric field that induces it. χ is the electric susceptibility of the material.

In the bulk of a homogeneous dielectric, the polarization does not result in a finite net charge density. As the dashed box in the center of Figure 2.6a illustrates, the amount of, say, negative charge that leaves any closed control volume due to the polarization on one side is exactly the same as the amount that enters on the opposite side. This is different for a material with a heterogeneous polarization. In that case, the net charge balance in a control volume can be nonzero. Bearing in mind charge conservation, we find a polarization charge density:

$$\rho_{pol} = -\nabla \vec{P} \tag{2.20}$$

If we consider a control volume at the surface of a dielectric or an interface between two dielectric materials of different susceptibility, the same will apply. Polarization charge that enters the control volume on one side is not compensated by charge leaving on the opposite side. This leads to a net surface charge density σ_{pol}. Applying Gauss' law to the dashed box on the left of Figure 2.6a and the fact that the polarization vanishes outside of the material, we find $\sigma_{pol} = -P$. Moreover, the positive and the negative polarization charge layer on the two surfaces together produce an electric field:

$$E_{pol} = \frac{\sigma_{pol}}{\epsilon_0} = -\frac{P}{\epsilon_0} = -\chi E_i \tag{2.21}$$

that is oriented in the opposite direction as the internal (and hence also the external) field (Figure 2.6b). Hence, the total internal field is partially screened by the polarization charge and is reduced in vectorial form according to

$$\vec{E}_i = \vec{E}_o + \vec{E}_{pol} \tag{2.22}$$

External and internal field are related by $\vec{E}_0 = \vec{E}_i(1 + \chi)$. This implies that the polarization charge at the interface has the density $\sigma_p = -\epsilon_0 \chi/(1 + \chi)$.

Introducing the relative permeability of the dielectric $\epsilon = 1 + \chi$, we can rewrite this expression as

$$\frac{\sigma_p}{\epsilon_0} = -\frac{\epsilon - 1}{\epsilon} E_o \tag{2.23}$$

The polarization charge thus always counteracts the applied external field. Its magnitude increases with increasing polarizability. In the limit $\epsilon \to \infty$, it assumes the value $\sigma_p = -\epsilon_0 E_o$, which implies that the internal field vanishes. That is, the material behaves as a conductor.

Generalizing our considerations from Figure 2.6 to interfaces between two different dielectric media 1 and 2 with permeabilities ϵ_1 and ϵ_2, we find that the normal component of the electric field always displays a jump due to the polarization charge $\sigma_{pol} = P_1 - P_2$ at the interface. Using Gauss' law to integrate over a control volume that contains the interface (see Figure 2.7a), we find that the magnitude of this jump is such that the normal component of ϵE_n remains constant. If we define the displacement field

$$\vec{D} = \epsilon_0(1 + \chi)\vec{E} = \epsilon_0 \epsilon \vec{E} \tag{2.24}$$

in each material, the boundary condition for the normal component of the electric field thus reads

$$D_{n1} = \epsilon_0 \epsilon_1 E_{n1} = \epsilon_0 \epsilon_2 E_{n2} = D_{n2} \tag{2.25}$$

Similarly, calculating the line integral along the edge of the rectangle shown in Figure 2.7b for an electric field tangential to the interfaces in combination with Eq. (2.4) shows that the tangential component of the electric field is constant across the interface, i.e.

$$E_{t1} = E_{t2} \tag{2.26}$$

Together, Eqs. (2.25) and (2.26) provide the general boundary conditions for the electric and the displacement field at the interfaces of dielectrics.

We can illustrate the meaning of the displacement field by considering a dielectric plate inserted into the gap of a parallel plate capacitor, Figure 2.8. We will revisit this concept in the first section of Chapter 9 (dielectrophoresis, dielectrowetting). The plates of the capacitor are kept at a separation d and carry the

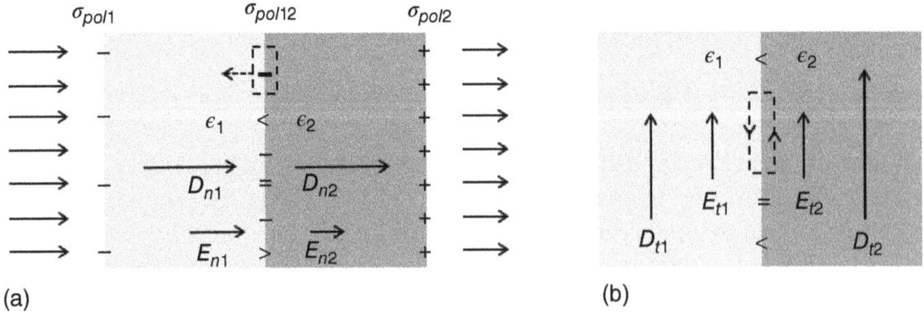

Figure 2.7 Illustration of the boundary conditions of the (a) normal component of the displacement field and (b) the tangential component of the electric field.

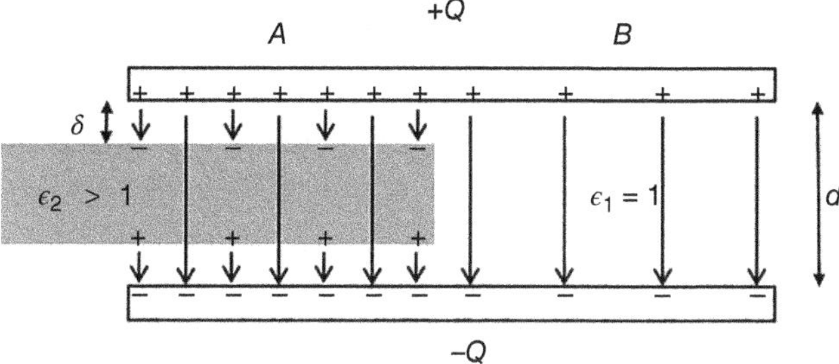

Figure 2.8 Schematic distribution of charges and fields for a dielectric (gray) partially inserted in a parallel plate capacitor with a constant charge Q. (The illustrated charge and field distribution refers to locations A and B far away from the edge of the dielectric. Close to that edge, local stray fields lead to a more complex distribution.)

charge $\pm Q$. At position B on the right far away from the edge of the dielectric, the electric field is given by $E_B = \sigma_B/\epsilon_1$, where σ_B is the surface charge density on the plates at location B. This results in a potential difference, or voltage, $U = E_B d$. Because the plates are conductors, the potential on each plate is constant, and hence, the voltage at locations facing the dielectric, such as A, is the same, i.e. $U = E_B d = \overline{E}_A d$, where $\overline{E}_A = (2 \cdot E_{A1}\delta + E_{A2}(d - 2\delta))/d$ is the average electric field at location A. The subscripts 1 and 2 indicate the electric fields within the gaps of width δ and within the dielectric, respectively. E_{A2} is smaller than E_B because the field inside the dielectric is partially screened by the polarization charges. Hence, E_{A1} must be larger than E_B. This implies that the charge on the plates is not homogeneously distributed. To guarantee a constant potential drop between the capacitor plates, we musthave $\sigma_A > \sigma_B$. From Eq. (2.25), we know that $E_{A1} = \epsilon_2 E_{A2}$, and from Eq. (2.26) we know that $D_{A1} = D_{A2}$ which is equal to $\epsilon_0 E_{A1}$. Because E_{A1} is the electric field generated by the *free* charge on the conductor surfaces, the latter conclusion implies that D is the field that is exclusively generated by the *free* charge density ρ_f. It does not contain any contribution from polarization charges. More generally, we can rewrite Eq. (2.1) as

$$\nabla \cdot \vec{D} = \rho_f \tag{2.27}$$

Figure 2.8 also illustrates another interesting aspect: If $\epsilon_2 \gg 1$, E_2 can become very small. This is particularly relevant if we consider a dielectric such as water that has a dielectric constant of 80. This means that E_2 is only somewhat more than 1% of E_1. This is rather close to the extreme case of inserting a metal plate instead of a dielectric. In that case, E_2 would strictly vanish, and all the potential would drop across the two narrow gaps of δ. This basic electrostatic effect explains why water as well as other strongly polarizable liquids can frequently be actuated in electrowetting-based lab-on-a-chip systems even if they do not contain any mobile ions and hence display a macroscopic contact reduction upon applying a voltage. (See also Problem 2.7.)

It is worthwhile to spend a few remarks on the particularly high value of the dielectric constant of water. Most other common solid and liquid materials used in electrowetting have much lower dielectric constants. The reason for the exceptionally high dielectric constant lies in the molecular properties of water, in particular its strong permanent dipole moment. To understand this, we consider an isolated nonpolar molecule, such as methane. If we expose such a molecule to an electric field, the electron cloud is pulled in one direction, whereas the nucleus is pulled in the opposite one. This results in an induced dipole moment $\vec{p} = \epsilon_0 \alpha \vec{E}$, provided that the field is not too strong. Here α is the polarizability of the molecule. The polarizability of molecules typically depends on their size. Larger molecules with more electrons are generally more easily polarizable. This is most easily seen for the noble gases. Their polarizability increases with size: He (1.32), Ne (2.38), Ar (10.77), Kr (16.47), and Xe (26.97 Å^3). Molecules with a permanent dipole moment $\vec{p}_0$ such as water have an additional degree of freedom. In addition to polarizing their electron cloud, they can also reorient to align their permanent dipole moment with the electric field. The energy of a permanent dipole in an external electric field is given by $W = -\vec{p}_0 \vec{E} = -p_0 E \cos\theta$, where θ is the angle between the dipole moment and the electric field. The dipole moment of water is particularly large. It amounts to $p_0 = 1.85$ D, where 1 D = 1 Debye = 3.3×10^{-30} m. For a field of, say, 10^8 V m^{-1}, the energy gain upon orienting the dipoles is thus of order $p_0 E \approx 6 \times 10^{-22}$ J, which is slightly less than thermal energies. (Remember that $k_B T = 4 \times 10^{-21}$ J at room temperature.) We expect the orientation of the dipoles to follow a Boltzmann distribution $\propto \exp(-W/k_B T)$. For $W < k_B T$, we can linearize the Boltzmann factor and achieve after averaging over all possible orientations an average dipole moment $\overline{p}_0 = p_0^2 E/3k_B T$. Note that this value is quadratic in p_0 because p_0 appears both in the expression of the quantity to be averaged and in the energy W (see Problem 2.4). Inserting numbers reveals that the dipole moment caused by the reorientation of permanent dipoles is typically much larger than the one due to induced dipole moments.

If there are c molecules per unit volume, the resulting net polarization is $\vec{P} = c\vec{p}$, in case the density is sufficiently low such that the local electric field experienced by one molecule is not affected by the local fields due to neighboring molecules. For the bulk response of dense fluids, one has to consider that the local electric field experienced by a molecule is not simply given by the macroscopic internal electric field E_i that we calculated above. Instead it is given by a superposition of E_i and the field generated by the molecules in its immediate vicinity. For common isotropic liquids (as well as isotropic solids), this leads to an effective reduction of the local field experienced by the molecule by a factor $1/(1 - c\,\alpha/3)$ for a molecule of polarizability α. This correction factor is known as Clausius–Mossotti factor (see, e.g. [1]).

We conclude with a remark on the dynamic response of dielectric media to time-dependent electric fields. While the DC electrical conductivity vanishes by definition, both electronic and orientational polarization give rise to a transient polarization current

$$\vec{j}_{pol} = \epsilon\epsilon_0 \partial_t \vec{E} \tag{2.28}$$

if a material is exposed to a time-dependent electric field. The characteristic response time of the polarization response is typically very fast in the ps range for orientational polarization and even faster for the electronic polarization.

For the purposes of EW, such time scales are typically not relevant. The observed behavior is given by the equilibrium polarization in the final state.

2.2.3 Dielectric Liquids and Leaky Dielectrics

While most solids can be easily identified either as conductors or as dielectrics, many fluids, in particular the widely used aqueous electrolyte solutions in electrowetting experiments, display an intermediate behavior. The water molecules by themselves have a very strong dipole moment that leads to a very strong polarizability due to reorientation of the molecules, as discussed above. This strong dielectric response leads to a strong enhancement of the capacitance if the space between two electrodes is filled with water. At the same time, ions dissolved in water are freely mobile and respond to any applied electric field inside the liquid similar to electrons in metals: They migrate until the external field is shielded and the internal field vanishes: Aqueous electrolyte solutions behave as conductors. This also implies that only nonequilibrium electrical currents can be imposed on electrolytes. DC currents imply the occurrence of electrochemical reactions at the electrodes that must be immersed into the fluids to drive the current. Transient currents and AC currents, however, can arise from relative displacements of the ions in the liquid. This situation is similar to metals, yet both the mobility and the density of ions in electrolytes are typically several orders of magnitude lower. Unlike the dynamics of electrons that is governed by the band structure of the solid, the net drift motion of dissolved ions in a liquid is governed by the balance of the electrical driving force F_{el} and the viscous *Stokes* drag force F_d experienced due to viscous friction with the ambient medium:

$$\vec{F}_{el} = q\vec{E} = \mu_v \vec{v} = \vec{F}_d \tag{2.29}$$

here, $\mu_v = 6\pi\eta_v a$ is the mobility of a particle or ion of radius a moving through an ambient medium of viscosity η_v. The motion of all ions with charge q_i in a given volume element of solvent gives rise to a net electrical current density:

$$\vec{j}_{el} = \sum_i c_i q_i \vec{v}_i = \lambda \vec{E} \tag{2.30}$$

here, λ is the electrical conductivity of the solution. It depends on the mobility and – in particular – on the concentrations c_i of all ions in the solution. For (very) low salt concentrations, the conductivity is linear in the mobility of the individual ions. In this case, we can write

$$\lambda = \sum_i \frac{c_i q_i^2}{6\pi\eta_v a_i} \tag{2.31}$$

For concentrations beyond 1 mM, collective effects arising from the interaction between ions give rise to corrections that lead to a weaker than linear increase of conductivity with concentration. Table 2.1 lists the so-called limiting mobilities in water of a variety of ions typically encountered in electrowetting along with the resulting conductivities for several concentrations.

Table 2.1 Ionic mobility μ_v and limiting ionic conductivity λ at 298 K for a selection of typical anions and cations.

Anions	μ_v (10^{-8} m² s⁻¹ V⁻¹)	λ (mS m² mol⁻¹)	Cations	μ_v (10^{-8}m² s⁻¹ V⁻¹)	λ (mS m² mol⁻¹)
F^-	5.70	5.54	H^+	36.23	34.96
Cl^-	7.91	7.64	Li^+	4.01	3.87
Br^-	8.09	7.81	Na^+	5.19	5.01
NO_3^-	7.40	7.15	K^+	7.62	7.35
OH^-	20.64	19.91	Rb^+	7.92	7.78
CO_3^{2-}	7.46	13.86	Mg^{2+}		10.6
CH_3COO^-	4.24	4.09	Ca^{2+}	6.17	11.90
SO_4^{2-}	8.29	16.00	NH_4^+	7.63	7.35

Next to this ohmic response to the electric field, electrolyte solutions also display a dielectric response, as discussed in the preceding section. For transient and/or time-periodic electric fields, both the ohmic and the dielectric response take place in parallel. The resulting total current in response to a time-dependent electric field is given by the sum of the expressions in Eqs. (2.28) and (2.30):

$$\vec{j}_{tot} = \lambda\vec{E} + \epsilon\epsilon_0\partial_t\vec{E} \tag{2.32}$$

Writing the total response of the material in this way as the sum of two components implies that we can represent the electrical properties of the liquid in an equivalent circuit model by an effective resistor in parallel with a capacitor. For plausible reasons, materials displaying this specific type of response are denoted leaky dielectrics.

For time-periodic electric fields $E(t) = E_0 \exp(i\omega t)$, the leaky dielectric response results in a combination of a dissipative in-phase response and a capacitive out-of-phase response of the system with a complex current $\vec{j}_{tot} = \lambda^*\vec{E}$ and complex conductivity

$$\lambda^* = \lambda + i\omega\epsilon\epsilon_0 \tag{2.33}$$

Alternatively, a complex dielectric response function $\epsilon^* = \epsilon\epsilon_0 - i\lambda/\omega$ can be defined with $\vec{j}_{tot} = \epsilon^*\partial_t\vec{E}$. In both cases, we can identify a characteristic relaxation time $\tau_{el} = \epsilon\epsilon_0/\lambda$. For AC frequencies $\omega \ll \tau_{el}^{-1}$, the response is governed by the ohmic response, and the electrolyte behaves as a perfect conductor in the bulk. Vice versa, for $\omega \gg \tau_{el}^{-1}$, the electrolyte behaves as a dielectric, in line with the expectations based on the equivalent circuit model. Inserting typical numbers from Table 2.1, we find 1 ns $< \tau_{el} <$ 100 μs corresponding to characteristic frequencies of 0.1 … 1 GHz for salt concentrations ranging between zero (i.e. deionized water) and approximately 1M. The *bulk* response of aqueous electrolytes may therefore be considered as predominantly ohmic in the vast majority of electrowetting applications.

To illustrate the consequences of this behavior, we consider the generic situation of a thin layer of a leaky dielectric layer in series with a dielectric between two electrodes; see Figure 2.9. In this situation, we have two separate domains,

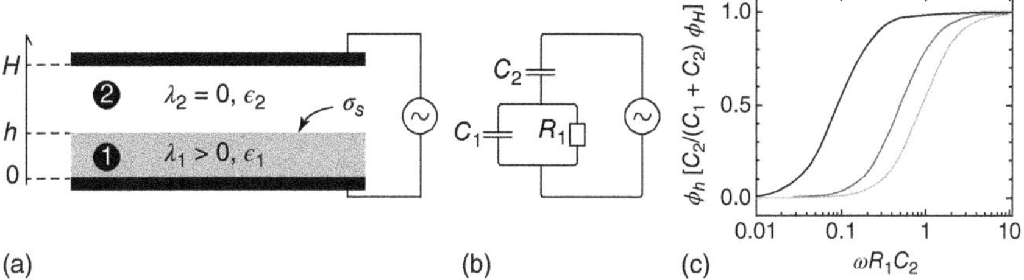

Figure 2.9 Schematic setup of a leaky dielectric layer in series with a pure dielectric. (a) Geometry with fluid properties. (b) Equivalent circuit diagram. (c) Normalized potential at the interface versus normalized frequency for values of the parameter $C_2/C_1 = 0.1,\ 1, 10$ (left to right curves).

$i = 1, 2$, with different material properties. In both domains, the divergence of the displacement field vanishes, i.e. $\nabla \cdot \vec{D}_i = 0$, and in both domains we can introduce an electrostatic potential with $\vec{E}_i = -\nabla\phi_i$. The material response, however, is different. In domain 1, we have $\vec{j}_1 = \lambda_1\vec{E} + \epsilon_0\epsilon_1\partial_t\vec{E}_1$, whereas in domain 2, we have $\vec{D}_2 = \epsilon_0\epsilon_2\vec{E}_2$. Inserting the definition of the electrostatic potential, we directly find for domain 2 the Laplace equation $\nabla^2\phi_2 = \partial_{xx}\phi_2 = 0$, where we have simplified the system to a one-dimensional problem in the last step. For domain 1, our starting point is the continuity equation of the electric charge, i.e. $\nabla\vec{j}_1=0$. (This implies that the material remains electroneutral in the bulk, as it should.) Inserting here the definition of the electrostatic potential, we find the governing equation $\nabla(\lambda_1 + \epsilon_0\epsilon_1\partial_t)\nabla\phi_1 = 0$. For homogeneous material properties, this equally results in the Laplace equation $\nabla^2\phi_1 = 0$. The absence of space charges thus guarantees a linear evolution of the electric potential in both media. Imposing the boundary conditions $\phi(0) = 0$ and $\phi(H) = \phi_H = U_0$, we thus have $\phi_1(x) = \phi_h\, x/h$ and $\phi_2(x) = \phi_h + (\phi_H - \phi_h)(x - h)/(H - h)$, where $\phi_h = \phi_1(h) = \phi_2(h)$ is the potential at the interface that needs to be determined from the boundary condition. Continuity of the total flux of charge at the interface guarantees that $\vec{j}_1(h) = \lambda_1\vec{E}_1 + \epsilon_0\epsilon_1\partial_t\vec{E}_1 = \epsilon_0\epsilon_1\partial_t\vec{E}_2 = \vec{j}_2(h)$. Assuming a periodic excitation $U = U_0e^{i\omega t}$ results in complex potential $\tilde{\phi}(x,t) = \phi(x)e^{i\omega t}$. Assembling all terms, we find

$$\phi_h = Re\{\tilde{\phi}_h\} = Re\left\{\frac{-\dfrac{i\omega\epsilon_2}{H-h}}{\dfrac{\lambda_1}{h} - i\omega\left(\dfrac{\epsilon_1}{h} + \dfrac{\epsilon_2}{H-h}\right)}\right\}\phi_H = \frac{\omega^2R_1^2C_2(C_1+C_2)}{1+(\omega R_1(C_1+C_2))^2}U_0 \tag{2.34}$$

For the right-hand side of the equation, we made use of the fact that the Ohmic resistance of the leaky dielectric layer is given by $R_1 = h/(A\ \lambda_1)$ and the capacitances of the two layers are $C_1 = A\epsilon_0\epsilon_1/h$ and $C_2 = A\ \epsilon_0\epsilon_2/(H-h)$, where A is the cross-sectional area. The same result can be obtained using directly the equivalent circuit model shown in Figure 2.9b (Problem 2.5). For low frequencies $\phi_h = 0$, whereas at high frequency it assumes the value $\phi_h = \phi_H\, C_2/(C_1 + C_2)$ as expected for a capacitive voltage divider; see Figure 2.9c. The system is characterized by a characteristic time scale $\tau = R_1 C_2 = \epsilon_0 \epsilon_2 h/((H-h)\lambda_1)$ and the dimensionless parameter $C_2/C_1 = \epsilon_2 h/(\epsilon_1(H-h))$. Note that both parameters do contain not only material properties but also geometric dimensions of the system. The reason is that the charge is injected into the system at the electrodes, and this charge needs to be transported over macroscopic distances to the interface between the two materials where it accumulates. In typical electrowetting applications, the dielectric layer is very thin compared with the drop such that $H - h \ll h$. As a consequence the characteristic cutoff time scale τ of the device will be much longer than expected based on the bulk properties of the materials, and hence the dielectric response of the materials will become more important at lower AC frequencies. This effect is essential for the actuation of dielectric liquids in electrowetting and similar devices.

2.3 Electrostatic Energy

The contact angle variation in electrowetting is governed by the distribution of electrostatic energy. Having summarized the basic laws of electrostatics and material response, we now focus on the calculation of the electrostatic energy of a system of conductors kept either at a constant charge or at constant potential.

2.3.1 Energy of Charges, Conductors, and Electric Fields

Two electric charges q_1 and q_2 at distance r exert a Coulomb force $F_c = q_1 q_2/4\pi\epsilon_0 r^2$ onto each other. If we bring two such charges from infinity to a finite separation r_{12}, we can calculate the mechanical work to do so by integrating Coulomb's law as $W_{el} = \int_{\infty}^{r_{12}} F_c dr$. This amount of work is stored as electrostatic energy E_{el} in the system. Hence, we have

$$E_{el} = \frac{1}{4\pi\epsilon_0}\frac{q_1 q_2}{r_{12}} \tag{2.35}$$

For a system of many charges, Eq. (2.35) can be generalized to

$$E_{el} = \frac{1}{2}\sum_{i\neq j}\frac{q_i q_j}{4\pi\epsilon_0}\frac{1}{r_{ij}} = \frac{1}{2}\iint dV_1\, dV_2 \frac{\rho(r_1)\rho(r_2)}{4\pi\epsilon_0}\frac{1}{r_{12}} \tag{2.36}$$

where the factor $\frac{1}{2}$ is introduced to avoid double counting of pairs of charges. For the right-hand side of the equation, we replaced the discrete sum by a volume integral over a continuous charge distribution $\rho(r)$ with r_{12} being the distance between the locations r_1 and r_2. Because the electrostatic potential $\phi(r)$ caused by a charge $\delta Q = \rho\, dV$ at a distance r is given by $\phi(r) = \delta Q/4\pi\epsilon_0 r$ we can rewrite Eq. (2.36) as

$$E_{el} = \frac{1}{2}\int \rho(r)\phi(r) dV \tag{2.37}$$

Equation (2.37) states that the electrostatic energy of a distribution of charges is given by half of the volume integral of the local charge density times the local electrostatic potential. We can also use Eq. (2.37) to calculate the electrostatic energy of an isolated charged conductor placed in an uncharged ambient dielectric medium. In this case, ρ vanishes everywhere except for the surface of the conductor. Moreover, the potential of the conductor is constant by definition. As a consequence, Eq. (2.37) reduces to

$$E_{el} = \frac{1}{2} Q \phi_0 \tag{2.38}$$

where Q is the total charge of the conductor and ϕ_0 is its potential. The equation can be generalized to i conductors with charges Q_i and potentials ϕ_i. By applying the same argument as above for each conductor separately, we find

$$E_{el} = \frac{1}{2} \sum_i Q_i \phi_i \tag{2.39}$$

A different interpretation of Eq. (2.37) can be obtained if we rewrite it using Poisson's equation (Eq. (2.7)) as

$$E_{el} = -\frac{\epsilon \epsilon_0}{2} \int \phi(r) \nabla^2 \phi(r) \, dV \tag{2.40}$$

Making use of the identity $\nabla(\phi(\nabla\phi)) = (\nabla\phi)^2 + \phi\nabla^2\phi$, we find

$$E_{el} = \frac{\epsilon_0}{2} \left\{ \int (\nabla\phi)^2 \, dV - \int \nabla(\phi(\nabla\phi)) \, dV \right\} \tag{2.41}$$

The second integral is a volume integral over a divergence and can hence be rewritten as a surface integral over $\phi(\nabla\phi)$. For a localized charge distribution, ϕ will decay as $1/r$, and hence the integrand vanishes as $\propto 1/r^3$, which is faster than the r^2 growth of the integration surface for $r \to \infty$. Hence, the second integral vanishes, and we can rewrite Eq. (2.41) as

$$E_{el} = \frac{\epsilon_0}{2} \int (\nabla\phi)^2 \, dV = \frac{\epsilon_0}{2} \int \vec{E}^2 \, dV \tag{2.42}$$

This means that the electrostatic energy contained in a charge and potential distribution as expressed in Eq. (2.37) can be interpreted as being localized in the electric field distribution. In other words, the electrostatic field itself has an energy density

$$e_{el}(\vec{r}) = \frac{\epsilon_0}{2} E^2(\vec{r}) \tag{2.43}$$

Equations (2.39) and (2.42) are derived here for conductors in vacuum. Calculating the corresponding relations in the presence of dielectric media involves some subtle considerations regarding the polarization charge. It turns out that the final equations for linear dielectric media look very similar. For the energy density, we find

$$e_{el}(\vec{r}) = \frac{\epsilon_0}{2} \vec{E}(\vec{r}) \vec{D}(\vec{r}) = \frac{\epsilon \epsilon_0}{2} \vec{E}^2(\vec{r}) \tag{2.44}$$

along with a corresponding volume integral analogue to Eq. (2.42). The expression for Eq. (2.39) remains unaltered. For details, the reader is referred to, e.g. [2].

2.3.2 Capacitance Coefficients and Capacitance

Equations (2.39) and (2.42) are conceptually simple, yet they both have the disadvantage that knowing the charge Q_i on each conductor, we still need to calculate the potential ϕ_i or the distribution of the electric field. For arbitrary shapes and distributions of the conducting surfaces, this is a complex task that requires in general numerical calculations. However, we can gain an additional step of insight by introducing the notion of capacitance or more generally speaking capacitance coefficients. The treatment in this subsection is somewhat formal. Yet, the key point is essential: Once we have understood why and how we can express the charge on a pair of conductors as a (linear) function of the potential, Eq. (2.39) turns into a quadratic form $E_{el} = CU^2/2$, where C is the capacitance and $U = \Delta\phi$ the potential difference between the two conductors. Electrostatic driving forces that generate mechanical motion can then be calculated as the gradients of the electrostatic energy, which turn out to be gradients of the capacity of the system. This notion is the basis of the frequently very convenient use of equivalent circuit models to analyze problems in electrowetting.

The idea of the calculation is the following: We consider a system of n isolated conductors. Each conductor carries a charge Q_j. Thanks to the linearity of the basic equations of electrostatics, there must be a linear relation between the electrostatic potentials and the charges of each conductor. Therefore, we can write for the potential ϕ_i on each of the, say, n conductors

$$\phi_i = \sum_{j=1}^{n} p_{ij} Q_j \tag{2.45}$$

where p_{ij} are known as the coefficients of potential. They indicate how the potential on conductor i changes upon depositing the charge Q_j on conductor j. The coefficients are symmetric, $p_{ij} = p_{ji}$. If we insert this expression into Eq. (2.39), we can write for the electrostatic energy

$$E_{el} = \frac{1}{2} \sum_{i,j} p_{ij} Q_i Q_j \tag{2.46}$$

To understand the origin of this relation and the manner to calculate the coefficients, we recall that the potential generated by an infinitesimal charge δQ and a distance r is given by $\delta Q/4\pi\epsilon_0 r$. If we consider an arbitrary location $\vec{r}$ in space, the local potential can thus be obtained by integrating the contribution of each infinitesimal charge on the surfaces of the conductors. Writing the infinitesimal charge $\delta Q = \sigma(\vec{r})dA$ in terms of the surface charge density σ and the surface element dA, we can express the potential as a sum of integrals over the surfaces of all conductors:

$$\phi(\vec{r}) = \sum_j \int \frac{\sigma_j(\vec{r}'_j)}{4\pi\epsilon_0} \frac{1}{|\vec{r} - \vec{r}'_j|} dA'_j = \sum_j Q_j \int \frac{\tilde{\sigma}_j(\vec{r}'_j)}{4\pi\epsilon_0} \frac{1}{|\vec{r} - \vec{r}'_j|} dA'_j \tag{2.47}$$

The charge distribution $\sigma_j(\vec{r}_j)$ on each conductor depends on the total charge Q_j on this conductor, on its shape, and on polarization effects due to the charges on the other conductors, as illustrated in Figure 2.10. For the right-hand equation, we normalized the charge density by the total charge on each conductor and wrote it

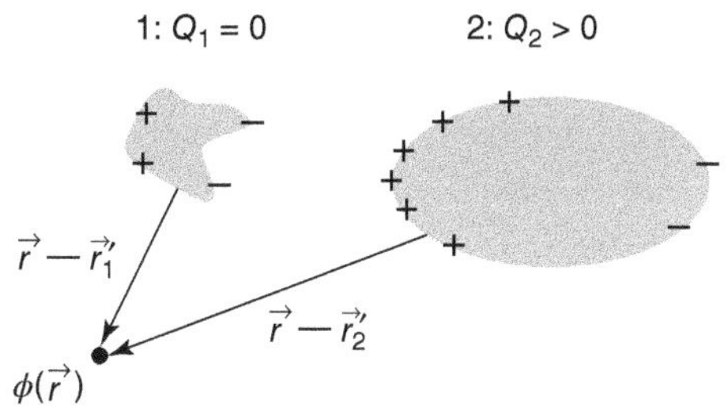

Figure 2.10 Charge distribution on two conductors with net charges Q_1 (taken to be 0) and Q_2 (taken positive). To calculate the potential at position $\vec{r}$, the contributions from both conductors need to be integrated over the respective surface. Note that the charge density on both surfaces is heterogeneous due to polarization.

as $\tilde{\sigma}_j = \sigma_j/Q_j$. The integral on the right then provides an expression for the potential at the location $\vec{r}$ as a function of the charges Q_j on all the conductors. If we want to calculate the potential of conductor i, we choose an arbitrary position $\vec{r}$ on the surface of that conductor. Because the potential on the surface of the conductor is constant, it does not matter which location $\vec{r}_i$ on the surface we choose. Doing so, we obtain

$$\phi_i = \sum_j Q_j \int \frac{\tilde{\sigma}_j(\vec{r}_j')}{4\pi\epsilon_0} \frac{1}{|\vec{r}_i - \vec{r}_j'|} dA_j' = \sum_j p_{ij} Q_j \tag{2.48}$$

Thus, we see that the integral in the equation provides the expression for the coefficients p_{ij} relating the potential on conductor i to the charge Q_j and its distribution $\tilde{\sigma}_j$ on all other conductors j. To calculate the integrals, we still need to solve the actual electrostatic problem using Poisson's equation and the appropriate boundary conditions. Nevertheless, this Eq. (2.48) provides us with a convenient general expression and with a feeling how the coefficients relating potential and charge in an arbitrary system of conductors arise from the geometry and the distribution of charges in the system. As mentioned above, for any nontrivial geometry, the explicit calculation of these coefficients is usually very complex and mostly requires numerical approaches. Nevertheless, we can gain additional insights by simply exploiting the linear relation between charge and potential without explicitly calculating the coefficients.

In capacitors, we usually consider only a pair of two conductors, as sketched in Figure 2.10. If we disregard the complexity of actually calculating the coefficients of potential, Eq. (2.45) simplifies to just two equations:

$$\begin{aligned} \phi_1 &= p_{11}Q_1 + p_{12}Q_2 \\ \phi_2 &= p_{21}Q_1 + p_{22}Q_2 \end{aligned} \tag{2.49}$$

If we make use of the symmetry of the coefficients ($p_{12} = p_{21}$) and of the fact that the two conductors on a capacitor typically carry opposite charge ($Q_1 = -Q_2 = Q$), we can express the potential difference $U = \phi_1 - \phi_2$ as

$$U = (p_{11} + p_{22} - 2p_{12})Q \tag{2.50}$$

Following the conventional definition for the capacitance of a capacitor $C = Q/U$, we have thus expressed the capacitance of the system in terms of the coefficients of potential as $C = 1/(p_{11} + p_{22} - 2p_{12})$. Now, we can write the electrostatic energy of a system of two conductors simply as the well-known expression

$$E_{el} = \frac{1}{2}\frac{Q^2}{C} = \frac{1}{2}CU^2 \tag{2.51}$$

The result can also be rewritten for the more general case of n conductors. If we invert the set of linear equations (Eq. (2.45)), we can express the charges on each conductor in terms of the potential as

$$Q_i = \sum_{j=1}^{n} c_{ij}\phi_j \tag{2.52}$$

where c_{ij} are called coefficients of capacitance. They can be obtained by inverting the matrix of coefficients of potential p_{ij}. With this expression, we can write the electrostatic energy of the system of conductors as a quadratic form:

$$E_{el} = \frac{1}{2}\sum_{i,j} c_{ij}\phi_i\phi_j \tag{2.53}$$

This analysis is the basis of equivalent circuit models that are frequently used to analyze electrowetting problems. In many practical applications of electrowetting, it is sufficient to consider only the two conductors and a single expression for the capacitance C. Yet, this is a simplification that ignores, for instance, the effect of the electric double layer that we will discuss in Chapter 4. Novel approaches, such as nanoscale electrowetting devices, require more detailed considerations, in which the more general expression (Eq. (2.53)) can become relevant.

2.3.3 Thermodynamic Energy of Charged Systems: Constant Charge Versus Constant Potential

So far, we have seen a number of different manners of writing the electrostatic energy of a system of conductors. For a given fixed configuration of conductors, we can express the electric energy as a function of either the charges or the potentials of the conductors. Equations (2.39) and (2.42) hold in both cases. However, if we want to calculate thermodynamic driving forces that are responsible – say – for the actuation of drops in an electrowetting experiment, it is important to consider whether the conductors in the system are kept at constant charge or at constant potential.

In deriving Eq. (2.39) we started out by calculating the work required to assemble a system of charges by bringing them together from infinity. During that process, each charge was assumed to remain constant. Correspondingly, the potential difference between them changed during the process. In this case, the electric energy can be obtained by the integral of the Coulomb force. Let us now consider instead a situation, in which we keep the conductors at fixed potentials. In this case, the charge on them would change (increase) as we bring them together. Changing the charge, however, requires us to connect the conductors to a reservoir of charge at fixed potential, i.e. to a battery. Taking charge out of the battery to keep the potentials constant implies that the electrostatic energy of the battery is reduced. Because the potential of the battery is constant, taking out the amount Q (> 0) means that the energy of the battery decreases by $\Delta E_b = -\phi_b Q$, where ϕ_b is the fixed potential of the battery. The fact that this amount is negative is consistent with the fact that the charge spontaneously flows from the battery to the conductors upon bringing them together. Because the battery is an integral component of a system of conductors at fixed potentials, the energy of the battery has to be included in the thermodynamic expression for the *free* electrostatic energy of the system. This free energy is the one that is minimized in equilibrium and the one that gives rise to the mechanical forces in electrowetting. Thus, we introduce the new *free* electrostatic energy as

$$\tilde{E}_{el} = E_{el} - \sum_i Q_i \phi_i = -E_{el} \tag{2.54}$$

This definition is of utmost importance to electrowetting. The minus sign that appears in this equation is responsible for the spreading of the fluid upon applying a voltage.

To illustrate the origin of this minus sign from a slightly different perspective, we consider explicitly the specific example of the electrostatic energy involved in charging a capacitor for the two complementary thermodynamic conditions of constant charge and constant voltage U_b, as sketched in Figure 2.11.

For the isolated capacitor, the work required to transfer a unit charge δQ from one capacitor plate to the other is given by

$$\delta W_c = U_c\, \delta Q \tag{2.55}$$

where the voltage $U_c = Q_c/C$ on the capacitor depends on the amount of charge Q_c that is already present. The total amount of charge required to charge the capacitor from zero to a finite charge Q is given by the integral over Eq. (2.55). It is equal to the total amount of electrostatic energy E_{el} that is stored in the system in the final state:

$$E_{el} = \Delta W_c = \int_0^Q U_c\, dQ = \frac{1}{2}\frac{Q^2}{C} \tag{2.56}$$

In the final state, the voltage across the capacitor is $U = Q/C$. Hence, we can write the electrostatic energy as $E_{el} = CU^2/2$. This is a positive amount of energy that can be recovered, e.g. by discharging the capacitor through some external circuit. Both expressions are equivalent to the volume integral over the electrostatic field energy, as the reader can easily verify using Eq. (2.42).

If we charge the same capacitor by connecting it to a battery, charge flows spontaneously from the battery to the capacitor. Being a spontaneous process, it is obvious that the charging of the battery must reduce the overall free energy of the total system. We can imagine the process of connecting the battery to the capacitor gradually by transferring discrete amounts of charge δQ across the switch in Figure 2.11b until the voltage across the capacitor is the same as U_b. The amount of work required to charge the capacitor is the same as in the case of the isolated capacitor. However, there is simultaneously a gain in electrostatic energy

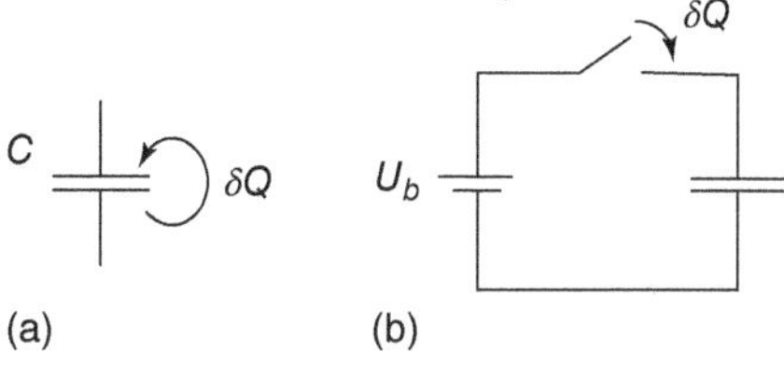

Figure 2.11 Progressive charging of an isolated capacitor (a) and a capacitor connected to a battery providing a constant voltage (b).

as charge is taken out of the battery. The amount of work done on the battery is given by the same expression (Eq. (2.56)) as for the capacitor, i.e. $\delta W_b = U_b \delta Q_b$. In contrast to the capacitor, however, U_b is constant. Moreover, taking into account charge conservation, we have $\delta Q_b = -\delta Q$. Hence, we can write the change in electrostatic energy of the battery as

$$\Delta E_{el}^b = \Delta W_b = \int_0^{Q_b} U_b \, dQ_b = -U_b Q = -CU_b^2 \tag{2.57}$$

Hence, the change in free electrostatic energy of the system is

$$\tilde{E}_{el} = E_{el}^c + \Delta E_{el}^b = -\frac{1}{2} CU^2 \tag{2.58}$$

here, the tilde indicates that we are considering the electric energy of a system where the conductors are kept at constant potential. The gain in free energy upon connecting a capacitor to a battery thus increases with the capacitance of the system. This simple thermodynamic fact is at the origin of all electrowetting-driven actuation of liquids. Because drops of conductive liquids function as deformable electrodes, electrowetting systems can reduce their total free energy by deforming or translating drops in such a manner that the capacitance of the system increases. To predict how an electrowetting system will react upon applying a voltage, it is thus sufficient to consider how the capacitance between the liquid(s) involved and the activated electrodes can be increased by deforming or moving the drops. For a quantitative description of the effect, we will obviously have to consider the competition between the gain in electrostatic energy and the costs in surface energy, as discussed in Chapter 1, and other forms of energy and dissipation, if applicable.

2.4 Electrostatic Stresses and Forces

2.4.1 Global Forces Acting on Rigid Bodies

One of the key goals in electrowetting is to transport liquids. Therefore, we need to calculate the electrostatic forces upon exposing the system to electric fields. According to the general principles of mechanics, we can calculate forces by taking the gradients of the proper thermodynamic energy with respect to the position of the object experiencing the force. Equations (2.56) and (2.58) provide expressions for the free electrostatic energy of a system of two conductors at fixed charge and at fixed potentials. The position-dependent quantity in both expressions is the capacitance C. For constant charge, we thus find that the force is given by

$$F_Q = -\nabla E_{el} = -\nabla \left(\frac{Q^2}{2C} \right) = \frac{Q^2}{2C^2} \nabla C \tag{2.59}$$

Similarly, in the case of constant potentials, we find

$$F_\phi = -\nabla \tilde{E}_{el} = -\nabla \left(-\frac{1}{2} CU^2 \right) = \frac{U^2}{2} \nabla C \tag{2.60}$$

That is, in both cases the force is directed along the gradient of the capacitance. If left to evolve freely, charged conductors will thus move (and/or deform in case of liquids) in such a manner that the capacitance of the system increases. In general, this is equivalent to saying that the oppositely charged conductors will attract each other, as they should. Problem 2.6 is devoted to an explicit calculation of the force on the plates of a parallel plate capacitor for the two conditions. Note, however, that the statements of Eqs. (2.59) and (2.60) go beyond forces acting on the conductors. The expressions equally hold for the forces experienced by dielectric bodies exposed to an electric field generated by fixed electrodes.

2.4.2 Local Forces: The Maxwell Stress Tensor

Equations (2.59) and (2.60) describe the total force on a solid body that arises from a gradient of the electrostatic energy. Liquids, however, are deformable. Therefore it is not sufficient to know the total force. In order to understand field-induced deformation of the liquid, it is important to know the local distribution of electrostatic fields and forces. To this end, we introduce in this section the so-called Maxwell stress tensor.

We know that the electrostatic force experienced by a charge Q in an externally imposed electric field $\vec{E}$ is given by $\vec{F}_{el} = Q\vec{E}$. Similarly, an arbitrary distribution of charge carriers (in vacuum) with a charge density $\rho(r)$ experiences an electrostatic body force per unit volume $\vec{f}^{\,v}_{el} = \rho\vec{E}$. Here, the charge density represents only the free charge carriers. In a dielectric medium, additional local charge density can arise from polarization charges. As discussed in Section 2.2, gradients of the polarizability χ (and correspondingly of the dielectric permeability ϵ) in combination with an electric field give rise to a polarization charge density $\rho_{pol} \propto \nabla\chi$. As a result, the net body force density in a dielectric medium reads

$$\vec{f}_{el} = \rho_f\,\vec{E} - \frac{1}{2}\epsilon_0 E^2\,\nabla\epsilon \tag{2.61}$$

This expression arises from the thermodynamic analysis by Korteweg and Helmholtz (see e.g. [2]). The factor ½ arises from the fact that the polarization charge is generated by the same electric field that exerts the force. Correspondingly, the net electrostatic force experienced by a finite volume is given by the integral

$$\vec{F}_{el} = \int\left(\rho_f\vec{E} - \frac{\epsilon_0}{2}E^2\nabla\epsilon\right)dV = \int\left((\nabla\vec{D})\vec{E} - \frac{\epsilon_0}{2}E^2\nabla\epsilon\right)dV \tag{2.62}$$

where we inserted Gauss' law for the right-hand equation. Making use of the identities $\nabla(\vec{D}\vec{E}) = (\nabla\vec{D})\vec{E} + (\nabla\vec{E})\vec{D}$ and $\nabla(\epsilon E^2) = E^2\nabla\epsilon + 2\epsilon\vec{E}\,(\nabla\vec{E})$, we find that the integrand on the right-hand side of the equation can be rewritten as the divergence of a tensor, the so-called Maxwell stress tensor:

$$T_{ij} = \epsilon_0\left(D_iE_j - \frac{1}{2}\delta_{ij}\vec{E}\vec{D}\right) \tag{2.63}$$

here, we introduce the tensor notation where T_{ij} is the ith component of the force onto a control volume dV arising from fields in the jth direction; see

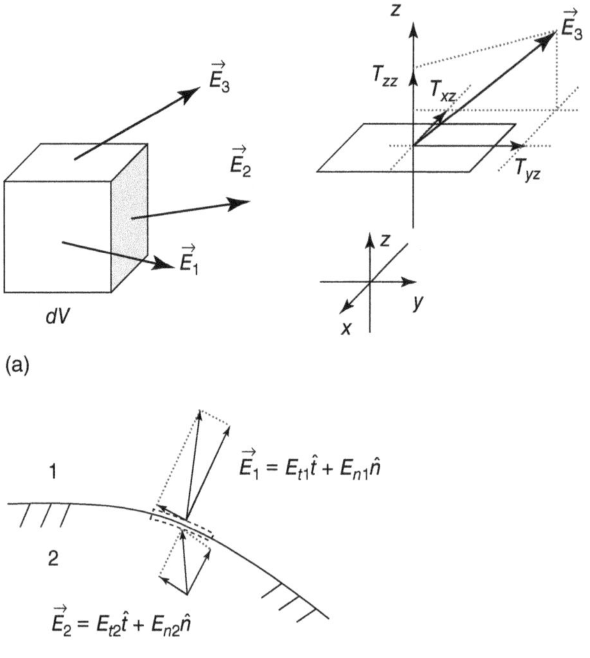

Figure 2.12 (a) Illustration of electrostatic stresses acting on a (cubic) control volume dV as described by the Maxwell stress tensor. For each of the six surface elements of the control volume, the electric field generates normal and tangential stresses as illustrated in the inset for the top facet. (b) Maxwell stress at an arbitrary interface between two media with dielectric permeabilities $\epsilon_1 < \epsilon_2$.

Figure 2.12. For the ith component of the total force on a volume V, we can write

$$F_{el,i} = \int \sum_j \frac{\partial}{\partial x_j} T_{ij}\, dV = \oiint_{\partial V} \sum_j T_{ij} \hat{n}_j\, dA \tag{2.64}$$

The reader may verify this equation by writing down explicitly the x, y, and z components of the force. For the second equation, we made use of Gauss' integral law. $\hat{n}_j$ is the jth component of the outward normal vector of the surface A enclosing the volume V. If we consider a control volume V as sketched in Figure 2.12, T_{ij} describes the stresses acting on the surface of V due the electric field.

The fact that we can write the electrostatic force as a volume integral over the divergence of the Maxwell stress tensor or equivalently as a surface integral of the tensor itself over the surface of the same probe volume implies that T_{ij} describes the flux of a conserved quantity. Equation (2.64) indicates that a finite divergence of the Maxwell stress tensor leads to a force, i.e. a change of mechanical momentum of the charges contained in the control volume. Assuming global momentum conservation, we can conclude that T_{ij} represents the tensor of the momentum flux density of the electric field. We note here that the expression given in Eq. (2.63) is the electrostatic version of the full Maxwell stress tensor of the electromagnetic field, which includes magnetic fields, too. The full electromagnetic stress tensor, for instance, also allows for the calculation of the radiation pressure due to electromagnetic light waves. The interested reader is referred to textbooks on electromagnetism. For the purposes of electrowetting, the effect of induced magnetic fields is negligible.

2.4.3 Stress Boundary Condition at Interfaces

In addition to the fundamental interpretation given above, the Maxwell stress tensor is of practical use for the calculation of electrostatic forces on localized distributions of charges. This is particularly useful if a control volume can be chosen such that the electric field is known by symmetry arguments or specific limits along the surface. The most important application of this principle is the calculation of the electric stresses acting on interfaces between two adjacent media, such as a conductor and a dielectric or two dielectrics of different permeabilities. Considering again an interface and a control volume as sketched in Figure 2.5a, Eq. (2.64) now allows for calculating the electrostatic force acting on the interface. Because the small sides of the control volume are of negligible width, the net force is obtained simply as the difference between the Maxwell stress tensor in the medium on the left side of the interface minus its value in the medium on the right times the corresponding surface area of the control volume. For a conductor in contact with a dielectric, we know that the electric field is oriented normal to the interface and vanishes inside the conductor, as discussed in Section 2.1. The force is oriented along outward normal of the conductor and has the magnitude $\epsilon\epsilon_0 E^2 A/2$ if A is the surface area and ϵ the permeability of the ambient dielectric. The resulting Maxwell stress is given by

$$(T \cdot \hat{n}) \cdot \hat{n} = \pi_{el} = \frac{\epsilon\epsilon_0}{2} E^2 \tag{2.65}$$

We can rationalize the origin of this formula also without invoking the Maxwell stress tensor: A surface of a conductor exposed to an electric field E carries the surface charge $\sigma = \epsilon\epsilon_0 E$. An infinitesimal increment $E \to E + \delta E$ gives rise to an increment of the stress by $\delta\pi_{el} = \sigma\,\delta E = \epsilon\epsilon_0 E\,\delta E$. Integrating this expression from zero to a finite value produces the total Maxwell stress given in Eq. (2.65). It describes the force by which the external electric field pulls on the screening charge that it has created by itself.

Similarly, we can derive expressions for the stress acting on the interface between two dielectric media. In contrast to the surface of a conductor, the electric field can contain components both normal and parallel to the interface. Correspondingly, the functional form of the expressions is slightly more involved. Yet, by considering a suitably oriented coordinate system, we can read from Eq. (2.64) for the normal direction $\hat{n}$ and for the tangential direction $\hat{t}$

$$\begin{aligned}(T \cdot \hat{n}) \cdot \hat{n} &= \frac{\varepsilon_0}{2}(\varepsilon_1 E_{n1}^2 - \varepsilon_2 E_{n2}^2 - (\varepsilon_1 E_{t1}^2 - \varepsilon_2 E_{t2}^2)) \\ (T \cdot \hat{t}) \cdot \hat{t} &= \sigma E_t\end{aligned} \tag{2.66}$$

If medium 2 is a conductor, the normal contribution of Eq. (2.66) reduces to the result given in Eq. (2.65), as expected, because $E_{t1} = E_{t2} = E_{n2} = 0$. To calculate the tangential component, we made use of $(\epsilon_1 E_{n1} - \epsilon_2 E_{n2}) = \sigma/\epsilon_0$ and the usual boundary conditions (Eqs. (2.24) and (2.25)) for E_n and E_t. It simply implies that the tangential stress at the interface is given by the product of the polarization charge density at the interface times the tangential electric field.

In mechanical equilibrium, the stresses described by Eqs. (2.65) and (2.66) must be balanced by other forces acting on the interface. The normal components give rise to a discontinuity in the pressure across the interface. For interfaces involving solid media, the electric stresses are typically compensated by elastic stresses that lead – in case of conventional hard solids – to negligible elastic strains. For instance, an electric field with a strength of $10^8\,\mathrm{V\,m^{-1}}$ at a metal–vacuum interface gives rise to a Maxwell stress $\pi_{el} \approx 4 \times 10^5$ Pa. For a typical metal with an elastic modulus of 10–100 GPa, this gives rise to a strain of order 10^{-5}. Hence, a 1 mm thick sheet exposed to such a stress would deform by no more than 10 nm. For liquid–liquid or liquid–air interfaces, however, the situation is different. In this case, the electric normal stress enters as an additional term in the Young–Laplace equation (Eq. (1.6)), which then reads

$$\gamma\,\kappa = \Delta p - (T \cdot \hat{n}) \cdot \hat{n} \tag{2.67}$$

In case of two dielectric liquids, there is also tangential component of the stress. That component can obviously not be balanced by a discontinuity of the pressure. Rather, it needs to be balanced by other stresses in the tangential direction that can arise either from surfaces stresses due to Marangoni effects (i.e. lateral gradients of the surface tension) or dynamically by viscous stresses [3]. We will come back to this extension of Eq. (2.67) in Section 6.1.

In electrowetting, one of the liquid phases typically behaves as a perfect conductor. In this case, the relevant boundary condition at the liquid–liquid or liquid–vapor interface reads

$$\gamma\kappa = \Delta p - \frac{\epsilon\epsilon_0}{2}E^2 \tag{2.68}$$

Like the preceding ones, Eq. (2.68) is a local equation. For arbitrarily shaped surfaces and arbitrary external fields, both the mean curvature κ and the Maxwell stress π_{el} in Eq. (2.68) can vary along the interface. In mechanical equilibrium, κ and π_{el} adjust each other in a self-consistent process such that Eq. (2.68) is fulfilled everywhere along the surface. This force balance equation is essential for the description of the liquid surface profile close to the contact line in electrowetting as we will discuss in detail in Section 5.1.

Equations (2.64)–(2.68) also allow for calculating the total force acting on a conductive or a dielectric body exposed to an external electric field. The total force is obtained by choosing the volume of the body as control volume in Eq. (2.64) and by subsequently integrating the Maxwell stress tensor over the entire surface. While conceptually simple and intuitive, this method is practically often of limited use because it requires knowledge of the electric field distribution everywhere along the surface. Except for cases of high symmetry, however, this usually requires numerical calculations.

2.5 Two Generic Case Studies

To conclude this chapter, we discuss two specific electrostatic problems of direct relevance to many electrowetting experiments. In the first one, we calculate electrostatic forces in a parallel plate geometry that is very similar to electrowetting-driven lab-on-a-chip devices. The second one shows an analysis of the charge and energy distribution of two capacitors in series in the presence of a finite net charge in a lumped parameter description. This configuration represents the situation of a dielectric layer and an electric double layer with a finite surface charge that we will analyze from a microscopic perspective in Chapter 4.

2.5.1 Parallel Plate Capacitor

The physical situation that we want to consider is very similar to Figure 2.8, except that we assume the inserted material to be a perfect conductor and that we connect the electrodes to a battery, as it is usually the case in electrowetting experiments. Again, the plate is separated from the electrodes by a gap δ that is thin compared with the separation d of the capacitor plates; see Figure 2.13a. The metallic plate is electrically floating. For symmetry reasons it adopts an electrostatic potential $U_b/2$. To calculate the force acting on the plate, we first use the method of calculating the gradient of the free electrostatic energy as a function of the position of the plate. To this end, we replace the system by an electrical equivalent circuit consisting of the battery and three capacitors representing the empty parallel plate capacitor on the right, C_d, and the two identical gaps between the metal plate and the electrodes, C_δ, on the left.

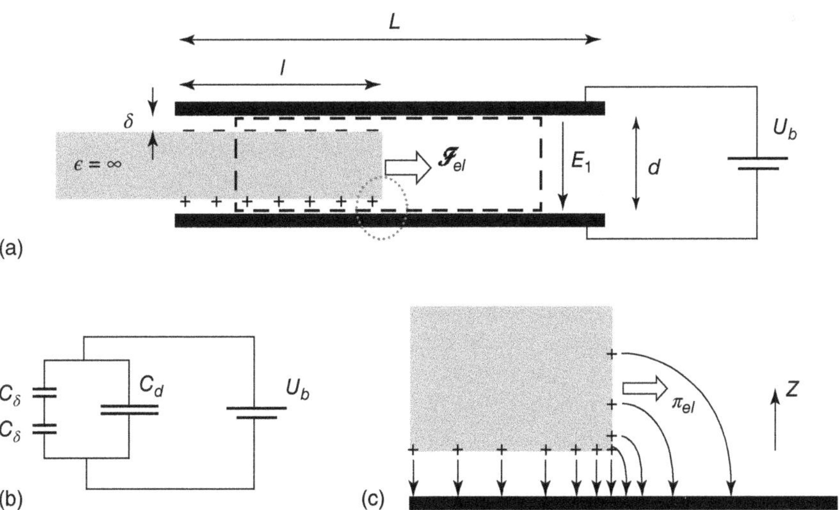

Figure 2.13 (a) Conductive plate inserted into a parallel plate capacitor with a small insulating gap of thickness δ. (b) Equivalent circuit. (c) Local distribution of electric field and screening charges at the corner of the metal plate (see encircled region in (a)).

The two latter are connected in series to each other and parallel to C_d, as sketched in Figure 2.13b. Assuming the capacitor to be air-filled, we thus have $C_d = (L - l)\epsilon_0/d$ and $C_\delta = l\epsilon_0/\delta$. Here, L is the total width of the capacitor, and l is the length by which the plate is inserted. The resulting total capacitance of the equivalent circuit is thus $C(l) = C_d + C_\delta/2$. This yields a free electrostatic energy:

$$\tilde{E}_{el}(l) = -\frac{1}{2}C(l)U_b^2 \tag{2.69}$$

Strictly speaking, this expression is incomplete, because we should have added a third capacitance in parallel to C_d and the two C_δ's to represent the more complex local field and charge distribution around the sharp corners at the front end of the metal plate. That capacitance, however, does not depend on the position of the plate and therefore merely represents a constant offset of the electrostatic energy. When we take the gradient of the energy to calculate the force, this contribution drops out and we find

$$F_{el} = -\partial_l \tilde{E}_{el} = \frac{1}{2}\epsilon_0 \left(\frac{1}{2\delta} - \frac{1}{d}\right) U_b^2 \approx \frac{\epsilon_0}{4\delta} U_b^2 \tag{2.70}$$

The latter approximation holds for $\delta \ll d$. There is thus a force that metal plate pulls the plate into the capacitor because this increases the total capacitance of the system.

Before discussing the implications of the result, let us consider the problem from the perspective of electrostatic forces. As mentioned above, the most intuitive way to calculate the total force is to integrate the Maxwell stress tensor (Eq. (2.65)) over the surface of the plate. Most of the charges are facing each other on opposite sides of the narrow gap δ. The forces acting on these charges do not contribute to the force in the horizontal direction. Close to the corners of the metal plate, however, screening charges spill over to the vertical edge of the plate, as illustrated in Figure 2.13c. The surface normal of this edge is oriented along the horizontal direction. Hence, the electric field pulling on these screening charges does give rise to a force that pulls the plate into the capacitor. In contrast to the lumped parameter picture, which was based on the gradient of the total free energy, we learn from the force picture that the origin and the microscopic distribution of the force depend on the details of the geometry near the edge of the plate. Since both methods must yield the same net force, we expect that the integral of the Maxwell stress should result again in the expression Eq. (2.70).

To calculate this integral, we need to know the distribution $\vec{E}(\vec{r})$ of the electric field and (equivalently) the screening charges along the surface. For the simple geometry of a corner with straight edges, as sketched in Figure 2.13c, it is possible to calculate field and charge distribution analytically using the method of conformal mapping [4, 5], which we will not discuss here. The calculation shows that both electric field and charge density display a weak divergence upon approaching the corner, as we already discussed in the context of Figure 2.4a. It turns out, however, that the divergence can be integrated. For a system with $d \gg \delta$, the two corners can be considered as separate. Per corner, the integrated force yields

$$F_{el,x} = \int_0^\infty \frac{1}{2}\sigma E \, dz = \frac{\epsilon_0}{2\delta} U^2 \tag{2.71}$$

where $U = U_b/2$ is the voltage between the electrode and the metal plate. Inserting this and multiplying by 2 for the two corners of our plate, we recover the same result as Eq. (2.70), as expected.

While it is obviously comforting to obtain the same result from the integration of the microscopic force as from the energy gradient, the agreement may seem somewhat fortuitous. Why does a complex local field distribution that depends on the details of the shape of the edge always result in a well-defined value independent of the local geometry? The Maxwell stress tensor formalism provides an answer to rationalize this result. In the spirit of Eq. (2.64), we can choose arbitrary control volumes and calculate the net electrostatic force acting on the material inside. By integrating the Maxwell stress along the surface of the metal plate, as we just did, we implicitly chose a control volume that tightly encloses the plate. If we make a different choice, however, we can avoid the calculation of the complex field distribution around the corners of the metal plate. Let us consider specifically the dashed square box in Figure 2.13a as our control volume. In this case, we can indeed calculate $F_{el,x}$ in a straightforward manner using Eq. (2.64). To be specific, we choose the control volume such that the vertical parts on the left and on the right are taken sufficiently far from the edge of the plate not to be affected by the local stray fields. The horizontal parts are taken just outside the surface of the electrodes at the top and at the bottom, where the electric field is oriented perpendicular to the surfaces. On the right-hand side of the metal plate, the relevant components of the stress tensor are

$$\begin{aligned} T_{xx} &= \varepsilon_0 \left(E_x^2 - \frac{1}{2}(E_x^2 + E_y^2 + E_z^2) \right) = -\frac{\varepsilon_0}{2}\left(\frac{U_b}{d}\right)^2 \\ T_{xy} &= T_{xz} = 0 \end{aligned} \tag{2.72}$$

Inside the metal plate, the electric field and thus the Maxwell stress tensor vanish. Within the small gap δ, the tensor components read

$$\begin{aligned} T_{xx} &= -\frac{\varepsilon_0}{2}\left(\frac{U_b}{2\delta}\right)^2 \\ T_{xy} &= T_{xz} = 0 \end{aligned} \tag{2.73}$$

Because the off-diagonal elements vanish everywhere and because of the orientation of the surface normal around our control volume, only the integrals along the right vertical edge and across the two gaps of width δ along the left vertical edge contribute to the integral. The reader can easily verify that this leads to the final result:

$$\vec{F}_{el,x} = 2\left(-\frac{\epsilon_0}{2}\left(\frac{U_b}{2\delta}\right)^2\right)(-\vec{e}_x) + \left(-\frac{\epsilon_0}{2}\left(\frac{U_b}{d}\right)^2\right)\vec{e}_x = \frac{\epsilon_0}{2}U_b^2\left(\frac{1}{2\delta} - \frac{1}{d}\right)\vec{e}_x \tag{2.74}$$

Thus, we recover exactly the same result as Eq. (2.70). By making this convenient choice of the control volume, we could calculate the force without the need to solve for the local distribution of stray fields at the edge of the plate. This result also illustrates that the net force exerted by the electric field does not depend on its actual microscopic distribution within the control volume. All that matters is the Maxwell stress along the surface. This very powerful result is a general consequence of momentum conservation.

Let us note a few more aspects of this result. Firstly, we discussed here the specific case of a metallic plate that is inserted into the capacitor. The same treatment can be generalized to dielectric plates, as the one sketched in Figure 2.6. The fact that the electric field then penetrates into the material reduces the electrostatic force, but the qualitative behavior remains the same. For the purposes of electrowetting-driven applications, this implies that electromotive or *ponderomotive* forces are not limited to perfectly conductive liquids but equally apply to dielectric liquids. The reader is invited to generalize the results obtained here to this situation (see Problem 2.7). The relation between dielectrophoretic and general ponderomotive forces in electrowetting was discussed in great detail by Jones [6]. It will be revisited in practical applications in Chapter 9. Secondly, we discussed all our considerations assuming DC voltages and electric fields. Yet all the conclusions presented can be extended to AC electric fields. As long as the AC frequencies are not too high, the conductors keep behaving as conductors, and dielectrics keep behaving as dielectrics with the same dielectric constant. If at the same time the AC frequency is sufficiently faster than the fastest relevant mechanical eigenfrequency of the system, the only difference is that the system will experience time-averaged electric energies and forces. Measurable energies and forces are then obtained by using the corresponding root-mean-square values.

2.5.2 Charge and Energy Distribution for Two Capacitors in Series

To complete this chapter, we turn back to the elementary problem of two capacitors in series, as shown in Figure 2.14. We will apply our analysis from the end of Section 2.3 to identify the distribution of charge and the free electrostatic energy in this system. This equivalent circuit is frequently used in EW to describe the dielectric insulating layer acting as a capacitor C_1 and the electric double layer at the solid-electrolyte interface acting as a second capacitor C_2. In practice, the solid-electrolyte interface typically carries a finite surface charge density due to a variety of spontaneous chemical processes, as we will discuss in Chapter 4. Translated to our simple equivalent circuit, this means that the electrically isolated conductor between the two capacitors carries a fixed finite charge, which we denote as Q_s.

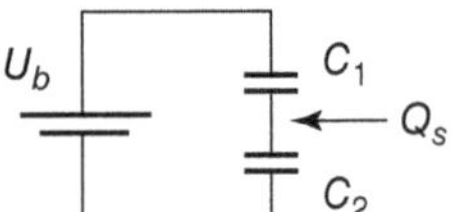

Figure 2.14 Two capacitors in series carrying a finite net charge Q_s.

The variation of the total electrostatic energy as we transfer charge from the battery to the system of the two capacitors is given by

$$\delta E_{el} = \delta E_b + \delta E_{C_1} + \delta E_{C_2} \tag{2.75}$$

where $\delta E_b = U_b \delta Q_b$ is the variation of the energy of the battery and $\delta E_{C_1} = U_1 \delta Q_1$ and $\delta E_{C_2} = U_2 \delta Q_2$ are the variations of the electric energy of capacitors 1 and 2.

The U_i and the δQ_i with $i = 1, 2$ are the voltages across and the charges on the two capacitors. To proceed, it is convenient to express the variation of the energy in terms of the variation of the charge, say, on capacitor 1. As already discussed in Section 2.3, charge conservation implies $\delta Q_b = -\delta Q_1$. If the charge on the first capacitor increases, then obviously the charge on the second capacitor increases, too. Hence, charge conversation imposes again $\delta Q_2 = \delta Q_1$. (This may not seem fully obvious at first glance, because each of the capacitors has two plates and the charge on the two plates is equal and opposite. If we speak about *the charge* on the capacitor, what we mean is the absolute value. Alternatively, we may choose to specify the charge on the terminal with the higher potential.) Finally, we need to fix the additional boundary condition that the total charge is Q_s on the central part connecting the middle terminals of the two capacitors. This constraint implies $Q_1 - Q_2 = Q_s$. Assembling all the terms, we find

$$\delta E_{el} = \left\{ -U_b + \frac{Q_1}{C_1} + \frac{Q_1 - Q_s}{C_2} \right\} \delta Q_1 = 0 \tag{2.76}$$

Equating the curly brackets to zero, we find the equilibrium charges on the two capacitors as

$$Q_{1/2} = C_{eff} \left(U_b \pm \frac{Q_s}{C_{2/1}} \right) \tag{2.77}$$

where $C_{eff} = \frac{C_1 C_2}{C_1 + C_2}$. For $Q_s = 0$, this formula indeed implies that the charge on the two capacitors is the same, as it should be. For zero applied voltage, we find $Q_1/Q_2 = -C_1/C_2$, again as expected. Inserting Eq. (2.77) into the expression for the total electric energy $E_{el} = Q_1^2/2C_1 + Q_2^2/2C_2$, we finally obtain

$$E_{el} = \frac{1}{2}\frac{C_{eff}^2}{C_1}\left(U_b + \frac{Q_s}{C_2}\right)^2 + \frac{1}{2}\frac{C_{eff}^2}{C_2}\left(U_b - \frac{Q_s}{C_1}\right)^2 = \frac{1}{2}C_{eff}U_b^2 + \frac{Q_s^2}{2(C_1 + C_2)} \tag{2.78}$$

The expression shows that the energy due to the applied voltage is stored in the effective capacitor formed by the two capacitors in series. The contribution due to the constant charge Q_s appears to be stored in a capacitor with capacitance $C_1 + C_2$. Electrowetting experiments are controlled by the variation of the energy due to the applied voltage and are thus controlled by the first term. For typical EW experiments, the capacitance of the electric double layer is much larger than the one of the dielectric layer, i.e. $C_2 \gg C_1$. In this case, the electrostatic energy is primarily stored in the capacitor with the smaller capacitance, i.e. $E_{el} \approx C_1 U_b^2/2 + \text{const}$. As will also be seen later in Chapter 7 for multilayer dielectrics, a detailed understanding of the charge and electric fields across each capacitor layer is critically important for achieving reliable electrowetting actuation.

Problems

2.1 Solve the Poisson equation in spherical coordinates to calculate the potential distribution (a) outside and inside a spherical shell of radius R with a constant surface charge density σ. (b) Repeat the same calculation for a homogeneously charged sphere of the same radius with bulk charge density ρ.

2.2 Calculate the electric potential distribution for a row of alternating positively and negatively charged electrodes as in Figure 2.4c but with an electrically grounded horizontal electrode at a distance h above the alternating electrodes. Discuss separately the limiting cases with $h \ll L$ and $h \gg L$ (L, periodicity of alternating electrodes), and relate this to electrowetting experiments with interdigitated electrodes.

2.3 Calculate the polarization charge density σ_p at the interface between two media with dielectric constants ϵ_1 and ϵ_2.

2.4 Calculate the average orientation of a permanent dipole moment p_0 in an external electric field E. Use the Boltzmann distribution and average over all possible orientations of the dipole moment relative to the electric field. Assume that $\frac{p_0 E}{k_B T} \ll 1$.

2.5 Consider the equivalent circuit model shown in Figure 2.9b. Use the impedances $Z_R = R$ and $Z_C = 1/i\omega C$ of the resistor and the two capacitors to verify Eq. (2.34).

2.6 Consider a capacitor consisting of two parallel plates of diameter R at a distance $d \ll R$ such that the effect of fringe fields at the edge can be neglected. (a) Assume that the two plates of the capacitor are electrically insulated and carry the charge $\pm Q$. Calculate the force between the plates and the potential difference as a function of d. Confirm that the gradient of the energy of the capacitor plates, of the electrostatic field energy inside the capacitor, and of the integral of the stress tensor all yield the same expression for the force. (b) Repeat the calculation of the force while keeping the potential difference between the plates constant. Calculate for this case also the charge as a function of the distance and the change in energy of the battery.

2.7 Consider a parallel plate capacitor as discussed in Figure 2.13. Replace the mobile metallic plate by a mobile dielectric plate with dielectric constant ϵ_d, and assume that the medium surrounding the plate between the electrodes is an oil with dielectric constant ϵ_{oil}. (a) Repeat the analysis of Section 2.3 to calculate the force. (b) Replace the straight right edge of the dielectric plate by some other shape, say, a circular arc. How does this affect the force experienced by the plate? Explain the physical origin of your result.

2.8 In Figure 2.13 we assumed $\delta \ll d$. If we miniaturize the device, we can reduce the plate distance d, yet the little gap δ needs to remain finite to guarantee electrical insulation between the electrodes and the inserted plate. How does the force experienced by the plate evolve if its thickness h is gradually reduced, i.e. in the limit $h = d - 2\delta \rightarrow 0$?

References

1 Jackson, J. D. (1983). *Klassische Elektrodynamik*, 2e. Berlin: Walter de Gruyter.

2 Landau, L. D. and Lifschitz, E. M. (1985). *Lehrbuch der Theoretischen Physik VIII: Elektrodynamik der Kontinua*, vol. VIII. Berlin: Akademie Verlag.

3 Saville, D. A. (1997). Electrohydrodynamics: The Taylor–Melcher leaky dielectric model. *Ann. Rev. Fluid Mech.* 29: 27–64.

4 Kang, K. H. (2002). How electrostatic fields change contact angle in electrowetting. *Langmuir* 18 (26): 10318–10322.

5 Vallet, M., Vallade, M., and Berge, B. (1999). Limiting phenomena for the spreading of water on polymer films by electrowetting. *Eur. Phys. J. B* 11 (4): 583–591.

6 Jones, T. B. (2002). On the relationship of dielectrophoresis and electrowetting. *Langmuir* 18 (11): 4437–4443.

3

Adsorption at Interfaces

The liquids used in electrowetting (EW) experiments are usually not pure but contain solutes. In many cases, manipulating the solutes is essential for the functioning of the device, or, as in the case of biomicrofluidics, manipulating the dissolved material is the goal of the experiment. The solutes cover a wide range of species including small inorganic ions, surfactant molecules, dye molecules, and biological molecules such as proteins and DNA, as well as colloidal particles and living cells. Figure 3.1 illustrates a variety of solutes that vary considerably in size and physical and chemical properties. In general, all these solutes interact with the interfaces present in the system and can either adsorb or become depleted from the interface. Attraction and repulsion of small ions at the drop–substrate interface are essential to the basic principle of EW. In contrast, adsorption of large charged or uncharged solutes such as proteins and particles to interfaces is often detrimental to the operation of EW devices because it compromises the homogeneity and intrinsic hydrophobicity of EW surfaces. Understanding the principles of solute adsorption to interfaces is therefore essential to make proper choices of materials and material combinations in EW experiments such that devices can be operated successfully. The goal of this chapter is to provide an overview of the basic physical principles governing interfacial adsorption of uncharged solutes. We consider adsorption equilibria (Section 3.1), kinetic aspects of adsorption (Section 3.2), and various aspects of specific types of solutes that are frequently encountered in EW experiments (Section 3.3). Readers interested in more details can find comprehensive overviews in many excellent textbooks of physical chemistry of interfaces, such as [1–3]. In the appendix, we present a generic statistical mechanics model of adsorption equilibria that provides additional insights for more theoretically interested readers and a derivation of some of the general formula discussed throughout the chapter. In Chapter 4, we will generalize the concepts introduced here to charged solutes, i.e. to ions, in order to understand the basic physics of electric double layers.

Electrowetting: Fundamental Principles and Practical Applications,
First Edition. Frieder Mugele and Jason Heikenfeld.

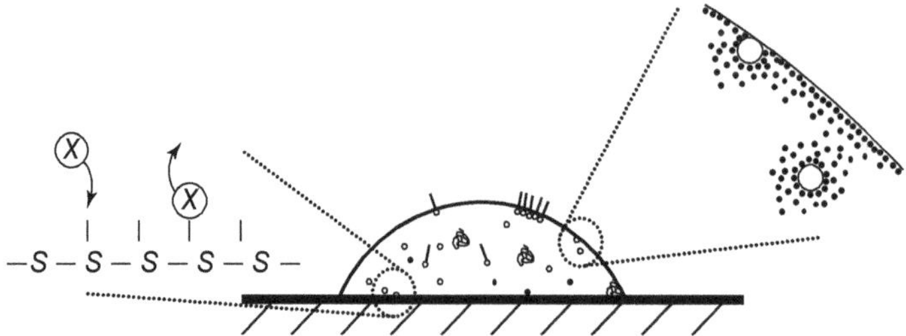

Figure 3.1 Illustration of a drop containing various types of solutes that adsorb to both the drop-surface and the drop–substrate interface. Left zoom: Adsorption of solute X to specific surface sites S. Right zoom: Solute molecules along with their solvation shells that become distorted upon adsorption to an interface.

3.1 Adsorption Equilibrium

3.1.1 General Principles

The physical origin of interfacial adsorption on the atomic scale is very similar to the origin of the interfacial tension of a pure fluid that we discussed in Chapter 1 (see Figure 1.2). A solute molecule in the bulk interacts with neighboring solvent molecules within the range of the dominant molecular interaction forces, i.e. typically within a few molecular diameters. Within this range, the presence of the solute molecule alters the structure of the solvent leading to a so-called solvation shell, as sketched in the right-hand inset of Figure 3.1. The formation of this solvation shell involves a certain solvation energy that is characteristic for the specific combination of solute and solvent. Similarly, interfaces are accompanied by solvation layers with a fluid structure that is different from the bulk. Like interfacial tensions, solvation energies are free energies that contain both enthalpic and entropic contributions. If a solute molecule approaches an interface, the solvation shell of the molecule and the solvation layer of the interface overlap. As a consequence, the solvation structure and hence the total solvation energy change. If the free energy decreases, the molecule adsorbs spontaneously to the interface and interfacial tension, i.e. the free energy per unit area of the interface decreases. Vice versa, if the combined free energy of solvated solute and interface increases with decreasing separation, the solute will be depleted from the interface and the interfacial tension increases. This general principle applies to all types of interfaces, i.e. to solid–liquid, liquid–liquid, and liquid–air interfaces that we encounter in EW experiments.

Two basic questions arise: How is the fraction of adsorbed molecules related to the bulk concentration? How is the variation of the interfacial tension related to the amount of adsorbed molecules?

3.1.2 Langmuir Adsorption

Conceptually, it is useful to consider the system as being split in two parts, the solute molecules that are dissolved in the bulk of the liquid and those that are adsorbed to the interface. In a typical experiment, the solute concentration in the bulk is fixed, and we are interested in the amount of adsorbed solute as a function of the bulk concentration. Crystalline solids typically display a certain density Γ_0 of surface sites per unit area to which the solute species can adsorb; see Figure 3.1. The adsorption and desorption process can then be seen as a chemical reaction

$$SX \leftrightarrow S + X \tag{3.1}$$

in which the solute X and the surface site S can either be separated or form a surface complex SX. If Γ solute molecules per unit area adsorb to the interface, a fraction $\alpha = \Gamma/\Gamma_0$ of the surface sites is occupied. The goal of our calculation is to determine that fraction. In the case of amorphous solids and liquid–liquid or liquid–air interfaces, the definition of the site density is less obvious. Yet, even if the surface does not provide any density of sites, there is still a certain maximum density of molecules per unit area at the interface that is determined by the maximum packing density. For instance, for surfactant molecules with a straight hydrocarbon chain as hydrophobic tail, the maximum packing density is $\Gamma_0 \approx 5\,\text{molecules/nm}^2$, corresponding to $\approx 10^{-5}\,\text{mol m}^{-2}$.

For the derivation of the so-called Langmuir model of adsorption, we assume that the adsorption of each solute molecule follows a reaction as in Eq. (3.1) completely independent of all the other molecules. This assumption implies that we neglect possible lateral interaction among adsorbed molecules. The arguably easiest way to derive the desired relation between surface coverage and bulk concentration is to consider the system from the perspective of a dynamic equilibrium between adsorption and desorption. We assume that Γ solute molecules per unit area are adsorbed. If each adsorbed molecule desorbs with a desorption probability k_d per time unit, the total desorption rate per unit area and time can be written as

$$v_d = k_d \Gamma \tag{3.2}$$

Conversely, the adsorption rate is controlled by an adsorption rate constant k_a and by the bulk concentration c of the solute. In this case, however, we must take into account the finite adsorption capacity of the interface: An arriving solute molecule can only adsorb if it encounters a vacant adsorption site. The density of the latter is given by $\Gamma_0 - \Gamma$. Hence, the adsorption rate is given by

$$v_a = k_a c\,(\Gamma_0 - \Gamma) = k_a c\,\Gamma_0(1-\alpha) \tag{3.3}$$

The term $(1-\alpha)$ guarantees that α cannot exceed unity. In equilibrium, adsorption and desorption balance each other to guarantee a steady surface coverage. Equating the expressions Eqs. (3.2) and (3.3), we find

$$\alpha = \frac{\Gamma}{\Gamma_0} = \frac{c/K}{1 + c/K} \tag{3.4}$$

where we introduced the equilibrium constant $K = k_d/k_a$. This expression is known as Langmuir adsorption isotherm. (Note that there is a certain arbitrariness in the definition of K. It would be equally possible to define K as the inverse of our definition here. Whenever you read an article or make use of tabulated equilibrium constants, make sure to verify how it is defined.) In the dilute limit,

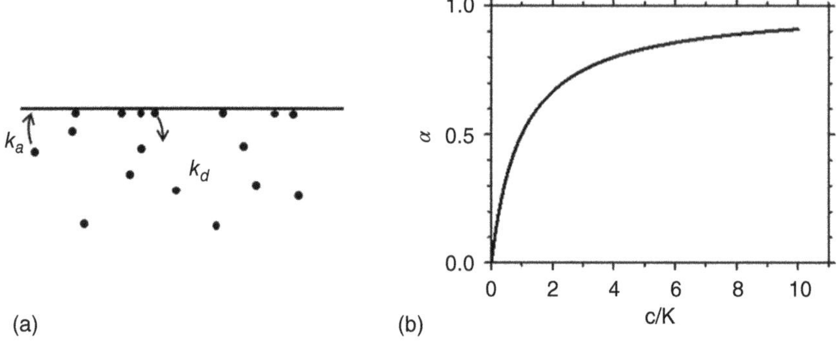

Figure 3.2 (a) Schematic illustration of adsorption and desorption at an interface. (b) Langmuir adsorption isotherm showing the fractional adsorption as a function of the ratio of the bulk solute concentration c and the equilibrium constant K.

$c/K \ll 1$, the adsorbed molecules do not hinder each other upon adsorption. In this case, known as Henry's regime, α simply increases linearly with the bulk concentration. Conversely, for higher concentrations, $c/K \gg 1$, α saturates as empty sites on the surface become scarce (see Figure 3.2). The steepness of the initial increase depends on the equilibrium constant.

To elucidate some of the implicit assumptions in our treatment and to provide a more intuitive understanding of the equilibrium constant K, it is instructive to re-derive Eq. (3.4) from slightly more general perspective involving equilibrium arguments rather than kinetics. We will do so in two different manners.

First, we return to the adsorption reaction equation, Eq. (3.1). Like in every equilibrated chemical reaction, the location of the equilibrium is determined by the corresponding equilibrium constant K and the concentrations of the species involved. Equilibrium concentrations and surface coverages follow the law of mass action:

$$K\{SX\} = \{S\}[X] \tag{3.5}$$

Here, square brackets indicate bulk concentrations and curly brackets indicate surface coverages. Because every surface site is either empty or occupied, Eq. (3.5) is complemented by a conservation law:

$$\Gamma_0 = \{SX\} + \{S\} \tag{3.6}$$

Combining Eq. (3.5) and Eq. (3.6) and identifying $\{SX\} = \Gamma$ and $[X] = c$, we recover Eq. (3.4). In this case, however, we did not make use of adsorption or desorption rates that require conceptual ideas about nonequilibrium processes, but we arrive at our result purely based on equilibrium arguments. From Eq. (3.5), we can see directly that the equilibrium constant corresponds to the specific value of the bulk concentration $[X]_{1/2}$, for which half of the surface sites are occupied. This relation provides an intuitive interpretation of the equilibrium constant K, and the corresponding pK value $pK = -\log_{10}K$ that is frequently found in chemistry reference tables and databases. A high pK value thus indicates that adsorption takes place already at very low concentrations. We will discuss specific examples of adsorption reactions in the context of ion adsorption in the next chapter.

Alternatively to using the law of mass action, we can consider the problem from the perspective of statistical physics and thermodynamics. If the bulk and the surface are two equilibrated subsystems of a bigger thermodynamic ensemble that can exchange solute molecules, the chemical potential of the solute molecules in the bulk μ_b and of those at the surface μ_s must be equal. For dilute solutions, we can write the bulk chemical potential as

$$\mu_b = \mu_b^0 + k_B T \ln c/c_0 \tag{3.7}$$

where μ_b^0 is the reference chemical potential involved in the solvation of an individual solute molecule at the reference concentration c_0 and the second term is the configurational entropy. Similarly, we can write for the surface contribution

$$\mu_s = \mu_s^0 + k_B T \ln \frac{\alpha}{1-\alpha} \tag{3.8}$$

with the surface reference chemical potential μ_s^0. Here, the second term due to the configurational entropy is not only ln α, but ln $\alpha/(1-\alpha)$. This correction term represents the *mixing* entropy of occupied and empty surface sites. (We elaborate on the derivation of this expression and other aspects of the statistical mechanics approach in the appendix of this chapter.) By equating Eqs. (3.7) and (3.8), we recover once again the Langmuir isotherm, Eq. (3.4).

Comparison to the preceding derivations shows that the equilibrium constant is related to the difference in chemical potentials $\Delta\mu^0 = \mu_s^0 - \mu_b^0$ between an individual dissolved solute molecule in the adsorbed state on the surface and the reference state in the bulk:

$$K = c_0 e^{\Delta\mu^0/k_B T} \tag{3.9}$$

Strong surface affinities correspond to large negative values of μ_s^0 and hence large negative values of $\Delta\mu^0$. This leads to small values of K and thus a substantial surface coverage even at low bulk concentrations, as expected. This expression also provides an intuitive link to our original definition of K in terms of adsorption and desorption rate constants, $K = k_d/k_a$. If the desorption rate constant is small compared to the adsorption rate constant, we obtain a small equilibrium constant and hence strong adsorption. In contrast to the equilibrium analysis, however, the kinetic analysis requires knowledge about the pathway of the adsorption (and desorption) process that may involve an energy barrier ε, as sketched in Figure 3.3. Kinetically, the Arrhenius-type adsorption and desorption rates are given by $k_a \propto \exp(-\varepsilon/k_B T)$ and $k_d \propto \exp(-(\varepsilon - \Delta\mu^0)/k_B T)$, respectively. Only upon calculating the equilibrium constant the barrier ε drops out again.

3.1.3 Reduction of Surface Tension

Once we know the surface coverage, we need to determine the resulting variation of the interfacial tension. The result is obtained using the Gibbs thermodynamic adsorption equation:

$$d\gamma = -\Gamma\, d\mu \tag{3.10}$$

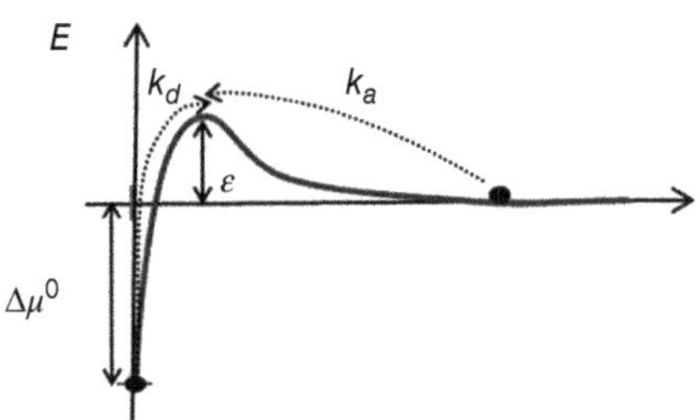

Figure 3.3 Adsorption energy and adsorption barrier experienced by a solute molecule in the vicinity of a surface. The abscissa is a general reaction coordinate.

The equation states that any increase of the chemical potential of a solute species by $d\mu$, e.g. due to an increase in the bulk concentration, gives rise to a reduction of the interfacial tension by a proportional amount for each solute molecule (per unit area) that is adsorbed to the interface. Using the relation between Γ and μ that we derived above, the Gibbs adsorption equation allows us to calculate the decrease of the interfacial tension as a function of the bulk concentration.

In equilibrium, μ is the same in the bulk and for the adsorbed solute molecules. It is thus fixed by the bulk concentration. For ideal solutions, we can use Eq. (3.7) to find

$$d\gamma = -\Gamma k_B T \, d\ln c = -\frac{k_B T \Gamma}{c} dc \tag{3.11}$$

Rearranging this expression, we find

$$\Gamma = -\frac{c}{k_B T} \frac{d\gamma}{dc} \tag{3.12}$$

Because we already know $\Gamma(c)$ from the Langmuir adsorption isotherm, Eq. (3.4), we can integrate Eq. (3.12) and obtain

$$\gamma(c) = \gamma_0 - k_B T \Gamma_0 \ln(1 + c/K) \tag{3.13}$$

where we denote the intrinsic surface tension of the pure solvent as γ_0. For $c/K \ll 1$, this leads to a linear decrease of the surface tension with increasing bulk concentration. We can also rewrite Eq. (3.13) as

$$\gamma(\alpha) = \gamma_0 + k_B T \Gamma_0 \ln(1 - \alpha) \tag{3.14}$$

which relates the interfacial tension directly to the fractional coverage α. Such a relation is known as the two-dimensional equation of state of the adsorbate, in analogy with three-dimensional equations of state, such as the ideal gas law relating the density to pressure. Similarly, we can interpret Eqs. (3.13) and (3.14) in a mechanistic manner: In Section 1.1, we discussed how the interfacial tension arises from the interaction amongst the solvent molecules at the interface. An ensemble of adsorbed solute molecules behaves like a two-dimensional gas at the interface. These adsorbed molecules tend to spread across the available interfacial area. This results in a (positive) two-dimensional pressure, the so-called surface pressure Π that can be measured, e.g. by placing a mobile barrier on the liquid surface and measuring the force acting on it. The corresponding stress is oriented in the opposite direction as the tensile surface tension forces. Hence, adsorption

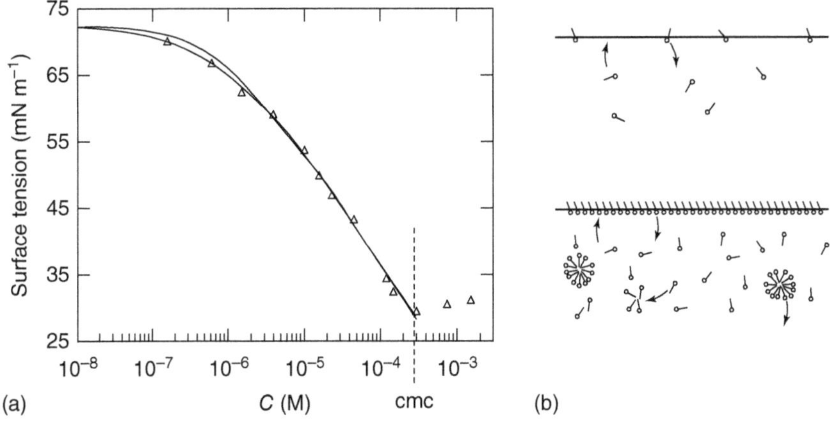

Figure 3.4 (a) Surface tension γ of water versus bulk concentration of a common uncharged surfactant (Triton X-100) along with a fit to Eq. (3.13). Note that γ seizes to decrease beyond the critical micelle concentration (cmc) of $\approx 3 \times 10^{-4}$M. Source: Chang and Franses 1995 [4]. Reproduced with permission of Elsevier. (b) Illustration of dissolved and surface-adsorbed surfactant molecules. Top: At concentration below the cmc. Bottom: Above the cmc, a dense layer of adsorbed surfactant coexists with a dense solution and micelles, spherical aggregates of surfactant molecules within the bulk solution.

of solute molecules to an interface reduces the net tension. The positive surface pressure is therefore defined as $\Pi = \gamma_0 - \gamma(c)$. For small α this definition results in $\pi = k_B T\Gamma$, as we can easily verify by expanding Eq. (3.14). This equation is the two-dimensional analog to the equation of state $p = \rho k_B T$ of an ideal gas of density ρ in three dimensions.

Figure 3.4 shows the dependence of the surface tension of water as a function of the bulk concentration for a common strongly adsorbing surfactant. Note that γ decreases by a factor of 2–3 for surfactant concentrations well below 1 mM. Provided that the concentration is not too high, the decrease in γ is indeed well represented by the prediction of Eq. (3.13) over several orders of magnitude. We will address the deviations occurring at concentrations beyond the so-called critical micelle concentration (cmc) briefly in Section 3.3. They are related to the formation of micelles, the globular aggregates of surfactant molecules in the bulk solution sketched in the lower part of Figure 3.4b.

3.2 Adsorption Kinetics

So far, we considered equilibrium aspects of interfacial adsorption. In EW devices, however, it is common to generate and displace drops or to generate new species by chemical reactions within a drop. The question that arises then is how long it takes before solutes reach the interfaces within the system and assume the equilibrium distribution discussed in the preceding section. Macroscopic flow fields involved in drop generation and displacement can play a role in bringing, *advecting*, solutes toward the interface. We will discuss this

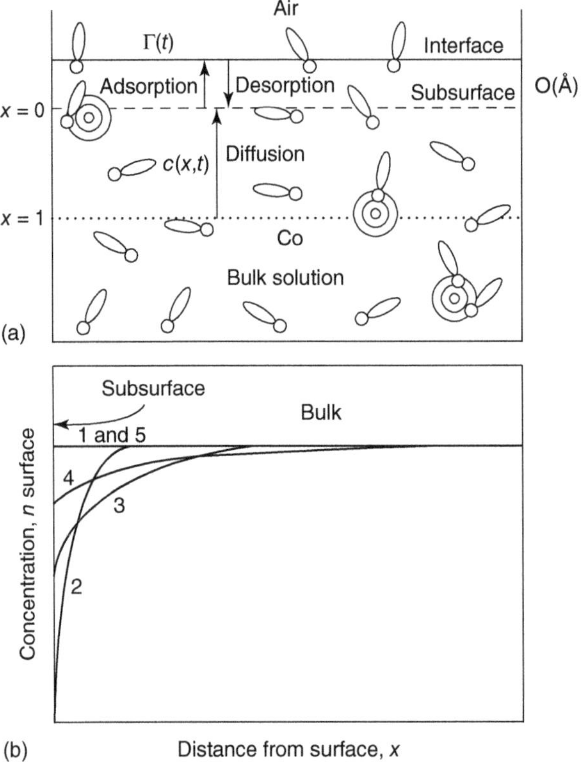

Figure 3.5 (a) Illustration of kinetic adsorption involving bulk solution, diffusive boundary layer, subsurface layer, and interface. Source: Chang and Franses 1995 [4]. Reproduced with permission of Elsevier. (b) Concentration profile $c(x, t)$ as a function of the distance from the interface for various time steps (1: initial condition; 2: immediately after interface creation; 3, 4: intermediate times; and 5: final state).

aspect in some detail in Section 8.4 in the context of microfluidic mixing. Yet, close to interfaces, the macroscopic flows generally cease, as expressed by the hydrodynamic no-slip boundary condition (see Section 6.1). As a consequence, there is always a so-called boundary layer, in which molecular diffusion is the dominant transport mechanism by which the solute molecules have to reach the interface. This situation is sketched in Figure 3.5a. The time it takes for solute molecules to equilibrate with a freshly formed interface is therefore governed by a combination of diffusive transport toward the surface and the specific processes involved in the final actual adsorption process on a molecular scale. The latter may involve adsorption barriers (see Figure 3.3) caused by microscopic phenomena, such as molecular interaction forces, rearrangement of solvation shell, conformational changes, etc. If the bulk concentration is low or if the solute molecules diffuse slowly (e.g. because they are large), the total flux toward the interface and hence the overall adsorption kinetics is called diffusion limited. Vice versa, if diffusion is fast and adsorption barriers become relevant, the adsorption kinetics can become *barrier limited* or *adsorption limited*.

If we create a new drop or displace an existing drop to a new location in an EW device, the new interface is initially free of adsorbates, $\Gamma(t = 0) = 0$, and the solute concentration in the liquid is c_0 everywhere. This situation is described by a flat horizontal concentration profile, curve 1 in Figure 3.5b. If we neglect adsorption barriers, all solute molecules in the *subsurface* layer in the immediate vicinity of the interface adsorb instantaneously to the interface. Therefore, the solute becomes depleted in the subsurface layer (curve 2). This leads to a concentration gradient between the subsurface and the rest of the solution that results in a diffusive flux. As a consequence, the solute concentration decreases in a depletion zone close to the surface, while the concentration in the subsurface layer directly at the surface gradually increases again (curves 3 and 4). As time goes by, the depletion zone gets wider and wider and the concentration gradients decrease. For an infinite bulk reservoir, the solute concentration in the final state equals the bulk concentration again, and the surface coverage Γ and hence the interfacial tension γ have reached their equilibrium values corresponding to the adsorption isotherm derived above. (In practice, molecular interactions may lead to a slight concentration gradient in the immediate vicinity of the interface. The thickness of such a zone will be determined by the range of the dominant molecular interaction.)

Mathematical treatment of the full diffusion problem leads to a general solution for the time-dependent surface coverage:

$$\Gamma(t) = 2c_0\sqrt{\frac{Dt}{\pi}} - \sqrt{\frac{D}{\pi}}\int_0^t \frac{c(0,\tau)}{\sqrt{t-\tau}}d\tau \tag{3.15}$$

Here, D is the diffusion coefficient of the solute and $c(0, \tau)$ is the concentration in the subsurface at time τ. For short times, the integral is negligible, and we find that the coverage increases as

$$\Gamma(t) = 2c_0\sqrt{\frac{Dt}{\pi}} \tag{3.16}$$

The coverage thus increases initially very quickly as $t^{1/2}$. The quicker the initial increase, the higher the bulk concentration c_0, and it increases with the diffusion coefficient. Note that the diffusion length $\sqrt{Dt}$ is the characteristic distance explored by a diffusing solute molecule during the time t. It provides a measure for the time-dependent characteristic thickness of the depletion zone in Figure 3.5b.

The consequences of gradual surfactant adsorption can be observed directly in EW experiments. Short times after the formation of a new aqueous drop in ambient oil, the interface is bare and hence γ is rather high. As a consequence, the drop is rather stiff and reacts only weakly to the applied voltage in EW. With increasing drop age, Γ increases and hence γ decreases: The drop becomes *softer* and responds stronger to the applied voltage. Figure 3.6 shows an example of a measurement of dynamic interfacial tension $\gamma(t)$ as extracted from the drop age-dependent EW response of a water drop in an ambient surfactant solution in oil. Details of the procedure to extract γ from the EW response are discussed in Section 5.1.

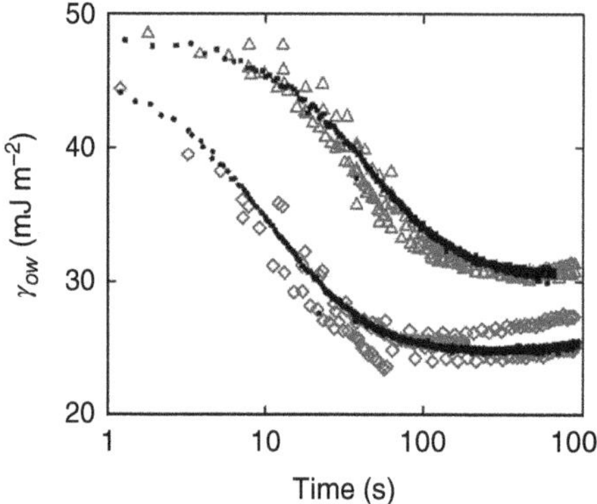

Figure 3.6 Oil–water interfacial tension versus elapsed time after formation of a water drop in an ambient solution of 0.001 wt% (bottom) and 0.0005 wt% (top) of Triton X-100 in heptane. Gray symbols: data extracted from EW measurements. Black symbols: reference measurement using pendant drop tensiometry. Source: de Ruiter et al. 2014 [5]. Reproduced with permission of Royal Society of Chemistry.

Table 3.1 Diffusion coefficient of common solutes in water at room temperature along with the characteristic times for diffusing 1 and 100 μm.

Solute	D (10^{-10} m^2 s^{-1})	τ_1 (s)	τ_{100} (s)
CO_2	19	0.000 5	5
Air	20	0.000 5	5
Ethanol	8	0.001 25	12.5
n-Decanol	1.7	0.006	60
Triton X-100	2.6	0.003 8	38
BSA (protein)	0.2	0.05	500
λ-DNA	0.0057	1.74	17 400
10 nm silica bead	0.2	0.05	500
1 μm PS bead	0.002	5	50 000

Characteristic values of diffusion coefficients for common solute molecules are provided in Table 3.1. The diffusion coefficient typically decreases with increasing size of the solute. For colloidal particles, D decreases inversely with the radius a according to the Stokes–Einstein relation $D = k_BT/6\pi\mu a$, where μ is the viscosity of the solvent. Randomly coiled polymers, such as DNA, typically diffuse like small colloidal particles with a radius corresponding to their radius of gyration $\langle r_g \rangle$. According to standard polymer physics, $\langle r_g \rangle$ scales with the number of monomers as $\propto N_m^{0.58}$; hence the diffusion coefficient scales as $D \propto N_m^{-0.58}$. For DNA, one finds, for instance, $D \cong 3 \times 10^{-10} N_m^{-0.58}$ m^2 s^{-1}. Table 3.1 also shows the characteristic time $\tau = x^2/D$ required to diffuse over distances of $x = 1$ and 100 μm. The table clearly shows that large molecules and particles clearly do not mix by diffusion on the time scale of typical experiments. Therefore, active measures need to be taken to reduce the distance that needs to be covered by diffusion in order to promote mixing and enhance the speed of a chemical reaction (see also Chapter 8).

3.3 Surface-Active Solutes: From Surfactants to Polymers, Proteins, and Particles

Throughout the preceding sections, we implicitly assumed that the solute molecules display a finite solubility in the liquid and a positive affinity to the interface. Moreover, we assumed that the solute molecules in the bulk and the ones adsorbed to the interface eventually reach thermodynamic equilibrium. These assumptions are reasonable and frequently fulfilled for small solute molecules. Yet, situations in which one or more of these conditions are not fulfilled are equally common, in particular for larger solute molecules, as we will illustrate in the following with a few examples.

The most generic example of strongly adsorbing surface-active solute molecules is surfactants. Surfactant molecules are amphiphilic, i.e. they consist of a polar head group and a nonpolar tail. The former provides affinity to the aqueous phase, and the latter to any nonpolar environment such as air, oil, and hydrophobic solids. The head group can be charged or simply contain a number of polar bonds, such as $—CH_2—O—CH_2—$. The tail typically consists of an aliphatic hydrocarbon chain that may or may not be branched. Typical examples are fatty acids and their salts, such as sodium stearate ($CH_3(CH_2)_{16}COO^-Na^+$), a basic ingredient of common soap. More recently, fluorinated tails have become more popular in droplet-based microfluidics as well.

The affinity of solute molecules to interfaces increases with increasing hydrophobicity of the tail and increasing hydrophilicity of the head group. As a consequence, surfactant molecules with the same head group but a longer hydrocarbon tail adsorb more strongly to interfaces. This generally leads to a more pronounced reduction of the corresponding interfacial tension. Figure 3.7 illustrates this. It shows the surface pressure $\Pi(c) = \gamma_0 - \gamma(c)$ for a series of homologous alcohols, i.e. for $HO—(CH_2)_n—CH_3$ with $n = 3, \ldots 7$, as a function of the bulk concentration of the alcohol in bulk water. As the graph shows, for a fixed bulk concentration, π increases with increasing hydrocarbon chain length. Conversely, this implies that the interfacial tension drops more strongly with

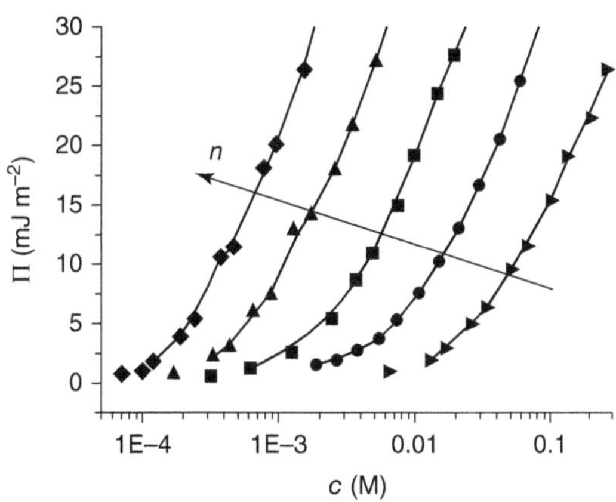

Figure 3.7 Surface pressure $\Pi = \gamma_0 - \gamma(c)$ versus bulk concentration for several n-alcohols adsorbed to an air–water interface. Butanol, pentanol, hexanol, heptanol, and octanol. Source: Posner et al. 1952 [6]. Data taken with permission of Elsevier.

increasing c. This reflects the stronger affinity of the longer molecules to the interface.

Increasing the length of the aliphatic chain, however, decreases the bulk solubility of the surfactants. Beyond a certain critical concentration, the surfactant molecules no longer dissolve as individual molecules in water. Instead, they spontaneously form aggregates called *micelles* of several tens or hundreds of molecules. The molecules in the micelles are arranged in such a manner that the hydrophobic tails point toward each other, while the hydrophilic head groups point toward the ambient aqueous phase, as sketched in the bottom of Figure 3.4b. The concentration, at which micelles first occur is called the cmc. It assumes a well-defined value for each surfactant that can vary from fractions of millimolars to many millimolars, depending on the specific properties of the molecules. Because the occurrence of micelles indicates the solubility limit of individual molecules in the bulk, it also limits the chemical potential of the solute in the bulk, μ_b^0. μ_b^0, however, controls the adsorption to the interface and the reduction of the interfacial tension, as discussed above. The fact that the interfacial tension becomes independent of c for $c >$ cmc, as shown in Figure 3.4, is thus a consequence of the fact that μ_b^0 cannot be increased any further by addition of more solute to the solution.

One important function of surfactants in microfluidics is to facilitate emulsification and to stabilize emulsions. The generation of emulsions, i.e. dispersions of small drops of oil in water (or vice versa), is facilitated by surfactants because they reduce the energetic costs of forming large areas of oil–water interfaces. Moreover, surfactants also stabilize emulsions by suppressing coalescence. To illustrate the mechanism, we consider two oil drops in water stabilized by a water-soluble surfactant with $c >$ cmc. If the two drops were to merge into a larger one, the total area available to each surfactant molecule at the oil–water interface would decrease because of the decreasing surface-to-volume ratio upon coalescence. Coalescence would thus imply that the surfactant molecules would have to be forced into an energetically less favorable, highly compressed state, from which the molecules cannot easily escape, because the bulk solution is already saturated with molecules. The only escape would be to form additional micelles. This requires a series of complex steps involving a cooperative rearrangement of many molecules, which takes time. While coalescence remains thermodynamically favorable and the emulsion remains intrinsically unstable, surfactants thus provide kinetic stabilization by preventing coalescence within a finite time. In addition to the unfavorable compression, molecular-scale forces such as electrostatic repulsion between charged head groups can further enhance the surfactant-induced stabilization of emulsions. Similar factors are at play when creating dispersed particles as well (e.g. a stable pigment dispersion).

Larger solutes such as polymer molecules and colloidal particles often display strong interfacial activity as well. Given their larger size, their affinity to interfaces is generally much stronger than for simple small molecules, unless they have a very strong preference for one of the two liquid phases. (For colloidal suspensions the latter case corresponds to complete wetting of the colloidal particles by one of the liquids.) For a colloidal particle with a finite contact angle, we can easily estimate the energy gain upon adsorption to an oil–water interface.

On dimensional grounds, it is clear that the adsorption energy must be of order $E_{ads} \approx \gamma a^2$. Considering a particle with $a = 20$ nm and a typical interfacial tension of $\gamma = 20\ \mathrm{mJm^{-2}}$, we find $E_{ads} = 10^{-19} J \gg k_B T$. This is very different from the case of surfactant molecules where thermal energy is sufficient to establish a thermodynamic equilibrium between adsorbed and dissolved molecules. For particles with surface affinities much larger than $k_B T$, adsorption is irreversible, and equilibration does not take place. As a consequence, adsorbed colloidal particles are very good stabilizers for emulsions. Particle-stabilized emulsions are known as Pickering emulsions. In Problem 3.5 we calculate the equilibrium configuration of a colloidal particle at a liquid interface as a function of the contact angle θ_Y.

The adsorption of polymer molecules, proteins, and soft particles is even more complex than the one of colloids because of their internal degrees of freedom. Protein molecules, for instance, typically consist of hydrophobic parts and hydrophilic parts with many functional groups, of which some are charged. In their natural aqueous environment, the hydrophobic parts are *hidden* from the ambient water deep inside the protein molecule, and the hydrophilic parts are exposed to the ambient similar to the configuration of a surfactant micelle. The underlying hydrophobic interactions give rise to the complex shapes that many proteins take in solution and to their responsiveness to variations of pH and salt concentration. Upon adsorption to a hydrophobic surface, this structure can rearrange as well such that the hydrophobic parts face the surface. At the same time, the hydrophilic parts again guarantee a low *interfacial energy* with the ambient water. As a result of this freedom to adapt their molecular configuration, proteins generally display a strong affinity to hydrophobic surfaces, which can lead to serious problems in many practical applications including EW-based microfluidic devices. In fact, the combination of intrinsic heterogeneity and conformational flexibility allows proteins to adhere to almost any type of surface. This is the primary cause of biofouling, the formation of layers of biological material on many natural and man-made surfaces.

3.A A Statistical Mechanics Model of Interfacial Adsorption

The conditions describing the thermodynamic equilibrium of surfactant adsorption can be derived in a consistent manner from the perspective of statistical mechanics based on a general expression for free energy density of the system. While leading to the same results, namely, the Langmuir adsorption isotherm, Eq. (3.4), and the corresponding reduction of the interfacial tension, Eq. (3.14), this approach reveals some of the implicit assumptions of our earlier derivations in Section 3.1.

First of all, we consider again the splitting of our total system into the bulk solution and the adsorbed species on the surface. If we consider the total number of N surfactant molecules in the system at some fixed temperature T and volume V, we can write the total free energy F as a sum of a bulk contribution F_b and a surface contribution F_s:

$$F(V,T,N) = F_b(V,T,N_b) + F_s(A,T,N_s) = F_b(V,T,N-N_s) + F_s(A,T,N_s) \tag{A.1}$$

where A is the surface area. For the second equation, we made use of the conservation of the number of surfactant molecules $N = N_b + N_s$. In thermodynamic equilibrium the solute molecules distribute themselves between bulk and surface to minimize F, i.e. we require

$$\left.\frac{\partial F}{\partial N_s}\right|_{V,T} = \left.\frac{\partial F_b}{\partial N_s}\right|_{V,T} + \left.\frac{\partial F_s}{\partial N_s}\right|_{A,T} \overset{!}{=} 0 \tag{A.2}$$

Using the definition of the chemical potential $\mu = \partial F/\partial N$, this leads to the condition that the chemical potential of the solute molecules in the bulk μ_b must equal the one in the surface μ_s that we used already in Section 3.1. Focusing on the surface contribution, we can introduce the free energy area density f_s:

$$F_s(A, T, N_s) = \int dA\, f_s(T, \Gamma) \tag{A.3}$$

Realizing that the bulk part of the system is very large compared to the surface, we can safely assume that the chemical potential of the solute is completely determined by its bulk value, Eq. (3.7). In this case, the quantity that is minimized in thermodynamic equilibrium is the corresponding grand potential:

$$\widetilde{f}_s(T, \mu) = f_s(T, \Gamma) - \mu\Gamma \tag{A.4}$$

where $f_s(T, \Gamma)$ contains several contributions. First, there are contributions due to interactions of each molecule with the surrounding solvent and with the interface that we described by μ_s^0 in Eq. (3.8). Second, there may be contributions to the interaction of the adsorbed molecules among each other and/or with an external field. Finally, there are entropic contributions due to the configurational arrangement of the adsorbed molecules at the interface. Ignoring the second contribution for the time being, we can write

$$f_s(T, \Gamma) = \Gamma\mu_s^0 - Ts \tag{A.5}$$

where s is the configurational entropy per unit area of the adsorbed molecules. The configurational entropy is determined by the number of possible arrangements Ω of the adsorbed molecules on the surface:

$$s = \frac{S}{A} = \frac{k_B}{A} \ln \Omega \tag{A.6}$$

To calculate Ω we resort to a two-dimensional lattice model. We assume that the surface is divided into a large number $M = A\Gamma_0$ of potential surface sites, where Γ_0 is the maximum coverage of adsorbed molecules, i.e. the inverse area per molecule at close packing, as defined before (see Section 3.1.2). Ω is given by the number of distinct possibilities to distribute N_s particles – and the corresponding $M - N_s$ vacant sites – over these M sites. According to standard statistical mechanics counting statistics, we find (Figure 3.A.1)

$$\Omega = \frac{M!}{N_s!(M - N_s)!} \tag{A.7}$$

Hence, the entropy per unit area reads

$$s = \frac{k_B}{A} \ln \frac{M!}{N_s!(M - N_s)!} = -k_B \frac{M}{A}(\alpha(\ln \alpha - 1) + (1 - \alpha)(\ln(1 - \alpha) - 1)) \tag{A.8}$$

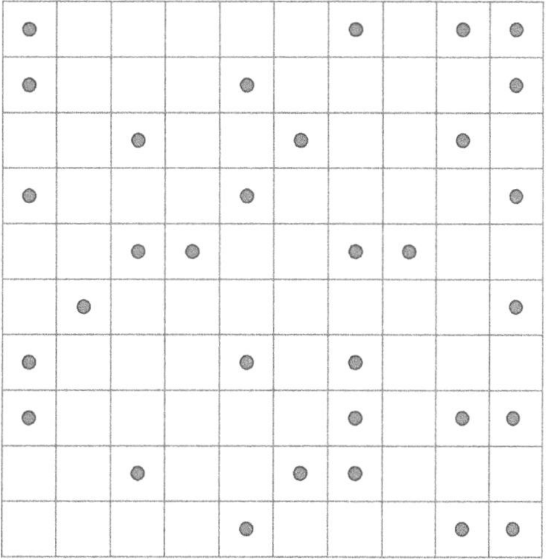

Figure 3.A.1 Schematic illustration of distribution N_s adsorbate molecules over M available surface sites in a two-dimensional lattice gas. Each lattice site has an area $1/\Gamma_0$.

For the last step, we made use of Stirling's formula $\ln M! = M(\ln M - 1)$ and the definition of the fractional coverage $\alpha = N_s/M$. This expression is the conventional mixing entropy. Inserting the result in Eq. (A.4), we find

$$\widetilde{f}_s(T, \mu) = \Gamma_0 \, [\alpha \mu_s^0 - k_B T[\alpha[\ln \alpha - 1] + [1 - \alpha][\ln[1 - \alpha] - 1]] - \alpha \mu] \quad \text{(A.9)}$$

For a given temperature and chemical potential μ, α adjusts itself such that $\widetilde{f}_s$ assumes a minimum, i.e. $\partial \widetilde{f}_s/\partial \alpha = 0$. This condition yields $\mu_s = \mu_s^0 + k_B T \ln \alpha/(1 - \alpha)$, the expression that we already gave in Eq. (3.8). To obtain the reduction of the free energy caused by the adsorption, we insert this result in Eq. (A.9) and obtain again $\gamma(\alpha) = \gamma_0 + k_B T \Gamma_0 \ln(1 - \alpha)$, our result from Eq. (3.14).

This derivation reveals an important implicit assumption in our earlier discussion: From Eqs. (A.5), (A.6), and the subsequent equations, it is very clear that our analysis treats the adsorbed molecules as a two-dimensional ideal gas without lateral interactions except for the fact that any surface site can only be either empty or occupied by one molecule. This assumption is frequently satisfied over a rather large range of concentrations, as illustrated in Figure 3.4. Yet, lateral attractions can also lead to the formation of adsorbate clusters at the interface that reduce surface chemical potential as compared with the ideal gas term. As a result, an increased surfactant coverage at the surface is observed at the same bulk concentration. Conversely, repulsive interactions reduce the coverage. Various extensions of the simple Langmuir isotherm have been studied and described in the physical chemistry literature, e.g. Frumkin isotherm, Davies isotherm, etc. (see, e.g. [4]), with corresponding two-dimensional equations of state. Equation (A.5) provides a clear pathway on how such additional interactions can be taken into account: They arise as interaction terms in addition to the entropy of the two-dimensional ideal gas. Furthermore, external fields, such as local electric fields, are equally readily included in Eq. (A.5), as we will see in the next chapter when we treat the electric double layer.

Problems

3.1 Consider the data in Figure 3.4. At which bulk concentration of Triton X-100 is the fractional coverage $\alpha = 1/2$? Assume a minimum area per adsorbed molecule of 0.17 nm^2. Estimate the surface coverage at the cmc.

3.2 Estimate the equilibrium constant K for the adsorption of Triton X-100 to the oil–water interface from the data shown in Figure 3.6. (The mass of Triton X-100 is 647 $g\,mol^{-1}$. Use again $1/\Gamma_0 = 0.17 nm^2$.)

3.3 Hexanol dissolved in water has an equilibrium constant of $K \cong 5 \times 10^{-3} mol\ l^{-1}$. At which bulk concentration is the surface coverage 10%, 50%, and 90%?

3.4 A drop of pure water with a volume of 40 nl is merged with a drop of an aqueous surfactant solution with a concentration of 5 mM and a volume of 10 nl. The equilibrium constant for adsorption to the air–water interface is $K = 1 \times 10^{-3} mol\,l^{-1}$, and the maximum packing density is $1/\Gamma_0 = 0.2\ nm^2$. What is the surface tension of the resulting mixed drop after equilibration? How long does it typically take for the drop to equilibrate if you assume that equilibration is limited by diffusion of the surfactant molecules across the drop (diffusion coefficient $D = 10^{-10} m^2 s^{-1}$)?

3.5 Consider a colloidal particle with a radius of 50 nm and a water contact angle of $\theta_Y = 30^{\circ}$. What is the equilibrium height of the particle with respect to the liquid surface? What is the minimum energy required to remove the particle from the interface?

3.6 Consider the kinetics of surfactant adsorption to a freshly created surface in contact with a bulk solution. (a) Show that the surface tension is expected to decrease $\propto \sqrt{t}$ at short times. (b) Estimate the diffusion coefficient D of Triton X-100 based on the data shown in Figure 3.6. (Use the surfactant properties as specified in Problem 3.2.)

References

1 Adamson, A.W. (1990). *Physical Chemistry of Surfaces*, 5e. New York: Wiley.
2 Hunter, R.J. (1993). *Introduction to Modern Colloid Science*. Oxford: Oxford University Press.
3 Butt, H.J. and Kappl, M. (2010). *Surface and Interfacial Forces*. Wiley.
4 Chang, C.H. and Franses, E.I. (1995). Adsorption dynamics of surfactants at the air/water interface – a critical-review of mathematical-models, data, and mechanisms. *Colloids Surf., A* 100: 1–45.

5 de Ruiter, R., Pit, A.M., de Oliviera, V.M. et al. (2014). Electrostatic potential wells for on-demand drop manipulation in microchannels. *Lab Chip* 14: 883.
6 Posner, A.M., Anderson, J.R., and Alexander, A.E. (1952). The surface tension and surface potential of aqueous solutions of normal aliphatic alcohols. *J. Colloid Sci.* 7: 623–644.

4

From Electric Double Layer Theory to Lippmann's Electrocapillary Equation

Having established the basic principles of capillarity, electrostatics, and interfacial adsorption, we are now in the position to address the electrical control of capillary phenomena, in particular the so-called electrocapillary effect, which was first discovered by Gabriel Lippmann in the late nineteenth century. After a short review of Lippmann's historic findings in Section 4.1, we extend our considerations of interfacial adsorption of solutes from the preceding chapter from neutral molecules to ions. Because ions carry charge, their distribution in the vicinity of the solid–electrolyte interface is coupled to the local electric field. This gives rise to the formation of the electric double layer (EDL) with a space-charge layer that screens any externally applied voltage. The energy of the EDL, which we calculate within the so-called Gouy–Chapman approximation, is essential for electrocapillarity and for all electrowetting (EW) experiments without dielectric. In Section 4.2, we discuss first the so-called diffuse part of the EDL within the framework of the Poisson–Boltzmann (PB) theory. This will lead us to the famous Young–Lippmann equation describing the contact angle reduction. In Section 4.3, we discuss some of the shortcomings of the PB model and provide an idea of how some of them are mitigated in more sophisticated models by considering a specific interactions and the formation of a *Stern layer* of adsorbed ions in analogy with Chapter 3. Finally, in Section 4.4 we discuss some aspects of ion adsorption and charge generation processes at hydrophobic–water interfaces as they are commonly used in EW.

4.1 Electrocapillarity: the Historic Origins

In Section 1.4, we discussed how the competition between capillary forces and gravity gives rise to capillary rise – or to capillary depression – depending on the value of the contact angle. Jurin's law, Eq. (1.26), describes that the observed height of rise is proportional to the interfacial tension at the meniscus. In the late nineteenth century, such measurements were first performed with liquid mercury. Mercury seemed to be a convenient liquid for such investigations because (i) being a metal it has a very large surface tension leading to a large height of rise and (ii) it displays – thanks to this large surface tension – almost universally a very large and constant contact angle close to 180°. Nevertheless, widely

Electrowetting: Fundamental Principles and Practical Applications,
First Edition. Frieder Mugele and Jason Heikenfeld.

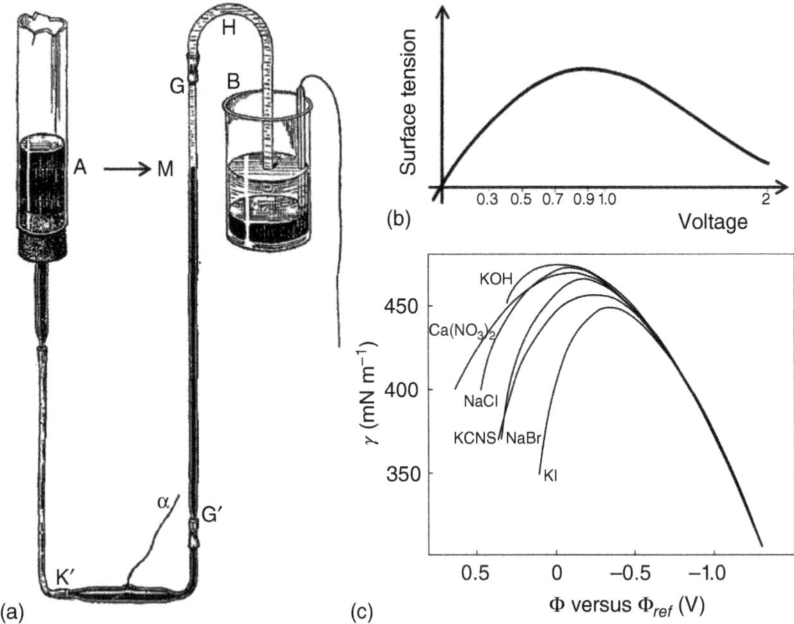

Figure 4.1 (a) Historic electrocapillarity setup by Lippmann for voltage-dependent capillary rise of liquid mercury–electrolyte interfaces. The *M* denotes the meniscus position; see arrow. Source: Adapted from [1]. (b) Surface tension versus applied voltage in units of *Daniels* as plotted by Lippmann for a mercury–sulfuric acid interface. (c) Set of more recent electrocapillary curves for various types of aqueous electrolytes. The right-hand side corresponds to negative potentials on the mercury. Source: Grahame 1947 [2]. Adapted with permission of American Chemical Society.

scattering results of the height of rise were found and tentatively attributed to surface contamination. In the 1870s, Gabriel Lippmann, who won the Nobel Prize in 1908 for the invention of an interferometry-based method for color photography, was interested in electrical phenomena. Motivated by the discrepancies in the capillary rise experiments with mercury, he performed a series of very systematic measurements with the liquid metal in contact with sulfuric acid in carefully cleaned glass capillaries; see Figure 4.1 [1]. As he applied bias voltages of the order of a few hundred millivolts between the mercury and the electrolyte solution, he found that the position of the meniscus varied. Analyzing his results using Jurin's law with the surface tension γ as a fit parameter, he found a characteristic voltage dependence of the mercury–electrolyte interfacial tension with a maximum at some specific voltage and a substantial decrease for both higher and lower voltages as seen in Figure 4.1b.

From his measurements and from an elegant thermodynamic analysis, he deduced the famous electrocapillary equation, nowadays also known as *Lippmann's equation*. It relates the charge density σ of the interface to the voltage dependence of the interfacial tension:

$$\sigma = -\frac{d\gamma}{dU} \tag{4.1}$$

This equation implies that the maximum of the interfacial tension in the experimental curves corresponds to the point of zero charge (PZC) of the interface. The fact that this maximum is generally found for nonzero voltage implies that chemical forces typically give rise to a finite interfacial charge even in the absence of any applied voltage. The negative sign in the equation is a consequence of the experimental observation that the height of rise decreases for voltages away from the PZC. Physically, it originates from the fact that the battery that keeps the voltage constant is part of the system. Therefore, the quantity to be minimized is the free energy, as discussed in Section 2.3.

In fact, Lippmann's equation became a widely used tool in electrochemistry because the resulting electrocapillary curves give direct access to the Gibbs free energy of the interface and therefore allow for detailed studies of adsorption processes at metal–electrolyte interfaces, in particular for liquid metals such as mercury. Measuring the dependence of the electrocapillarity response on the chemical nature of the ions was one of the most popular approaches for such investigations for decades during the twentieth century. Details on the thermodynamic derivations as well as the subtleties of the measurement and definition of electrostatic potentials and reference electrodes can be found in classical textbooks and review articles (e.g. [2, 3]).

In addition to this fundamental scientific impact, Lippmann's work also provides an impressive example of typical modern days' strive to develop practical applications based on novel scientific insight. In his groundbreaking paper from 1875, Lippmann describes an engine that used bundles of partially mercury-filled capillaries. Upon applying a suitable voltage, the mercury menisci moved up and down like pistons to convert electrical energy to mechanical motion. The engine achieved a maximum speed of 100 rpm. While little is known about the fate of this idea, Lippmann also developed a very sensitive electrometer based on electrically-driven capillary rise. Electrometers based on electrocapillarity remained an important measurement instrument for decades. Together with Étienne-Jules Marey, one of Lippmann's capillary electrometers was used to record the first human electrocardiogram ever. An English translation of Lippmann's seminal article, which was originally published in French, can be found as an appendix in Ref. [4].

In the following sections, we will discuss the properties of the EDL that forms at solid–electrolyte interfaces. We will focus in particular on metal–electrolyte interfaces, as used in the original electrocapillarity experiments. The key elements of the model, however, and the PB theory that we use to describe the distribution of ions are very general. It is the basis of many phenomena in colloid and interface science, including in particular EW on dielectrics.

4.2 The Electric Double Layer at Solid–Electrolyte Interfaces

Let us consider a solid surface immersed into an aqueous electrolyte. For simplicity, we assume that the solid is a metal such that we can control its electrostatic potential. Let us fix the potential of the metal at ϕ_0, and let us assume further that the bulk of the liquid is electrically grounded. Moreover, we exclude chemical reactions. Generally, this implies that the applied voltage should not exceed a few hundred millivolts, depending on the specific metal. Beyond this limit the onset of chemical *faradaic* reactions leads to electrical currents and to degradation of the system including in particular electrolysis of the aqueous solvent. Only in the absence of chemical reactions we can assume that the ions reach an equilibrium distribution. (In practice, guaranteeing the absence of faradaic reactions is *the* limiting factor of direct EW without dielectric.) Once a voltage is applied, the ions in the fluid redistribute to screen the electric field. Counterions with opposite sign of the applied potential are attracted toward the surface, and co-ions are repelled. The redistribution of ions leads to the formation of a space-charge layer, with a typical thickness of a few nanometers. This is called the diffuse part of the EDL. The redistribution of ions continues until the electric field in the bulk of the electrolyte vanishes. This is a basic consequence of the mobility of the charge carriers, as in any electric conductor. One important consequence of this screening effect is that the electric field vanishes in the bulk of the electrolyte. Thanks to the screening, neither the bulk electrolyte itself nor any dissolved molecule or suspended particle or biological cell experiences the presence of any externally applied voltage. This is crucial for many biological applications of EW.

4.2.1 Poisson–Boltzmann Theory and Gouy–Chapman Model of the EDL

The first goal of the present section is to calculate the thickness and the distribution of ions within the diffuse layer. Ideally, one might think that the counterions adsorb directly onto the solid surface. This, however, would lead to a very high – in fact diverging – local density of counterions and hence to a very strong (diverging) gradient of the counterion density between the surface and the bulk of the liquid. Such a strong gradient gives rise to a diffusive flux away from the surface that pushes some ions back into the liquid next to the surface. Thermodynamic equilibrium is reached when the electrically driven ionic fluxes due to the applied voltage and the diffusive flux balance each other. This balance determines the thickness of the diffuse part of the EDL.

For a quantitative mathematical analysis of the problem, we consider a laterally wide solid surface in contact with the fluid, as sketched in Figure 4.2. The distribution of the ions and of the electric field then simply depends on one geometric variable, the distance z from the surface. To be specific, let us assume we apply a negative potential to the electrode, and let us further assume that the electrolyte contains dissolved salt ions i with a bulk density c_i^∞ far away from the surface. For instance, we can think of $i = +, -$ representing Na^+ ions and Cl^- ions. In this case, we know that $c_+^\infty = c_-^\infty$. From statistical physics, we know that the ions will follow a Boltzmann distribution in thermal equilibrium. That is, we can write

$$c_i(z) = c_i^\infty \exp\left(\frac{-Z_i e\phi(z)}{k_B T}\right) \tag{4.2}$$

where Z_i is the valency of the ith ion species and $\phi(z)$ is the (unknown) electrostatic potential at a distance z from the surface. The condition that the ions

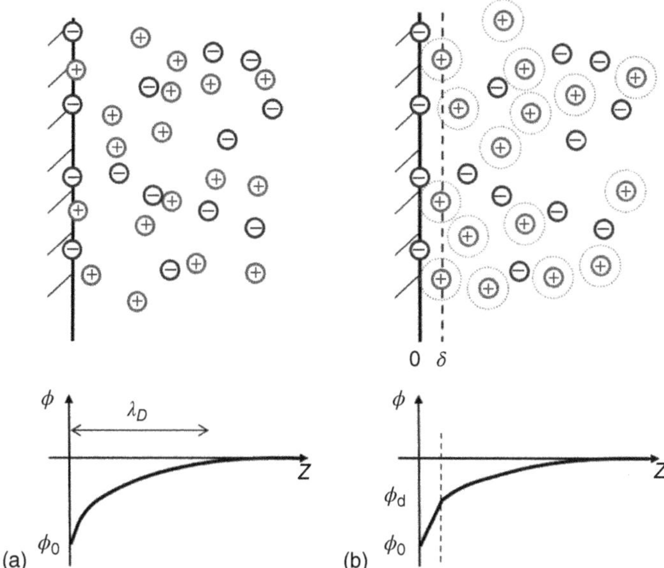

Figure 4.2 Schematic structure of the electric double layer including bare surface charges (top) and electrostatic potential distribution (bottom). (a) Gouy–Chapman model with diffuse layer (thickness: O ($\lambda_D \approx 0.1 - 100$ nm)) extending up to the surface. (b) Basic Stern model with one layer of adsorbed ions (here: cations) adsorbed at a distance $z = \delta$; dashed circles around cations indicate molecular details such as the finite ion radius or a hydration shell of solvating water molecules.

obey the Boltzmann distribution is equivalent to the statement that the electrochemical potential μ_i of each ionic species (Eq. (4.2)) is constant throughout the system, i.e.

$$\mu_i = k_B T \ln \frac{c_i(z)}{c_i^{\infty}} + Z_i e\, \phi(z) = \text{constant} \tag{4.3}$$

We know that the electrostatic potential $\phi(z)$ must become constant at infinity because the electric field inside the electrolyte must vanish. If we choose to define the potential $\phi(z)$ with respect to the bulk, i.e. if we choose $\phi(\infty) = 0$, the constant in Eq. (4.3) is also zero.

As a side remark, we note that we will use concentrations of ions and treat the electrolyte as an ideal diluted solution throughout this chapter. In practice, this approximation is only true at very low concentrations. At finite concentrations, electrolytes are nonideal, and the ions interact with each other. This is commonly taken into account by replacing the concentrations in the formulas by *activities*. Even at moderate concentrations of, say, 10 mM, activities can deviate from concentrations by 10% or 20%, depending on the specific salt.

So far, the distribution of $\phi(z)$ in the vicinity of the surface is unknown. It depends both on the potential (or charge density) at the solid surface and on the distribution of the ions according to Poisson's law of electrostatics (see Eq. (2.7)):

$$\frac{d^2\phi}{dz^2} = \phi''(z) = -\frac{\rho_{el}(z)}{\varepsilon\varepsilon_0} \tag{4.4}$$

For the electrolyte solution, the local electric charge density is determined by the distribution of the ions: $\rho_{el}(z) = \sum_i Z_i \, e \, c_i(z)$. In the following, we limit ourselves for the sake of simplicity to the case of a single species of a monovalent salt such as NaCl. In this case, we can write

$$\rho_{el}(z) = e(c^+(z) - c^-(z)) \tag{4.5}$$

(The formalism that we will discuss is easily extended to ions of arbitrary charge (see Problem 4.1), yet equations become somewhat more cumbersome.) The focus on one salt species implies that the concentration of these ions will be much larger than the ones of all other ions that may be present, including in particular free hydronium (H_3O^+) and hydroxyl (OH^-) ions. The ions of the salt are then the *potential-determining* ions. For the conditions of most EW experiments, it is indeed fair to assume that the concentration of the added salt ions is large compared with one of hydronium and hydroxyl ions and hence to neglect the contribution of the latter on the right-hand side of Eq. (4.4).

With these preliminaries discussed, we can combine Eqs. (4.5) and (4.4) to obtain the basic equation governing the electrostatics of electrolytes, the PB equation

$$\phi''(z) = \frac{2c_\infty e}{\varepsilon_0 \varepsilon} \sinh(e\,\phi(z)/k_B T) \tag{4.6}$$

As we can see, the PB equation is an ordinary but nonlinear second-order differential equation for the distribution of the electrostatic potential $\phi(z)$. It requires two boundary conditions to be solved. As already mentioned, we know that the electric field $E = -d\phi/dz$ must vanish for $z \to \infty$. Moreover, we typically either know the potential $\phi(0)$ on the surface (as in the case of a metal electrode), or the surface charge σ_0, which is then related to the electric field at the surface by Gauss' law. By solving the PB equation, we obtain a self-consistent solution for the electrostatic potential $\phi(z)$ and upon inserting the solution into the Boltzmann distribution, we find the ion distribution $c_\pm(z)$ within the diffuse part of the double layer.

Equation (4.6) can be rewritten more compactly if we introduce the nondimensional electrostatic potential $\Psi = e\phi/k_B T$ as

$$\frac{d^2\Psi}{dz^2} = \kappa^2 \sinh \Psi \tag{4.7}$$

where,

$$\kappa = (2c_\infty e^2/\varepsilon\varepsilon_0 k_B T)^{1/2} \tag{4.8}$$

is the Debye parameter. It has the dimensions of an inverse length. The corresponding length scale, the Debye length $\lambda_D = 1/\kappa$, is *the* characteristic length scale of most electrostatic phenomena in electrolytes. The thickness of the diffuse layer, the range of electric fields within the electrolyte, the range of electrostatic forces between charged objects, etc. all scale with λ_D. At a distance of a few times λ_D, they all vanish.

The expression in Eq. (4.8) contains all the ingredients representing the balance between electrostatics and (thermal) diffusion: $c_\infty e^2/\varepsilon\varepsilon_0$ measures the electrostatic energy per area between two unit charges in a dielectric medium at their average separation, and $k_B T$ is the thermal energy. Increasing salt concentration improves the screening and therefore reduces the characteristic length λ_D. The typical value of λ_D ranges from a few hundred nm in deionized water down to fractions of a nm for salt concentrations of 100 mM and beyond. (Deionized water in contact with air typically has a pH ≈ 6 due to dissolved CO_2 converting to bicarbonate and carbonate ions.) For monovalent salts in water at room temperature, it is convenient to remember the expression

$$\lambda_D = \frac{0.34\,\text{nm}}{\sqrt{c_\infty\,[M]}} \tag{4.9}$$

where c_∞ is the (bulk) salt concentration in M, i.e. moles per liter.

The character of electrostatic interactions is thus completely changed from a long-range force with a potential that decreases with the inverse of the distance in vacuum to a short-range exponentially decaying force in an electrolyte that is negligible at separations beyond a few nanometers for typical salt concentrations. This dramatic change is caused by the screening effect of the mobile ions that accumulate in the diffuse layer and prevent any (static) electric field from penetrating the solution. In contrast, in perfect dielectric liquids without free ions, electrostatic interactions retain their long-range character as in vacuum, only reduced by a constant overall factor ε due to *dielectric* screening. Interesting intermediate cases arise when small amounts of ionizable polar species are dissolved in dielectric liquids. In this case, electrostatics can become an exponentially decaying force with a rather long Debye screening length. Such situations can arise in EW devices if complex mixtures of liquids and solutes are used to match various requirements such as the matching of densities or refractive indices (see the materials in Chapter 7). Similarly, long-range penetrating electric fields can arise in the presence of high frequency AC fields. In that case, however, the system is not in equilibrium.

Despite its simple appearance, the PB equation, Eq. (4.7), is difficult to solve. Analytical solutions exist only for very few specific geometries, including the planar interface considered here. Yet, their derivation is mathematically somewhat tedious and does not offer much physical insight. We refer the interested reader to standard textbooks (e.g. [5]). The physical principles can be understood by considering the linearized version the PB equation, the so-called Debye–Hückel equation:

$$\Psi''(z) = \kappa^2 \Psi(z) \tag{4.10}$$

The Debye–Hückel equation is obtained by linearizing Eq. (4.7), i.e. by writing $\sinh\Psi \approx \Psi$. Hence, it is valid for $\Psi = e\phi/k_B T \ll 1$, i.e. for maximum values of the electrostatic potential $\phi < 25\,\text{mV}$. This voltage is rather small, even by the standards of EW without dielectric. Nevertheless, the Debye–Hückel approximation still provides qualitatively correct physical trends and thus valuable insights for understanding the behavior of the system.

The general solution of Eq. (4.10) is a superposition of exponential functions. Because we know that the electric field vanishes far away from the surface, i.e. $\phi' \to 0$ for $z \to \infty$, we know that ϕ decays exponentially:

$$\phi(z) = \phi_0 \exp(-\kappa z) \tag{4.11}$$

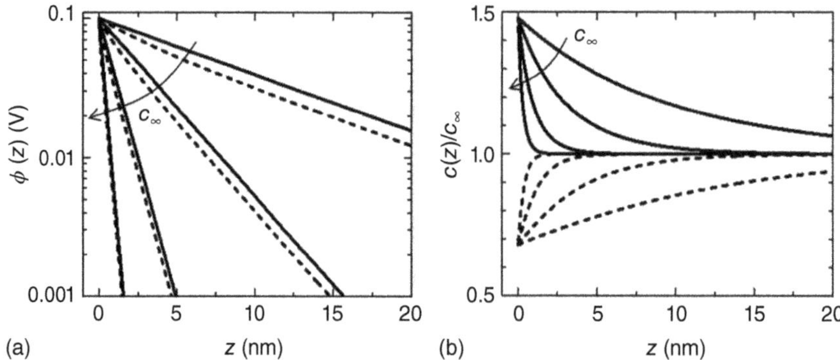

Figure 4.3 (a) Potential distribution for $\phi(0) = 0.1$ V (full lines: Debye-Hückel approximation; dashed lines: Poisson–Boltzmann theory) for variable bulk concentration (1, 10, 100, 1000 mM) of a monovalent salt. (b) Local ion concentration (for $\phi(0) = 0.01$ V) close to solid–liquid interface ; solid: counterions ($c/c_\infty > 1$); dashed ($c/c_\infty < 1$): co-ions. Same concentration range as in a).

Figure 4.3 shows potential and ion distribution within the diffuse part of the double layer for variable salt concentration. Note the distinct reduction of the range of the electric potential and hence the electric field with increasing salt concentration due to improved screening. Figure 4.3b) also shows that the ion distribution closely follows the potential distribution. The counterion density is enhanced close to the surface, and the co-ion density is reduced. This gives rise to a space-charge density in the diffuse layer that exactly compensates the charge on the electrode. As Figure 4.3a) shows, the solutions of the full PB equation look qualitatively similar to the linearized Debye–Hückel theory.

4.2.2 Total Charge and Capacitance of the Diffuse Layer

Global charge neutrality requires that the total charge in the diffuse layer exactly compensates whatever may be the charge on the surface σ_0. This implies that we can measure the charge on the surface by measuring the charge density σ_D in the diffuse layer. The solution of the PB equation provides a direct relation between the charge density in the diffuse layer and the potential drop across it. The total charge of the diffuse layer is obviously given by integrating the charge density

$$\sigma_D = \int_0^\infty \rho_{el}(z)dz \tag{4.12}$$

Here, we assumed that the diffuse layer stars directly at the surface, i.e. we assumed $\delta = 0$, according to the general idea of the Gouy–Chapman model (see Figure 4.2). The right-hand side depends on the potential distribution $\phi(z)$. Interestingly, it is possible to derive the relation between the charge density and the potential drop across the diffuse layer directly from the full PB equation without even knowing the exact solution (see Problem 4.2). Gouy and Chapman first obtained the expression

$$\sigma_D = -4c_\infty \lambda_D e \sinh \frac{e\varphi_D}{2k_BT} \tag{4.13}$$

(Sometimes, this equation is also denoted as Grahame's equation.) Here, ϕ_d is the potential at the edge of the diffuse layer next to the solid surface. For an ideal metal electrode as considered here, it is identical to the applied potential ϕ_0. Note that the prefactor in Eq. (4.13) provides an intuitive natural measure for the surface charge density: the bulk ion concentration times the characteristic thickness of the space-charge layer. Expanding the sinh function in Eq. (4.13) for small potentials, we find the corresponding solution for the Debye–Hückel equation, which we could have derived equally by integrating Eq. (4.12) using the exponentially decaying solution for $\phi(z)$ that we calculated above:

$$\sigma = \frac{\varepsilon\varepsilon_0}{\lambda_D}\phi_D \tag{4.14}$$

(Note that we dropped here the sign following standard conventions. When speaking about *the charge* on a capacitor, it is common to speak of the absolute value, understanding that there is usually the same amount of charge with opposite sign on the two electrodes of a capacitor. To be precise, $\sigma = \sigma_0 = -\sigma_D$ in Eq. (4.14).) This expression provides a linear relation between the potential difference ϕ_D and the charge density σ, like in a conventional capacitor ($Q = CU$). In Debye–Hückel approximation, the diffuse layer thus behaves like an ideal capacitor with a capacitance per unit area of $c_D = \varepsilon\varepsilon_0/\lambda_D$, i.e. like a parallel plate capacitor with a plate spacing λ_D and dielectric constant $\varepsilon\varepsilon_0$. Assuming, for instance, $\lambda_D = 10$ nm corresponding to a salt concentration of 1 mM for a salt consisting of monovalent ions, the resulting capacitance per unit area is $c_D \approx 0.07\ \mathrm{F\,m^{-2}}$. This gives rise to very large charge densities per unit area (see Figure 4.4). For the full PB theory, the relation between ϕ_d and σ_d is nonlinear. Analog to conventional electrical circuit theory, we can define a voltage-dependent differential capacitance $C_{diff} = dQ/dU$. From Eq. (4.13) we obtain the differential capacitance per unit area as

$$c_{diff} = \frac{d\sigma_0}{d\phi_d} = \frac{\varepsilon\varepsilon_0}{\lambda_D}\cosh\frac{e\phi_d}{2k_BT} \tag{4.15}$$

For $e\phi/k_BT \ll 1$, this approximation reduces to $c = \varepsilon\varepsilon_0/\lambda_D$, in line with Eq. (4.14), as expected.

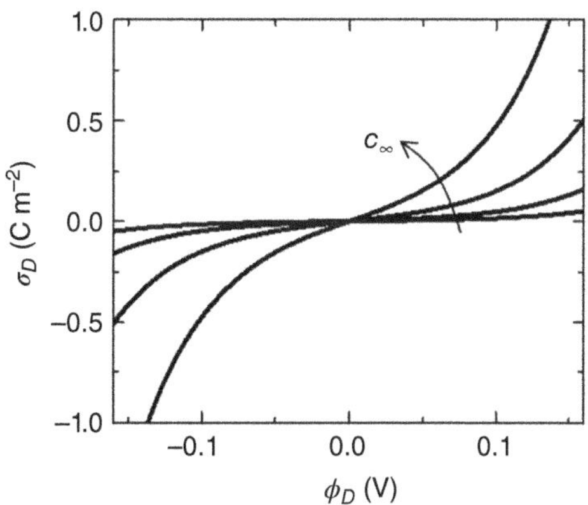

Figure 4.4 Diffuse layer charge as a function of diffuse layer potential according to Gouy–Chapman, Eq. (4.13), for concentrations of 1–1000 mM of a monovalent salt.

4.2.3 Voltage Dependence of the Free Energy: Electrowetting

To evaluate the consequences of these interfacial charging phenomena for wetting, we need to calculate the variation of the energy per unit area involved in building up the EDL as we apply the voltage. In general, this calculation starts from an expression of the free energy density that includes the electrostatic energy of the ions in the electric field calculated in the preceding section, their entropy, and the electric field energy. Integrating this free energy density over the thickness of the EDL leads to the desired expression for the (negative) excess free energy per unit area of the EDL. For the interested reader we provide the details of this calculation in the appendix of this chapter. Here, we make use of the results of the preceding section and insert them into Lippmann's general thermodynamic expression, Eq. (4.1). If we assume that the electrolyte is perfectly uniform all the way up to the interface at zero voltage, the variation of the interfacial energy upon applying a potential ϕ_d at the solid surface is given by

$$\gamma(\phi_d) = \gamma^0 - \int_0^{\phi_d} \sigma(\phi) d\phi \tag{4.16}$$

Inserting the Gouy–Chapman equation, Eq. (4.13), for the diffuse layer charge, we obtain

$$\gamma(\phi_d) = \gamma^0 - 8k_B T c_\infty \lambda_D \left(\cosh \frac{e\phi_d}{2k_B T} - 1 \right) \approx \gamma^0 - \frac{\varepsilon\varepsilon_0}{2\lambda_D} \phi_d^2 \tag{4.17}$$

The interfacial tension of the metal–electrolyte interface thus decreases with increasing potential difference across the EDL. Hereby, the PB description of the EDL provides a microscopic explanation of the global parabolic shape of the electrocapillary curves obtained by Lippmann (cf. Figure 4.1). In the linearized Debye–Hückel expression, the approximation on the right-hand side of Eq. (4.17), the free electrostatic energy, is simply given by the one of a parallel plate capacitor with capacitance c_D per unit area. Inserting typical numbers, we find $\Delta\gamma = -\frac{c_D \phi_d^2}{2} \approx -\frac{1}{2}\, 0.1 \frac{F}{m^2}\, 1\,\mathrm{V}^2 = -0.05\,\mathrm{J\,m^{-2}}$, which is qualitatively consistent with the electrocapillary curves shown in Figure 4.1c.

Lippmann, like most electrochemists following him in the early to mid-twentieth century was primarily interested in the properties of the EDL at metal–electrolyte interfaces and in the energy involved in forming such interfaces rather than wetting. Consequently, they designed their electrocapillarity experiments in such a way that the surfaces were always completely wetted by the aqueous phase even in the absence of any applied voltage. In this case, reducing the interfacial tension of the mercury–electrolyte interface upon applying a voltage only changed the height of rise in their capillary rise experiments but not the contact angle. Wetting experiments are conceptually more difficult because the contact angle is determined by the tensions of all three interfaces meeting at the contact line rather than just a single one, as expressed by Young's equation. They are also more difficult in a technical sense because they are hampered by heterogeneities on the interfaces, such as the roughness of a solid surface that can pin the contact line and prevent the contact angle from achieving its equilibrium value, as we discussed in Section 1.7. Only in the 1930s researchers mainly around the Russian (electro)chemist Frumkin (see [6]) performed systematic experiments exploring the consequences of electrocapillarity for wetting. Because solid metal surfaces were (and still are) difficult to prepare in a sufficiently clean and smooth manner for their measurements, they primarily studied air bubbles and oil drops on liquid mercury surfaces immersed in ambient electrolyte. Upon varying the potential between the mercury and the ambient electrolyte, they could produce variations of the contact angle of more than 150° for potential differences $\phi_D \approx 1$ V. The actual analysis of the experiments is slightly more complex than the wetting on a solid surface because both tangential and normal forces need to be balanced following Neumann's triangle (see Section 1.3).

Let us leave these complications aside and assume that equilibration of the contact line and thus the measurement of an equilibrium contact angle can be achieved directly at a (solid) metal–electrolyte in a bulk dielectric environment such as air or oil. To calculate the contact angle, we have to insert the interfacial tensions into Young's law. The external potential difference is applied between the electrolyte and the metal surface. Hence, it is the tension of this interface that decreases according to Eq. (4.17) as we apply the voltage. The tensions of the other interfaces are not affected because there is no substantial potential drop across them. Resorting to our convention from Chapter 1 to denote the drop phase by the letter "l," the ambient medium by a letter "v" no matter whether it is actually a vapor phase or another condensed immiscible dielectric such as an oil, and the metal electrode as a solid "s," we can write the voltage-dependent contact angle as

$$\cos\theta(U) = \frac{\gamma_{sv} - \gamma_{sl}(U)}{\gamma} = \cos\theta_Y + \frac{1}{\gamma}\int_0^U \sigma(\phi)d\phi \tag{4.18}$$

Here, we ignored some subtleties regarding the exact value of the electrostatic potential and its measurability with suitable reference electrodes and simply equated the applied voltage U to the potential difference ϕ_D across the diffuse layer. Readers interested in such aspects are referred to classical textbooks or review articles on electrochemistry [2, 3]. Upon inserting the approximate expression for $\sigma(\phi)$ at low voltage, Eq. (4.14), the equation assumes the characteristic form of an upward parabola as a function of the applied voltage, which is often denoted as Young–Lippmann equation:

$$\cos\theta(U) = \cos\theta_Y + \frac{\varepsilon\varepsilon_0}{2\lambda_D\gamma}U^2 \tag{4.19}$$

This key equation relates the electrostatic energy involved in the formation of the EDL on the scale of a few nanometers to the wettability, which is observed on macroscopic scales by the naked eye.

For EW on dielectric, we will recover the quadratic scaling of cosθ with the applied voltage. Yet, as we will see in Chapter 5, the relevant energy in that case is not the free energy of the EDL but the electrostatic energy in the macroscopic capacitor formed by the liquid, the electrode(s), and the dielectric layer separating them.

4.3 Shortcomings of Poisson–Boltzmann Theory and the Gouy–Chapman Model

The PB theory and Gouy–Chapman model of the EDL capture the overall decrease of the interfacial energy and the contact angle. Yet, our theory also misses out on a number of aspects. First of all, our model predicts both the contact angle and the interfacial tension to be maximum at zero applied voltage and to decrease symmetrically for both positive and negative applied bias voltages. The electrocapillary curves in Figure 4.1 clearly show that this is not correct. Moreover, these curves also display a considerable dependence on the type of ion, which is not included in the model in any way. In addition, a more detailed analysis of electrocapillary curves often reveals various *humps*, and differential capacity curves were found to deviate in a non-monotonic manner from Eq. (4.15); see ref. [2]. These shortcomings arise from the fact that the basic picture of continuum physics breaks down when the screening length λ_D approaches molecular scales. That is, the assumptions that we can describe charge and ion distributions as continuous functions $c_\pm(z)$ and $\rho(z)$ and that these functions are related to a meaningful *mean field* potential $\phi(z)$ according to the PB equation are no longer valid. In fact, the limitations of our model are reached rather quickly. For a bulk concentration $c_\infty = 0.1$ M, we have $\lambda_D = 1$ nm. Since the concentration of counterions can be enhanced by a factor of 10–100 upon approaching the surface, the EDL becomes very crowded. Upon approaching the solid surface, ions come very close together and – according to our Gouy–Chapman model – should even overlap (see also Problem 4.4). If neighboring ions come this close, the electrostatic energy is dominated by their direct Coulomb interaction with each other rather than the mean potential ϕ that appears in the PB theory. Moreover, additional short-range chemical interactions will start to play a role. Many models have been and are still being proposed to compensate various aspects of these problems, and in recent years, numerical simulations have become an increasingly popular tool to study such microscopic effects. Historically, one of the first approaches was to account for the finite size of the ions by considering them as hard spheres with some ion specific radius a, as schematically indicated by the dashed circles around the cations in Figure 4.2b. This gives rise to an excluded volume that prevents the ions from getting closer to the solid surface than a and closer to each other than $2a$. As a consequence, counterions accumulate in a layer, the *Stern layer*, at a distance δ from the surface with a maximum packing density corresponding to a two-dimensional close-packed lattice. In practice, δ can be the bare radius of the ion, or it may include a very tightly bound shell of hydration water.

In addition to the geometric exclusion because of finite size, ions may also display some chemical interactions with the solid surface even in the absence of an applied potential. In this case, ions spontaneously adsorb to the solid surface and generate a finite surface charge σ_0 at zero voltage. The adsorption process is governed by the same type of adsorption and desorption equilibria as described in Chapter 3, except that the local concentration of the adsorbing species, i.e. the ions, next to the interface is different from the bulk due to the electrostatic forces.

The finite spontaneous charge density and the corresponding potential give rise to a finite reduction of the interfacial tension as compared to the uncharged state. To compensate for this effect, a finite bias voltage, the so-called potential of zero charge U_{pzc} needs to be applied to discharge the interface. According to Eq. (4.1), this voltage of zero charge corresponds to the maximum of the interfacial tension. If spontaneous charging is properly taken into account, Eqs. (4.20) and (4.1) change to

$$\gamma(U) \approx \gamma^0_{pzc} - \frac{\varepsilon\varepsilon_0}{2\lambda_D}(U - U_{pzc})^2 \tag{4.20}$$

and

$$\cos\theta(U) = \cos\theta_Y + \frac{\varepsilon\varepsilon_0}{2\lambda_D\gamma}(U - U_{pzc})^2 \tag{4.21}$$

This explains the asymmetry of the electrocapillary curves. Taking into account that the affinity to the interface will be different for each specific ion, it is obvious that the value of U_{pzc} is also specific to each combination ion and surface. Hence, the model also includes an explanation for the observed ion specificity of electrocapillary curves. The quantitative calculation of electrocapillary curves and the corresponding contact angle variations requires implementation of a specific model of ion adsorption. An example of a generic treatment that captures the basic phenomenology is provided in Appendix 4.A of this chapter. A more detailed discussion of various types of Stern layer models is beyond the scope of this book. The interested reader can find discussions in specific textbooks and review articles (see e.g.[5, 7, 8]).

4.4 Teflon–Water Interfaces: a Case Study

Modern EW experiments involve interfaces between an electrolyte and a hydrophobic substrate, frequently an aqueous electrolyte and a fluorinated polymer. Similar charging processes as mentioned in the preceding section also take place at such hydrophobic–water interfaces. A large number of observations such as titration experiments and electrokinetic measurements (electrophoresis, streaming potential) show that such interfaces spontaneously assume a negative surface charge upon immersion in water at not too low pH (see Figure 4.5; the ζ potential is the electrostatic potential at the *shear plane* of electrokinetic experiments. It is often close to the δ plane of the adsorbed ions; see Figure 4.2b). The resulting net charge densities in the diffuse layer are of the order of $0.1 - 0.2\ e/\text{nm}^2$, as calculated from the Gouy–Chapman equation. For very low pH, σ_0 reverses sign at a *PZC*, which is typically located somewhere between pH 2 and 4. It is interesting to note that the value of these intrinsic charge densities is more than an order of magnitude larger than the charge densities on the dielectric layers that are induced by the applied voltage in typical EW experiments.

The precise origin of the interfacial charge at hydrophobic–water interfaces has been debated in the physical chemistry literature for quite some time (see, e.g. [10]). At the stage of writing of this book, the issue has not been

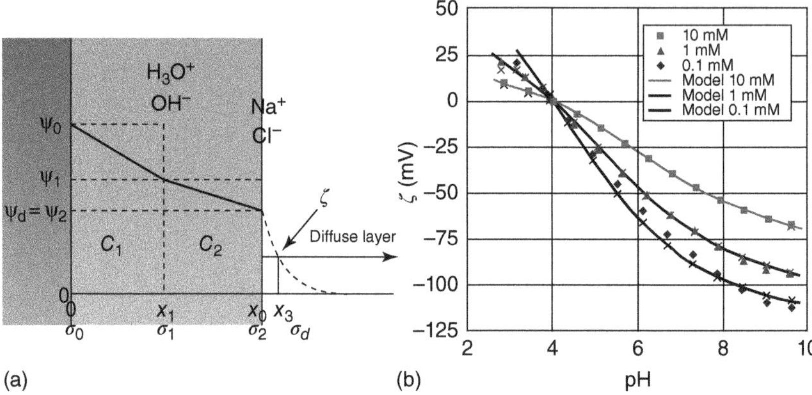

Figure 4.5 (a) Triple layer model of a Teflon–water interface with OH^-, H_3O^+, Na^+, Cl^- ions adsorbing at different distances from the solid surface in the Stern layer (central gray region). The planes denoted by *1* and *2* correspond to two different δ planes for the various ions in the spirit of Figure 4.2b. (b) Experimental surface potential of Teflon–water interfaces versus pH for KCl solutions of variable concentrations along with the theoretical description of an evolved surface complexation model. Source: Lutzenkirchen et al. [9]. Reprinted with permission of Royal Society of Chemistry.

settled. One widespread interpretation assumes that the charge originates from preferential adsorption of hydroxide (OH^-) ions to the interface. This is consistent with the observation that σ_0 becomes more and more negative with increasing pH. The primary charge generation process is believed to be the autolysis of water molecules at the interface, i.e. the reaction

$$(H_2O)_0 \rightleftharpoons (OH^-)_0 + H_b^+ \tag{4.22}$$

where the subscript 0 denotes surface-adsorbed species and b the bulk. The reversal of the surface charge at low pH is then naturally explained by the abundance of hydronium ions (H_3O^+) under these conditions. In addition to hydroxyl and hydronium ions, anions and cations from added salt can also accumulate near the interface. To describe the charging behavior of Teflon–water interfaces, rather complex models involving multiple planes to which different types of ions adsorb have been invoked (see Figure 4.5a). While some of the assumptions of these models seem to be ad hoc, they nevertheless provide a reasonable general idea of the processes occurring at these interfaces. Various aspects, albeit qualitatively, have indeed been reproduced by molecular simulations in recent years both for fluoropolymer surfaces and more generally for other hydrophobic–water interfaces that play a key role in many biological processes, including protein folding.

As we will see in the following chapter, unlike the electrocapillary curves discussed here, the contact angle response in modern EW on dielectric experiments is usually rather insensitive to these chemical processes. This is to some extent curious, given the fact that the charge densities caused by surface chemistry are much higher than the ones induced by the applied voltage. The reason is that the capacitance of the EDL is much larger than the one due to the dielectric layer. Hence the voltage drop and electrostatic energy, which is the relevant quantity for wetting, is dominated by the dielectric layer, as we saw in Section 2.5. Therefore, we refrain here from a deeper discussion of the physical chemistry of hydrophobic–water interfaces and refer the interested reader to the specific literature (see, e.g. [10]). Nevertheless it should be noted that chemical processes and ion adsorption may play an important role for the degradation of EW substrates upon long-term exposure to water and other polar electrolytes.

4.A Statistical Mechanics Derivation of the Governing Equations

Like in the preceding chapters, we can derive the governing equations for the equilibrium distribution of ions in the EDL on the basis of a microscopic mean field statistical mechanics model starting from a free energy density.

For our discussion of the adsorption of uncharged solutes in Chapter 3, we split the system into two subsystems, namely, the (unperturbed) bulk liquid and the surface. The two subsystems were characterized by the bulk density c_0 and the surface coverage Γ. To describe the properties of EDLs, we must include several extensions: First of all, we split the system in three subsystems, namely, the unperturbed bulk (b), the diffuse layer (d), and the surface (Stern) layer (s), i.e. we write

$$\begin{aligned}\mathcal{F}(V,T,N) &= \mathcal{F}_b + \mathcal{F}_d + \mathcal{F}_s \\ &= \int dV_b f_b(T,c) + \int dV_d f_d(T,c) + \int dA f_s(T,\Gamma)\end{aligned} \tag{A.1}$$

In practice, it is sufficient to consider only the free energy per unit area. Hence we can skip the lateral integration and rewrite Eq. (4.22) as

$$F(V,T,N) = F_b + F_d + F_s = \int_D^\infty dz f_b(T,c) + \int_0^D dz f_d(T,c) + f_s(T,\Gamma) \tag{A.2}$$

where F, F_b, F_d, and F_s are thus free energies per unit area. The exact location of the boundary D between the bulk volume V_b and the volume of the diffuse layer V_d is not crucial, as long as it is taken sufficiently far away from the surface, i.e. typically a few times λ_D. The unperturbed bulk has the same energy density

$$f_b(T,c_i) = \sum_i \mu_{b,i}^0 c_i + k_B T\, c_i \ln \frac{c_i}{c_0} \tag{A.3}$$

and chemical potential

$$\mu_{b,i}(T,c_i) = \mu_{b,i}^0 + k_B T \ln \frac{c_i}{c_0} \tag{A.4}$$

as described in Section 3.1.2, except that we need to sum over all ionic species i. Because the bulk remains unperturbed, all concentrations remain equal to the bulk values, i.e. $c_i = c_i^\infty$.

Within the diffuse layer, the ion concentrations are no longer simple fixed numbers, but they are position-dependent fields $c_i(z)$. To keep the notation simple, we will assume in the following that there is only one monovalent anionic and one monovalent cationic species. Hence, we have only two concentration fields $c_{\pm} = c_{\pm}(z)$. Moreover, we need to introduce the position-dependent electrostatic potential $\phi(z)$, which is finite both within the diffuse layer and at the surface. With these ingredients, we can write free energy density within the diffuse layer as

$$f_d(T, c_+(z), c_-(z), \phi(z)) = c_+\mu^0_{b+} + c_-\mu^0_{b-} + k_B T\, c_+\left(\ln\frac{c_+}{c_0} - 1\right) + \cdots$$
$$\cdots + k_B T\, c_-\left(\ln\frac{c_-}{c_0} - 1\right) + c_+ e\phi - c_- e\phi - \frac{\varepsilon_0\varepsilon}{2}\phi'^2 \tag{A.5}$$

The first two terms on the right-hand side represent the chemical potential of the individual solvated ions. Terms three and four are the contribution from the configurational entropy, as before. The fifth and sixth terms describe the electrostatic energy of the ions due to the electrostatic potential, and finally, the last term describes the energy density of the electrostatic field, with $\phi' = d\phi/dz$. The corresponding grand potential density is given by

$$\tilde{f}_d(T, \mu_+, \mu_-, \phi(z)) = f_d(T, c_+(z), c_-(z), \phi(z)) - \mu_+ c_+(z) - \mu_- c_-(z) \tag{A.6}$$

Assuming for simplicity that only positive ions adsorb to the interface, we can write for the contribution of the surface layer

$$f_s(T, \Gamma_+) = \Gamma_+\mu^0_s - k_B T\left(\Gamma_+ \ln\frac{\Gamma_+}{\Gamma_0} + (\Gamma_0 - \Gamma_+)\ln\frac{\Gamma_0 - \Gamma_+}{\Gamma_0}\right) + \Gamma_+ e\,\phi_s \tag{A.7}$$

Here, $\phi_s = \phi(0)$is the electrostatic potential at the surface. Again, the corresponding grand potential is

$$\tilde{f}_s(T, \mu_+) = f_s(T, \Gamma_+) - \mu_+\Gamma_+ \tag{A.8}$$

Thus, our model contains three position-dependent fields, $c_{\pm}(z)$ and $\phi(z)$. The free energy density is no longer a simple function but a functional of these three fields. Variational minimization of the functional with respect to the concentration fields yields the Euler–Lagrange equation

$$\frac{\partial}{\partial c_{\pm}}\tilde{f}_d[T, \mu_+, \mu_-, \phi(z)] = (\mu^0_{b\pm} \pm e\phi) + k_B T \ln\frac{c_{\pm}}{c_0} - \mu_{\pm} = 0 \tag{A.9}$$

From Eq. (A.7) we can recover two of our earlier statements: First of all, $\mu_{\pm}$ is (up to a constant) the *electrochemical* potential defined in Eq. (4.2). In the presence of a finite electric potential, it is this quantity rather than the chemical potential that is constant throughout the system. Moreover, equating $\mu_{\pm}$ to the bulk chemical potential (Eq. (A.2)) yields the Boltzmann distribution, Eq. (4.1), for the ions.

Variational minimization with respect to the electrostatic potential yields the Euler–Lagrange equation

$$\frac{d}{dz}\frac{\partial\tilde{f}_d}{\partial\phi'} - \frac{\partial}{\partial\phi}\tilde{f}_d = -\varepsilon\varepsilon_0\phi'' - e(c_+ - c_-) = 0 \tag{A.10}$$

Remembering that the charge density in the diffuse layer is given by $\rho_{el}(z) = e(c_+ - c_-)$, we see that we just recovered the Poisson equation, Eq. (4.3). Similarly, variation with respect to ϕ_s connects the electric field to the surface charge, i.e.

$$\left.\frac{d}{dz}\phi\right|_{z=0} = -\frac{e}{\varepsilon\varepsilon_0}\Gamma_+ \tag{A.11}$$

(The term on the right-hand side arises simply from the partial derivative of f_s with respect to ϕ_s. The one on the left-hand side is derived from the variation of f_d with respect to ϕ_s' taken at the lower boundary, i.e. from the integration by parts as in Eq. (1.35).) Like in the discussion of capillarity in Section 1.5, we thus recover the basic governing equations from variational energy minimization.

Minimization of Eq. (A.6) for the surface coverage does not require any variational calculation because Γ is still simply a scalar and not a field. The only addition compared with the non-charged case considered in Chapter 3 (see Eq. A.9 in the appendix) is the last term on the right-hand side of Eq. (A.5), i.e. the electrostatic energy of the adsorbed ions. Calculating $\partial\tilde{f}_s/\partial\Gamma_+ = 0$, we find

$$\mu_{s+} = \mu_{s+}^0 + k_B T \ln\frac{\alpha_+}{1-\alpha_+} + e\phi_s \tag{A.12}$$

which is the same as Eq. (3.8), extended by the last electrostatic term. Inserting Eq. (A.11) back into Eq. (A.7) yields the interesting result $\tilde{f}_s^{eq} = k_B T\,\Gamma_0 \ln(1-\alpha_+)$, which is identical with Eq. (3.14). This means that the additional electric field has no influence on the functional form of the energy per unit area. The effect of the electrostatic potential is completely incorporated in the variation of α_+. Equating $\mu_{s+} = \mu_{b+} = \mu_{b+}^0 + k_B T \ln c_{b+}/c_0$, we obtain the so-called Davies adsorption isotherm for surface-active cationic species:

$$\alpha_+ = \frac{\frac{c_{b+}}{c_0}}{\frac{c_{b+}}{c_0} + \exp\left[\frac{-(\Delta\mu^0 - e\phi_s)}{k_B T}\right]} \tag{A.13}$$

(Remember from our definition in Chapter 1 that positive values of $\Delta\mu^0$ correspond to a strong affinity of the solute to the substrate.) Having solved all the equations, we can now insert the solutions into Eq. (A.1) to obtain the free energy per unit area involved in the formation of the double layer. Subtracting the constant bulk contribution and inserting the equilibrium solutions calculated in the preceding equations, we find

$$\Delta F = \int_0^D \tilde{f}_d^{eq}\,dz + \tilde{f}_s^{eq} \tag{A.14}$$

The diffuse layer part consists of three contributions:

$$\begin{aligned}\Delta F_d &= \int_0^D dz\left\{\sum_i -k_B T(c_i(z) - c_{bi}) - \frac{\varepsilon\varepsilon_0}{2}\phi'^2\right\}\\ &= \int_0^D dz\left\{\sum_i -k_B T\, c_{bi}\left(\exp\left(-\frac{eZ_i\phi}{k_B T}\right) - 1\right) - \frac{\varepsilon\varepsilon_0}{2}\phi'^2\right\}\end{aligned} \tag{A.15}$$

Evaluating the integral using suitable integration by parts results in the same expression as Eq. (4.20).

Problems

4.1 (a) Rewrite the Poisson–Boltzmann equation for a salt consisting of divalent cations and anions, such as $CaSO_4$. (b) Rewrite the Poisson–Boltzmann equation for a salt containing a divalent cation and a monovalent anion such as $CaCl_2$. (c) Show that the corresponding version of the Debye–Hückel equation can still be written in terms of Debye screening parameter $\kappa = (2e^2 I/\varepsilon\varepsilon_0 k_B T)^{1/2}$ with an ionic strength $I = \sum_i Z_i^2 c_i^\infty/2$.

4.2 (a) Derive the Gouy–Chapman equation (Eq. (4.13)) by integrating the charge distribution in the diffuse layer using the Poisson–Boltzmann equation for a monovalent salt. (b) Show that the result can be written as $\sigma_D \approx -0.12\sqrt{c_\infty}\sinh(20\,\phi_d)$ for a monovalent salt at room temperature (c_∞ in moles per liter; ϕ_d in volts).

4.3 Total ion concentration close to a solid–electrolyte interface. (a) Consider a solid surface with a surface potential ϕ_D in contact with an aqueous electrolyte of NaCl of bulk concentration c_∞. Write down an expression for the total ion concentration in the diffuse layer. Show that the increase in counterion concentration and the decrease in co-ion concentration have the same magnitude to first order in $e\phi/k_B T$ such that the total ion concentration stays constant. (Hint: Show first that using Debye–Hückel solution for the potential gives consistent expansion in the small parameter, which is correct up to second order.) As a result, you will obtain the curves shown in Figure 4.3b). (b) Show that this is no longer true if second-order terms in $e\phi/k_B T$ are taken into account. Show that the total ion concentration under these conditions is given by $c_{tot}(z) = c_+(z) + c_-(z) = c_\infty(2 + \left(\frac{e\phi_0}{kT}\right)^2 \exp(-2\kappa z))$.

4.4 Use the last result of the preceding problem to estimate the density of counterions and co-ions at the solid surface for surface potentials of $\phi_0 = 0,\ 0.1, 1$ V and bulk concentrations of $c_\infty = 0.01, 0.1, 1$ M. Compare the average distance between adjacent ions (based on the concentration) to a typical radius of a hydrated ion (e.g. $r_{Na^+} = 0.116$ nm; $r_{Cl^-} = 0.167$ nm). What do you conclude about the validity of the Poisson–Boltzmann approximation?

References

1 Lippmann, G. (1875). Relations entre les phénomènes électriques et capillaires. *Ann. Chim. Phys.* 5: 494–549.
2 Grahame, D.C. (1947). The electrical double layer and the theory of electrocapillarity. *Chem. Rev.* 41: 441–501.
3 Adamson, A.W. (1990). *Physical Chemistry of Surfaces*, 5e. New York: Wiley.

4 Mugele, F. and Baret, J.C. (2005). Electrowetting: from basics to applications. *J. Phys. Condens. Matter* 17 (28): R705–R774.
5 Butt, H.J. and Kappl, M. (2010). *Surface and Interfacial Forces*. Wiley.
6 Frumkin, A. et al. (1932). Kapillarelectrische Erscheinungen und Benetzung von Metallen durch Elektrolytlösungen. *Phys. Z. Sowjetunion* 255–284.
7 Ben-Yaakov, D. and Andelman, D. (2010). Revisiting the Poisson-Boltzmann theory: charge surfaces, multivalent ions and inter-plate forces. *Physica A* 389 (15): 2956–2961.
8 Lyklema, J. (2000). *Fundamentals of Interface and Colloid Science*, vol. III. San Diego: Academic Press.
9 Lutzenkirchen, J., Preocanin, T., and Kallay, N. (2008). A macroscopic water structure based model for describing charging phenomena at inert hydrophobic surfaces in aqueous electrolyte solutions. *Phys. Chem. Chem. Phys.* 10 (32): 4946–4955.
10 Zimmermann, R., Freudenberg, U., Schweiß, R. et al. (2010). Hydroxide and hydronium ion adsorption – a survey. *Curr. Opin. Colloid Interface Sci.* 15 (3): 196–202.

5
Principles of Modern Electrowetting

In the preceding chapter, we discussed the derivation of Lippmann's equation for the voltage dependence of the interfacial tension of a metal–electrolyte interface and the resulting contact angle variation based on the free energy of the electric double layer at the interface. Modern electrowetting (EW) experiments almost exclusively use a dielectric layer to separate the conductive fluid from the electrode on the substrate. This layer makes the systems a lot more stable by preventing electrochemical reactions at the interface. In Section 5.1, we will derive the contact angle response in this so-called electrowetting-on-dielectric (EWOD) configuration. We will first consider the global response of the macroscopic contact angle, and subsequently we analyze the microscopic configuration of the drop surface in the vicinity of the contact line. Together these two approaches form what we will call the electromechanical standard model of EW. In Section 5.2, we discuss the link between the standard model and the electrocapillary model discussed in Chapter 4. Section 5.2.3 addresses some of the limitations of the present theory. Finally, Section 5.3 is devoted to differences that arise when the DC is replaced by AC voltage, as it is very commonly done in experiments.

5.1 The Standard Model of Electrowetting (on Dielectric)

5.1.1 Electrowetting Phenomenology

Modern EW experiments almost exclusively involve a dielectric layer to insulate the conductive fluid from the actuating electrode(s) on the substrate. By blocking the electrical current and thereby preventing chemical reactions, these layers contribute substantially to the robustness and reliability of EW. In basic experiments, the conductive fluid is frequently a drop of an aqueous salt solution, and the dielectric layer is a hydrophobic polymer film with a thickness d of order μm. The ambient medium surrounding the drop can be either air or a second immiscible dielectric liquid, which we will collectively denote as *oil*. The generic setup is shown in Figure 5.1. This configuration is often denoted as EWOD to emphasize the presence of the dielectric layer. Because EW *without* dielectric layers has become rather exceptional in recent decades, we will skip the explicit

Electrowetting: Fundamental Principles and Practical Applications, First Edition. Frieder Mugele and Jason Heikenfeld.

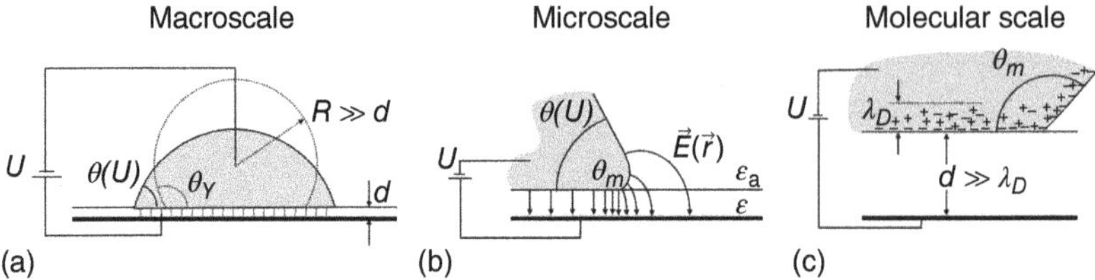

Figure 5.1 Basic setup of an electrowetting (on dielectric) experiment. (a) Reduction of the macroscopic contact angle from θ_Y at zero voltage (dashed drop contour) to $\theta(U) < \theta_Y$ at finite voltage (solid gray drop). *d*, thickness of insulating layer; *R*, drop size. (b) Electric field distribution and deformation of the surface profile on a microscopic length scale of order *d*. θ_m denotes the microscopic contact angle on a scale $\ll d$. ϵ and ϵ_a are the dielectric constants of the insulating layer and the ambient medium, respectively. (c) Distribution of ions at the nanometer scale of the Debye screening length $\lambda_D \ll d$.

reference to the dielectric and automatically imply its presence when we speak about *electrowetting* – unless explicitly stated differently. When a voltage U is applied to the system, the macroscopic contact angle decreases as in the case of direct metal–electrolyte contact in the preceding chapter. The voltage required to achieve the same contact angle reduction, however, is now much higher. As the snapshots of drops in Figure 5.2a illustrate, the variation of θ is – to a first approximation – independent of the polarity of the applied voltage, and the maximum of the contact angle is found at zero voltage. In the literature, there is not really a consistent sign convention regarding the polarity of U. For the purpose of this book, we will choose – unless explicitly noted otherwise – to keep the electrode on the substrate electrically grounded. Positive voltage thus means that the potential of the drop is higher than the one of the substrate.

To quantify the response of the EW system to the applied voltage, one commonly extracts the macroscopic contact angle from side-view images. The resulting curve $\theta(U)$ is known as *EW curve.* A typical example is shown in Figure 5.2b. Overall, the shape of EW curves can be divided in two regimes. In the first regime at low voltage, the contact angle progressively decreases at increasing rate, as the applied voltage is increased toward higher positive or

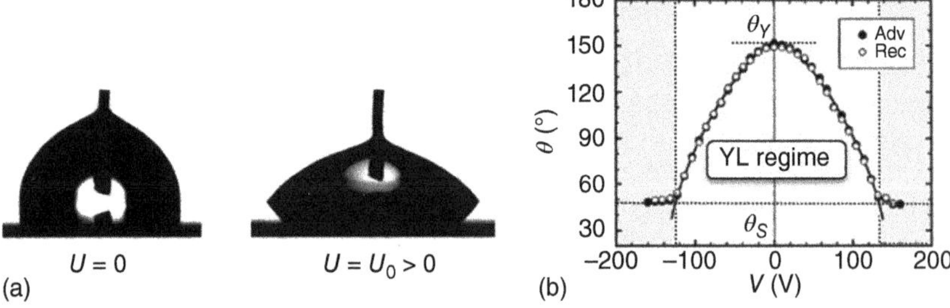

Figure 5.2 (a) Snapshots of a drop at zero voltage, and $U = \pm 100$ V. (b) Typical macroscopic response curve contact angle versus applied voltage, denoted as *electrowetting curve*. The white central region shows the Young–Lippmann (YL) regime, and the gray-shaded regions at $|U| > 130$ V corresponds to the saturation regime (specific system: ionic liquid in ambient oil). Source: Paneru et al. 2010 [1]. Adapted with permission of ACS.

negative values. This regime in the central region of Figure 5.2b is commonly denoted as *Young–Lippmann* (YL) regime. All presently known applications of EW operate in this regime. In the second regime at higher voltage, the slope of the EW curve decreases again, and θ eventually becomes independent of U beyond some more or less sharply defined transition voltage U_s. This phenomenon is known as contact angle saturation, and we denote the high voltage regime as *saturation regime*. Contact angle saturation and the mechanisms that determine the *saturation angle* θ_s are probably the most elusive problem in EW. Despite extensive efforts, at the time of writing of this book, the phenomenon is not completely understood. We will address some aspects of contact angle saturation in Section 5.3 and more extensively later in Section 7.3.

The decrease of the macroscopic contact angle in the YL regime is described by the EW or YL equation

$$\cos\theta(U) = \cos\theta_Y + \frac{c_d}{2\gamma}U^2 \tag{5.1}$$

where $c_d = \epsilon\epsilon_0/d$ is the capacitance per unit area between the drop and the substrate electrode. The solid line in Figure 5.2b is a fit of the YL equation to the experimental data, where the insulator thickness d was taken as a fit parameter.

Before we start to derive the YL equation and the more subtle aspects of the EW response of sessile drops on smaller length scales, it is useful to consider some basic assumptions and relevant length scales of the system. First of all, we will assume that the drop is perfectly conductive and that the ambient medium and the insulating layer behave as perfect dielectrics. This implies that the bulk liquid is kept at a fixed potential and that the electric field vanishes everywhere within the drop. From our discussion in the preceding chapter, we know that this is not completely true because the electric field is in fact screened within the electric double layer. Yet, the thickness of the double layer, λ_D, is typically orders of magnitude smaller than all other length scales in the problem (Figure 5.1c). Setting $\lambda_D \approx 0$ therefore seems to be a reasonable starting point for any model aiming to capture the macroscopic response. In the next section, we come back to this assumption when we establish the relation between standard EW and the insulator-free EW response discussed in Chapter 4.

The YL equation, as we said, describes the response of the drop on a macroscopic scale as we can perceive it by eye or by a moderately magnifying optical system. That scale is of the order of the drop size R or some fraction of it. Typically, this is much larger that the thickness of the insulator d (see Figure 5.1a). For many practical applications of EW, knowing and understanding the YL equation, i.e. the macroscopic response of the contact angle is sufficient. We will derive it in the subsequent subsection. To understand the microscopic origin of the EW response, to get some insight into the origin of contact angle saturation, and to become able to design advanced, say, nanoscale EW devices, it is necessary to consider also the response on scales of the order of the insulator thickness (see Figure 5.1b). From our considerations in Chapter 2, we know that there should be electric stray fields in the vicinity of the contact line. We also know that such stray fields exert a Maxwell stress that will pull on the free liquid surface. This gives rise to local deformations of the surface profile, similar to the effect of the disjoining pressure discussed in Section 1.6. The details of the local surface profiles will be addressed in the subsection on the microscopic model of EW. Together, the YL equation and the microscopic model form the current electromechanical standard model of EW.

5.1.2 Macroscopic EW Response

The easiest manner to derive the EW equation is to revisit our derivation of Young's equation based on energy minimization in Section 1.3 (Eq. (1.17)) and to include the contribution due to the electrostatic energy. In analogy with that derivation, we write down and express the total energy as a function of the drop shape and seek a minimum as a function of the contact line position. The total energy now consists of the sum of the interfacial energies and the electrostatic energy:

$$E_{tot}[A] = E_{surf} + E_{el} \tag{5.2}$$

Following Young, we can calculate the equilibrium contact angle by minimizing the surface energy upon varying the position of the contact line. A necessary condition for an energy minimum is that the variation of the total energy vanishes upon displacing the contact line by an infinitesimal distance dx. As in Section 1.3 the variation of the surface energy per unit length upon displacing the contact line is $\delta E_{surf} = dx\,(\gamma_{sl} - \gamma_{sv} + \gamma \cos\theta)$, Eq. (1.16) (Figure 5.3).

The contribution of the electric energy looks at first glance more evolved. As we discussed in Section 2.3, the (free) electrostatic energy between two conductors kept at a fixed potential difference U can be written as $E_{el} = -CU^2/2$, where C is the capacitance between the two conductors, i.e. the drop and the electrode on the substrate in the present case. In order to keep the notation simple, we drop here and in the following the tilde above $E_{el.}$, which we introduced in chapter 2 to emphasize that the relevant electric energy in EW is the *free* electrostatic energy including the negative sign. In general, calculating the capacitance requires a calculation of the field and charge distribution, which depends on the geometry of the electrodes, i.e. on the exact shape of the liquid surface. However, if we are only interested in the response of the drop on scales $\gg d$, we do not need to worry about such local effects. No matter how complex the local shape may look like, it will simply be translated along with the contact line in the course of our virtual displacement without any change in energy. Hence, the net contribution of the electric term to the energy variation is reduced to the generation of an extra piece of drop–substrate interface far away from the contact line. There, however, the dielectric layer and the electrode on the substrate form a parallel plate

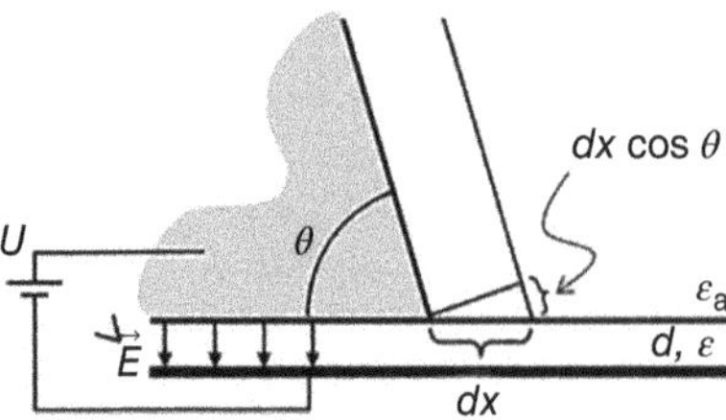

Figure 5.3 Illustration of the variation of the interfacial areas and the electric field upon displacing the contact line by *dx*. (The distribution of the electric field in the vicinity of the contact line is not drawn. It simply translates along with the contact line and therefore does not change.)

capacitor with a uniform electric field of strength U/d and a capacitance per unit area $c_d = \epsilon\epsilon_0/d$. This corresponds to an electrostatic energy per unit area of the drop–substrate interface $E_{el}/A_{sl} = -c_d U^2/2$. Hence, we have $\delta E_{el} = -dx\ c_d U^2/2$. Assembling all terms, we find

$$\delta E_{tot} = \delta E_{surf} + \delta E_{el} = dx\left(\gamma_{sl} - \gamma_{sv} + \gamma\cos(\theta) - \frac{c_d U^2}{2}\right) \tag{5.3}$$

Equating the bracket to zero as required for mechanical equilibrium yields the YL equation, Eq. (5.1). If we introduce the dimensionless EW number $\eta = c_d U^2/2\gamma$, we can abbreviate the expression and rewrite the YL equation as

$$\cos\theta(U) = \cos\theta_Y + \eta \tag{5.4}$$

The EW number measures the relative strength of the electrostatic energy and surface tension.

A few remarks are in place. First, the YL equation reproduces the observed independence of the EW response of the polarity. The solid line in Figure 5.2b illustrates that it properly captures the contact angle reduction in the YL regime (see also Figure 5.4). However, the YL equation fails to describe the saturation regime. Instead, it predicts in contrast to experiments that the voltage-dependent contact angle should vanish at a critical angle $U_c = 2\gamma\ d\,(1 - \cos\theta_Y)/\epsilon\epsilon_0$. There is no indication of saturation within the model, and it is therefore a priori unclear up to which maximum voltage the model can be expected to work in a given experiment.

Second, we can see from the YL equation that EW can only reduce the contact angle to values less than Young's angle. From a practical perspective, this means that ideal materials for EW should be chosen such that Young's angle is as high as possible in order to maximize the range of actuation that can be achieved by applying a voltage. Furthermore, the definition of the EW number also specifies how the contact angle reduction depends on capacitance $c_d \propto d^{-1}$ and on the surface tension γ. The same contact angle reduction can be achieved at lower voltage if either the thickness of the insulating layer and/or the surface tension

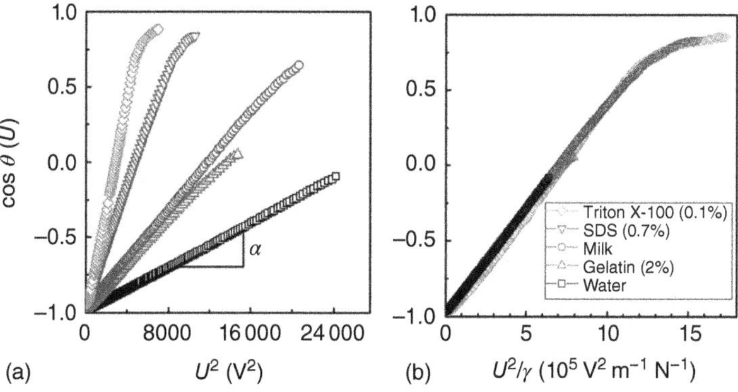

Figure 5.4 Electrowetting curves for various aqueous solutions in ambient silicone oil with interfacial tensions varying from 4.4 (top curve; Triton X-100) to 38 mJ m^{-2} (bottom curve; water). Source: Data taken from Banpurkar et al. 2008 [2].

of the drop is reduced. In the design of EW devices, it is therefore often a goal to minimize both d and γ within the constraints imposed by other aspects of the system. We will discuss various material-related issues in Chapter 7.

Third, we can see in Eq. (5.3) that the electrostatic energy gain scales with the drop–substrate interfacial just as the drop–substrate interfacial energy. From a macroscopic perspective, one may therefore combine the electrostatic contribution and the chemical solid–liquid interfacial tension γ_{sl} to a voltage-dependent *effective* interfacial tension:

$$\gamma_{sl}^{eff}(U) = \gamma_{sl} - \frac{c_d}{2}U^2 \tag{5.5}$$

This expression is convenient in many circumstances. It looks very similar to Lippmann's equation of electrocapillarity (see Eqs. (4.17) and (4.20)), except that c_d appears instead of the capacitance of the electric double layer c_{DL}. Yet, one should keep in mind that the electrostatic energy is in fact distributed over the entire volume of the dielectric layer, whereas the *true* chemical interfacial tension γ_{sl} that arises from the very localized molecular interaction forces at the solid–liquid interface remains constant. We will come back to a discussion of the similarities and differences in the next section.

To illustrate the success of the YL equation and the effect of reducing γ, we display in Figure 5.4a an example of EW curves for a variety of liquids with variable interfacial tension all obtained on the same substrate with the same thickness of the insulating layer, i.e. for the same value of c_d. Indeed, the weakest contact angle response is found for pure water (bottom curve – open squares), which displays the highest interfacial tension, whereas the solution containing the surfactant Triton X-100 has the lowest interfacial tension and displays the strongest EW response. Upon normalizing U^2 by the independently measured interfacial tension γ, all curves collapse into a single master curve (Figure 5.4b), as predicted by the YL equation.

5.1.3 Microscopic Structure of the Contact Line Region

In our derivation of the YL equation, we interpreted the term $-c_dU^2/2$ as an electrostatic energy per unit area. From our derivation of Young's equation in Section 1.3, we know that we can equally interpret the chemical interfacial energies as interfacial tensions, i.e. as force per unit length pulling on the contact line. Therefore, we may also expect that it should also be possible to interpret the electrostatic term in the YL equation as originating from an electrostatic force per unit length:

$$f_{el} = -c_dU^2/2 \tag{5.6}$$

which pulls radially outward on the contact line. How does this force arise? The electric field on the far left of the contact line that we used above to calculate the energy per unit area is oriented normal to the drop–substrate interface. Therefore it does not contribute to the force balance in the tangential direction. Hence, we need to consider the more complicated electric fringe fields in the vicinity of the contact line. The fringe fields are concentrated in a region of the order of the thickness d of the insulating layer around the contact line. Their distribution over a finite region above the solid surface implies that the free drop surface in the vicinity of the contact line experiences a Maxwell stress $\pi_{el} = \frac{\epsilon_a \epsilon_0 E(\vec{r})^2}{2}$, as we discussed in Section 2.4. (Here ϵ_a is the dielectric constant of the ambient medium.) This stress is oriented along the outward normal of the liquid surface (see Eq. (2.63) and Figure 5.1b) and hence contains a horizontal component, as needed to explain the origin of f_{el}. In order to calculate the total force exerted by the electric field, we need to integrate the Maxwell stress along the drop surface, exactly along the same lines as we did in our discussion of the force pulling a metallic plate into the gap between two parallel plate electrodes in Section 2.5.

To perform this integration, we need to know the distribution of the electric field $\vec{E}(\vec{r})$. This calculation is slightly more complex than for the metallic plate in Section 2.5, however, because the drop is a liquid and thus deforms under the influence of the normal stress. To guarantee mechanical equilibrium, the liquid surface adjusts its shape in such a manner that the Maxwell stress $\pi_{el}(\vec{r})$ is balanced at every location on the surface by the local Laplace pressure, i.e. by surface tension times the local curvature $\kappa(\vec{r})$. As we know from Section 2.4, the local force balance equation reads

$$\gamma\kappa(\vec{r}) = \Delta p - \frac{\epsilon_a \epsilon_0}{2} \vec{E}(\vec{r})^2 \tag{5.7}$$

(see Eqs. (2.66) and (2.67)). Because the liquid is a conductor, however, deformations of the liquid surface also affect the distribution of the electric field. Drop shape and field distribution are thus coupled and need to be calculated in a self-consistent numerical simulation. Figure 5.5a provides a qualitative sketch of the equilibrated drop shape and the corresponding field distribution.

Before discussing a few specific aspects and the results of the calculation, we can already deduce a number of qualitative conclusions. First, it is clear that the electric field vanishes sufficiently far away from the contact line. In that region,

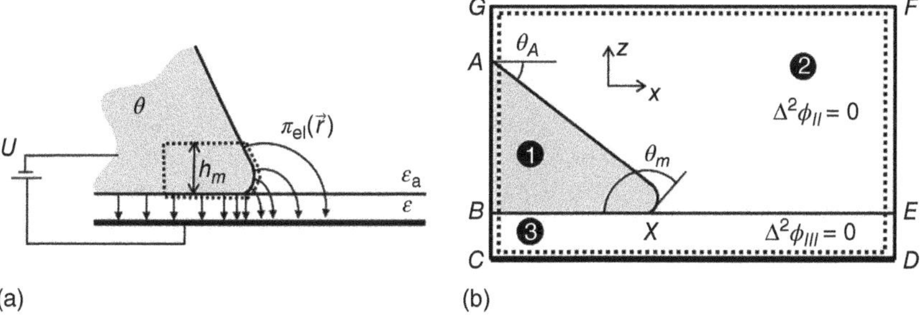

Figure 5.5 (a) Illustration of the drop shape and electric field distribution in EW geometry. Dotted line: boundary of control volume for the calculation of the force balance to calculate the local contact angle at the contact line in the limit $h_m \to 0$. (b) Setup of the physical system for a numerical simulation. External boundary conditions: $\phi = \phi_1$ on the drop (i.e. along *AB*, *AX*, *BX*), along *AG* and along *FG*. $\phi = 0$ along *CD*. Along *BC*, the electric field is U/d oriented in the *z*-direction. Along *ED* and *EF* $\partial_x \phi = 0$ is imposed and the field corresponds to the one of a parallel plate capacitor partially filled with a dielectric. Source: Buehrle et al. 2003 [3]. Adapted with permission of APS. Dotted line: control volume used to calculate the macroscopic contact angle θ_A.

the surface assumes its asymptotic shape, which can be a spherical cap shape or a flat configuration, depending on geometric boundary conditions including the imposed by the external pressure jump Δp. Second, the simple fact that the Maxwell stress and along with it the curvature of the surface increase upon approaching the contact line obviously implies that the local slope of the surface varies also, as we can calculate by integrating the curvature along the surface. Hence, the microscopic contact angle at the contact line, θ_m in Figure 5.5b, must be different from the asymptotic contact angle far away, θ_A.

From a more detailed analysis such as a numerical simulation, we expect to obtain quantitative answers to the following questions:

- What is the exact shape of the equilibrated surface profile?
- What is the relation between the asymptotic contact angle θ_A and the local contact angle θ_m?
- How does the net electrostatic force $f_{el} = -c_d U^2/2$ arise from the local field distribution?

Several implementations of numerical simulations are conceivable to address this problem. Here, we discuss only a few generic aspects and refer the interested reader to the more detailed original literature [3–5]. Common to all implementations is the general setup of the physical system, as schematically shown in Figure 5.5b, which can be implemented either in two or in three dimensions. The system typically consists of three different domains: the liquid drop (gray; domain 1), the ambient medium (domain 2), and the dielectric layer (domain 3). The drop is a perfect conductor; the ambient medium and the dielectric layer are perfect dielectrics with dielectric constants of ϵ_a and ϵ, respectively. The drop is kept at a fixed potential $\phi \equiv \phi_I$. Domains 2 and 3 are free of electric charge. Hence, the electrostatic potential in these domains fulfills the Laplace equation, $\nabla^2\phi = 0$. To calculate $\phi(\vec{r})$ everywhere, we solve the Laplace equation in the two domains with their constants $\epsilon_{II} = \epsilon_a$ and $\epsilon_{III} = \epsilon$ separately and patch them together along the section XE in Figure 5.5b with the standard boundary conditions for the normal and tangential electric field (see Eqs. (2.24) and (2.25)). Numerical finite element solvers such as COMSOL frequently implement this by solving Gauss' law $\nabla(\epsilon_0\epsilon(\vec{r})\,\nabla\phi(\vec{r})) = -\rho(\vec{r})$ instead, where $\epsilon(\vec{r})$ varies within a few computational grid cells of the interface, leading to a finite charge density $\rho(\vec{r})$ at the interface. The simulation box is chosen large enough to guarantee that the solutions within domains 2 and 3 have approached the asymptotic solution that is imposed along the edges of the simulation box. Typically, this will require lateral dimensions of a few tens of d. Along the drop surface AX, the potential is fixed to $\phi \equiv \phi_I$. In addition, Eq. (5.7) is imposed as a mechanical boundary condition along the free surface AX. Both the equilibrium shape of the surface and the contact line position X thus become a result of the calculation. For the specific two-dimensional configuration sketched in Figure 5.5b, it is assumed that the drop is infinitely large, which implies that $\Delta p = 0$ and hence a flat surface profile upon approaching the asymptotic point A. The local slope at point A corresponds to the asymptotic macroscopic contact angle θ_A that is observed in side-view images in experiments.

Figure 5.6a shows the resulting equilibrium profiles for increasing values of the EW number η, i.e. for increasing non-dimensional voltage, calculated for $\Delta p = 0$.

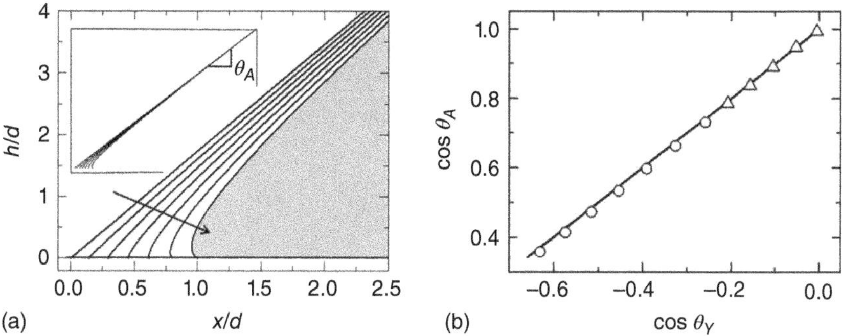

Figure 5.6 (a) Numerically calculated normalized surface profiles $h(x)/d$ for variable electrowetting number $\eta = 0, 0.2, \ldots, 1$ increasing along the arrow. Note the increasing curvature of the surface in the vicinity of the contact line (at $h = 0$) for higher η. The electrode on the substrate is located at $h/d = -1$. Inset: same profiles as in main panel but on larger scale to demonstrate that all profiles converge to the same macroscopic angle θ_A. (b) Symbols: cosine of macroscopic apparent contact angle $\cos\theta_A$ versus cosine of Young's angle $\cos\theta_Y$ from a series of simulations at variable η. Solid line: EW equation, Eq. (5.4). Source: Data taken from Buehrle et al. 2003 [3].

The profiles confirm the qualitative expectations discussed above. At zero voltage, the surface profile is straight all the way down to the contact line. For higher η, the profiles are still flat far away from the contact line but become progressively more curved upon approaching the substrate. The transition between the flat and the appreciably curved region occurs at a height of order d above the substrate. Throughout the specific simulations shown in Figure 5.6, θ_Y was increased along with η in such a manner that the asymptotic contact angle θ_A remained constant. The profiles directly confirm that the microscopic local contact angle θ_m at the contact line is always larger than the asymptotic macroscopic contact angle θ_A. Simulations performed at variable asymptotic contact angle θ_A showed that the relation between θ_A and the local contact angle θ_m follows the YL equation, i.e. $\cos\theta_A = \cos\theta_m + \eta$, as shown by the solid line in Figure 5.6b. Moreover, it turned out that θ_m is always equal to Young's angle θ_Y, independent of the applied voltage. This is arguably one of the most amazing results of the microscopic model: For any given physical system in an experiment, θ_Y is obviously fixed by the various interfacial tensions of the materials involved. The fact that $\theta_m \equiv \theta_Y$ thus implies that the local contact angle at the contact line does *not* vary in EW experiments despite the decrease of macroscopic angle θ_A with increasing voltage.

This at first glance surprising observation was indeed verified experimentally by imaging the drop profile on a scale smaller than the thickness of the insulator. For a common insulator thickness of just a few μm, this is difficult. Yet, by using very thick insulators of order 100 μm and beyond, local imaging is not a problem, as shown in Figure 5.7. The low magnification images in Figure 5.7A shows that the macroscopic contact angle of the drop indeed decreases from almost 180° at zero voltage to approximately 90° at high voltage corresponding to $\eta \approx 1$. The bottom row for $d = 150$ μm shows, however, that the microscopic contact angle indeed remains essentially constant. The zoomed views of the near contact line region at $\eta = 1$ in Figure 5.7B demonstrate that the transition from the invariant local

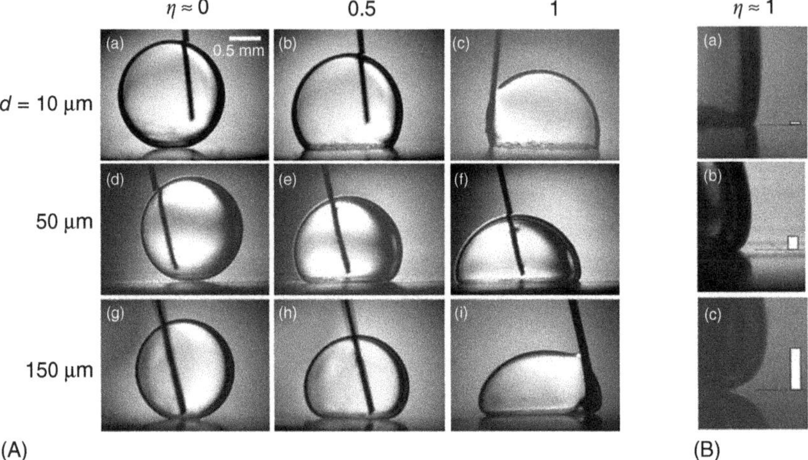

Figure 5.7 (A) Snapshots of drops on three different substrates with variable thickness of 10, 50, and 150 μm and for variable η, as indicated. The system consists of a drop of aqueous NaCl solution in ambient silicone oil and commercial Teflon films as substrates. (B) Zoomed view close to the contact line for $\eta = 1$. The scale bars indicate d in each subfigure. Source: Data taken from Mugele and Buehrle 2007 [4].

contact angle to the voltage-dependent macroscopic contact angle takes place within a region of order d of the substrate.

The origin of this remarkable voltage independence of θ_m becomes clear if we analyze the dependence of the Maxwell stress (which is equal to the local curvature of the surface for $\Delta p = 0$ 5.8) as a function of the distance from the contact line. Plotting these quantities as obtained from the simulations on a double logarithmic scale reveals the presence of two asymptotic regimes, as shown in

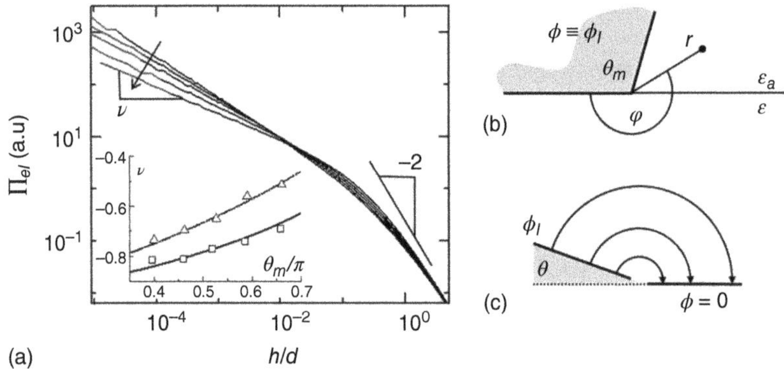

Figure 5.8 (a) $\pi_{el} \propto \kappa$ versus height of the liquid surface above the substrate for the same range of η and θ_Y as in Figure 5.6, both increasing along the arrow. Note the two different algebraic regimes with $\pi_{el} \propto (h/d)^{-\nu}$ for $h/d \ll 1$ and with $\pi_{el} \propto (h/d)^{-2}$ for $h/d \gg 1$ corresponding to the different asymptotic regimes illustrated in (b) and (c). Inset in (a): exponent ν versus θ_m from numerical calculations (triangles: $\epsilon = 1$; squares: $\epsilon = 2$; both: $\epsilon_a = 1$) and from the corresponding asymptotic analysis (solid line). Source: Data taken from Buehrle et al. Phys. Rev. Lett. 2003 [3, 4]. [6].

Figure 5.10. At distances $h/d \geq 1$, π_{el} decays with an exponent −2, which implies that the electric field decays as $1/h$. This behavior is universal, independent of the specific values of θ_m and θ_A. This scaling is not surprising. For $h \gg d$, the presence of the insulating layer can be neglected. Hence the electrostatic problem reduces to the problem of calculating the field distribution in a wedge with different potentials on the two legs that we already analyzed in Section 2.1 (see Figures 5.8c and 2.4b). In this geometry, the electric field decays linearly along the circular arc-shaped electric field lines. For distances $h/d \ll 1$, π_{el} decays with an exponent that depends on the local contact angle θ_m. Under these conditions, the electrode on the substrate is very far away. Asymptotically, we can therefore ignore its presence, as sketched in Figure 5.8b. The problem therefore comes very close to the situation of calculating the electric field around a simple metallic wedge at constant potential that we also considered in Section 2.1 (Figure 2.4a). In contrast to our previous calculation, we now need to take into account the presence of the two different dielectric media: the insulating layer and the ambient medium. This means that we need to calculate the solution of the Poisson equation separately for the two media and match the coefficients such that the conventional boundary conditions for the tangential and the normal component at the interface are met. The result of this analysis is that the Maxwell stress indeed diverges algebraically as $\pi_{el} \propto h^{\nu}$ with an exponent $-1 < \nu < 0$ (see Problem 5.1). The exponent found in the asymptotic analysis is consistent with the numerical result, as shown in the inset of Figure 5.8a.

Based on this weak algebraic divergence of the Maxwell stress, we can understand why the local contact angle remains constant upon applying the voltage. To this end, we analyze the horizontal component of the electrostatic force acting on the dashed control volume shown in Figure 5.5a. The net electric force acting on the control volume of height h_m is given by the integral over the Maxwell stress, i.e.

$$f_{el}(h_m) \propto \int_0^{h_m} \pi_{el}\, dh \propto h_m^{1+\nu} \tag{5.8}$$

The integrated electrostatic force acting on the control volume thus scales algebraically with h with a positive exponent $1 + \nu > 0$. Hence, $f_{el}(h_m) \to 0$ for $h_m \to 0$. Thus, the electrostatic force does not contribute to the force balance at the contact line. The only forces acting on the control volume in the limit $h_m \to 0$ are the surface tension forces, as in the absence of any applied voltage. As a consequence, we find that the local contact angle equals Young's angle, i.e. $\theta_m \equiv \theta_Y$.

Finally, we are left with the third question, namely, to explain why the apparent macroscopic contact angle decreases according to the YL equation. This statement is equivalent to the conclusion that the total electrostatic force arising from the integral over the local Maxwell stress results in $f_{el} = -c_d U^2/2$. To calculate f_{el}, we return to the electrostatic forces acting on the same control volume that we just analyzed. Except that in this case, we do now consider the limit $h_m \to \infty$ instead of $h_m \to 0$. If we make h_m large enough, we can expect that the electric field at the drop surface vanishes, i.e. $\pi_{el} \to 0$. At first glance, it seems only possible to calculate f_{el} by numerically integrating the electric field distribution obtained from a simulation, as described above. From that perspective, it is surprising that the calculation should yield exactly the fixed value that is required to match the macroscopic behavior expressed by the YL equation. To solve this apparent puzzle, we can consider first the simplified geometry of a perfect straight wedge in two dimensions with an opening angle α that is separated from a flat electrode by a thin insulating layer of thickness d. For this geometry, the Poisson equation can be solved analytically by the method of conformal mapping [include reference to ref. 7 (Kang) and ref. 9 (Vallet et al.) here.]. This calculation shows that the integrated force per unit length of the contact line is indeed exactly $f_{el} = -\epsilon_0 \epsilon U^2/2d$, independent of the opening angle α. As derived in Chapter 2 for the specific case of $\alpha = \pi/2$, the net force per unit length resulting from this calculation is indeed exactly $f_{el} = -\epsilon_0\epsilon\, U^2/2d$. This independence of the opening angle may seem counterintuitive given the fact that the exponent of the divergence of the local electric field does become more pronounced for smaller opening angles. Yet, returning to Figure 5.5 we notice that the specific choice of the control volume as drawn in Figure 5.5a is not the most convenient one. To calculate f_{el}, it is much more convenient to consider a very large control volume with boundaries far to the left and to the right of the three-phase contact line, as sketched by Figure 5.5b. In this case, the electric field can be assumed to vanish everywhere along the sections *BA*, *BG*, *GF*, *FE*, and *ED*. Only along *CD* and along *BC* the electric field is finite and – because the section *CD* is chosen just above the surface of the bottom electrode – oriented strictly along the z-direction. To calculate the net force, we must integrate the Maxwell stress tensor over the surface of the control volume, as described in Eq. (2.64). For our specific geometry here, the force is given by

$$f_{el,x} = \oiint \sum_j T_{xj}\hat{n}_j\, dA = -\vec{e}_x d\, T_{xx}^{CD} = -\vec{e}_x d \left(-\frac{\epsilon_0 \epsilon}{2} \left(\frac{U}{d} \right)^2 \right) = -\epsilon_0 \epsilon\, U^2/2d \tag{5.9}$$

Like in the example of the parallel plate capacitor in Chapter 2 (see Figure 2.13), only the section *CD* in Figure 5.5b contributes to the force in the x-direction, and we obtain indeed the expected result. From the derivation using this large control volume, it is mathematically immediately clear that the result must indeed be exactly equal to the expected value. Physically, this result is a direct consequence of the facts that (i) the Maxwell stress tensor T_{ij} describes the momentum flux density of the electric field and that (ii) the total momentum of the system electric field plus matter is conserved (see also Section 2.4). This general principle also implies that the value of $f_{el,x}$ must indeed not depend on the opening angle α, as is mathematically apparent from the absence of α in Eq. (5.9). Even more strongly, the global conservation argument implies that the net force is completely independent of the details of the microscopic shape of the surface profile. This means in particular that we do not need to know the details of the equilibrium surface profiles as shown in Figures 5.6 and 5.7 to calculate the correct voltage dependence of the macroscopic contact angle. Historically, this explains why the correct expression for the YL equation could be obtained in early calculations [7] despite the use of mechanically nonequilibrated straight wedge-shaped surface profiles.

5.2 Interpretation of the Standard Model of EW

5.2.1 The Electromechanical Interpretation

The preceding subsections describe the standard model of EW as it has emerged since the early 2000s. It clearly explains how the distribution of the electric field in the surrounding dielectric media and the screening by the ions in the conductive liquid give rise to electric stresses that pull on the liquid surface, leaving the microscopic contact angle unaffected. The stresses are thus the primary driver, and the reduction of the macroscopic apparent contact angle is a consequence of the localization of the electric field with a range of order d of the contact line. As long as $d \ll R$, the microscopic consequences and local deformations of the drop surface as shown in Figure 5.7 are typically not resolved, and only the reduction of the macroscopic apparent contact angle following the YL equation is observed.

This picture was denoted by T.B. Jones, one of the pioneers of the field in the early 2000s, as the *electromechanical model* of EW. The name differentiates the phenomenon from the more *electrochemical* phenomena that cause the reduction of the microscopic contact angle in the traditional EW experiments without dielectric layer. From the electromechanical model, it is clear that the reduction of the apparent contact angle in conventional EW (with dielectric) and the ponderomotive force that is used to transport drops along solid surface are strictly speaking two separated consequences of the same underlying microscopic electric stresses. The ability to actuate drops by inhomogeneous electric fields is thus independent of the reduction of the macroscopic contact angle. From our treatment of the problem in this book, this should be very clear: When we first introduced the electrostatic forces and calculated how they arise from the local field distributions in Chapter 2, we were considering undeformable solid objects. The resulting value of f_{el} was exactly the same as in Eq. (5.6). From Eq. (5.9), we know why this is the case.

From Chapter 2 we also know that the forces exerted by the electric field in an EW-like parallel plate capacitor geometry are not limited to perfectly conductive objects. Dielectric plates also experience a physically very similar force, except for a reduction in magnitude that depends on the dielectric contrast. This implies, for instance, that typical EW-based lab-on-a-chip devices, in which the drops are sandwiched between a top and a bottom plate, can not only actuate conductive liquids that display a reduction of the contact angle upon applying a voltage. The same type of device can also be used to actuate dielectric fluids, provided that the dielectric contrast between the liquid drop and the ambient medium is not too small. Such actuation of dielectric liquids has indeed been demonstrated, for polar liquids with a large dielectric constant above, say, 10 [8] even in the absence of a macroscopic contact angle reduction.

5.2.2 Standard Model of EW Versus Lippmann's Electrocapillarity

In our development of the standard model of EW, we considered the liquid as a perfect conductor and hence neglected the contribution of the distribution of ions in the electric double layer to the EW phenomenon. This is in contrast to our discussion in Chapter 4, where we analyzed the distribution of charge and energy at solid–electrolyte interfaces and derived Lippmann's electrocapillary equation:

$$\gamma_{sl}(U) = \gamma_{sl} - \frac{\epsilon_l \epsilon_0}{2\lambda_D}(U - U_{pzc})^2 \tag{5.10}$$

Formally, Eq. (5.10) looks very similar to the YL equation, Eq. (5.4), except for two important differences. First, the dielectric layer thickness d appears in the denominator of Eq. (5.4) instead of the Debye layer thickness λ_D. Second, the potential of zero charge, U_{pzc}, is absent in Eq. (5.4).

To understand the differences, we analyze the distribution of the electrostatic potential at the solid–liquid interface far away from the contact line. In the perfect conductor case, the electric potential ϕ drops from the value at the electrode, ϕ_e, to the value of the bulk liquid, ϕ_l, which we can set to zero for the present discussion (see Figure 5.9). In the experiment, the boundary condition that we apply is the potential difference $U = \phi_e - \phi_l$. The potential ϕ_s at the surface is not known. We know that the insulating layer behaves as a perfect dielectric. Hence, the electric field E_d decays linearly between $z = 0$ and $z = d$, i.e. we know $E_d = (\phi_e - \phi_s)/d$. We also know that the liquid behaves as an electrolyte with mobile free charges that adjust their distribution within the electric double layer according to the Poisson–Boltzmann equation. The potential thus decays in a near exponential fashion from its unknown value at $z = d$ to zero within a distance of a few times the Debye length λ_D. If we neglect for the moment the possible presence of some free charge density σ_s at the solid–liquid interface and consider that the externally applied electrostatic potential is screened exclusively by screening charges within the diffuse layer, we know from Gauss' law applied at the dielectric–liquid interface that the electric displacement field must be continuous, i.e. $\epsilon_d E_d = \epsilon_l E_l$. Making use of a first integral of the Poisson–Boltzmann equation (see Chapter 4), we can express E_l in terms of the unknown potential at the surface ϕ_s and obtain

$$\epsilon_d \frac{\phi_e - \phi_s}{d} = \epsilon_l \frac{2k_B T}{\lambda_D e} \sinh \frac{e\phi_s}{2k_B T} \tag{5.11}$$

The only parameters of the system that enter Eq. (5.11) are the dielectric constants ϵ_d and ϵ_l and the thicknesses d and λ_D. Solving the equation numerically for ϕ_s for typical values of these parameters, we find that ϕ_s does generally not exceed

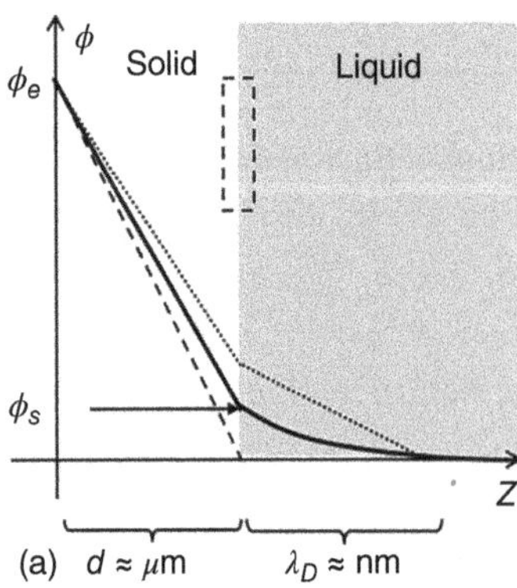

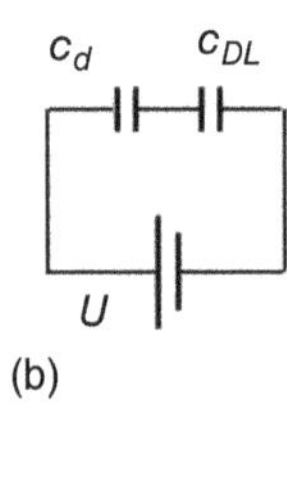

Figure 5.9 (a) Potential distribution at an electrowetting surface with the electrode at $z = 0$ and the dielectric–liquid interface at $z = d$. Solid line: full solution of Eq. (5.11). Dashed: perfect conductor solution ($\phi_s = 0$). Dotted: Debye–Hückel solution. Dashed square: box to apply Gauss' law to derive the boundary condition Eq. (5.11). (b) Schematic representation as two capacitors in series.

a few hundred mV. Hence, the vast majority of the electric potential that we apply in EW experiments indeed drops across the dielectric layer, as assumed in the perfect conductor approximation. We can very simply derive an upper limit for ϕ_s if we linearize the sinh function on the right-hand side of equation Eq. (5.11) (see Problem 5.2). Doing so, the equation simplifies to $\epsilon_d(\phi_e - \phi_s)/d = \epsilon_l \phi_s/\lambda_D$, which is equivalent to replacing the electric double layer by a capacitor with the capacitance per unit area $c_{DL} = \epsilon_0\epsilon_l/\lambda_D$, as it is usually done in the linearized Debye–Hückel theory (see Chapter 4). From this approximation, we obtain the upper estimate

$$\phi_s = \frac{\phi_e}{1 + \frac{\epsilon_l}{\lambda_D}\frac{d}{\epsilon_d}} \approx \frac{c_d}{c_{DL}}\phi_e \tag{5.12}$$

This relation tells us that neglecting the electric double layer is a good approximation as long as $c_d \ll c_{DL}$.

Knowing the surface potential, we can calculate the electrostatic energies per unit area of the two capacitors in series (see Figure 5.9b). In fact, we already discussed this problem at the end of Chapter 2 (see Figure 2.14). There, we realized that in a series circuit, the smaller capacitor, i.e. the dielectric layer in the present case, is not only the one across which the majority the electrostatic potential drops but also the one where majority of the electrostatic energy is stored. From Eq. (2.78) we can indeed conclude that the ratio between the energies stored in the dielectric layer and in the electric double layer scales is indeed $E_d/E_{DL} = c_{DL}/c_d$ (neglecting a possible small correction due to the intrinsic surface charge σ_s). This ratio provides the justification for neglecting the electric double layer in the standard model of EW in the preceding section.

Aside from simplifying modeling and the quantitative analysis of EW experiments, the dominance of the dielectric layer also implies that the properties of the electric double layer at the solid–electrolyte interface and hence chemistry of the materials involved hardly matter in standard EW experiments. As long as possible adsorption and desorption processes of ions at the dielectric surface are reversible, we do not need to care about them. They only contribute a small fraction to the total electrostatic energy of the system anyway. This is an important reason why modern EW on dielectric experiments are so much more robust and reliable than the historic experiments by the pioneers of electrocapillarity with direct metal (mercury)–electrolyte interfaces.

All the comments made above obviously only hold as long as the underlying assumption $c_d \ll c_{DL}$ is indeed fulfilled. Recently, intensive materials research is invested into the development of particularly thin and at the same time dielectrically stable insulating layers in order to reduce the required actuation voltage of EW devices. Advanced fabrication technologies nowadays enable the use of insulating with a thickness well below 100 nm. If the two length scales d and λ_d approach each other, the effect of the electric double layer can no longer be neglected. Hence, in such nano-EW applications, the role of the chemistry of the solid–liquid interface reappears. The first deviation that becomes apparent is that the EW curve in no longer symmetric around zero. That is, the finite value of U_{pzc} becomes apparent in the contact angle response. Figure 5.10 illustrates

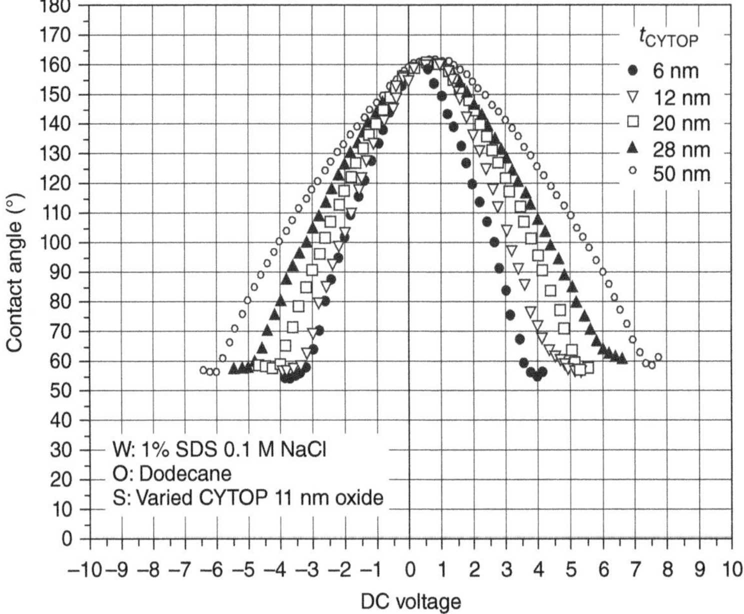

Figure 5.10 Asymmetric electrowetting curve for ultralow voltage EW on a composite nanoscale dielectric film of 11 nm SiOx and a variable thickness of fluoropolymer (6–50 nm) immersed in dodecane in the presence of surfactant. The voltage is applied to the drop and the substrate is kept at ground potential. Source: Berry et al. 2006 [6]. Reproduced with permission of Elsevier.

the effect for a specific system involving an aqueous electrolyte in ambient oil in the presence of a surfactant that reduces the interfacial tension to $\gamma \approx 5\ \text{mJ m}^{-2}$. More examples and specific material-related issues arising in the context of ultralow voltage EW will be discussed in Chapter 7.

Next to the practical consequences for the functional form of the proper EW equation and the symmetry of the EW curves, the dominance of the dielectric layer versus the electric double layer for the storage of electric energy is important for the interpretation of the origin of the EW effect: If the electrostatic energy is stored in the electric double layer, as it is the case in EW without dielectric, it is indeed very much localized at the interface. In this case, it seems natural to assign this electrostatic energy (per unit area) directly to the metal–electrolyte interface and to define a voltage-dependent (free) interfacial energy, as we did in Eq. (5.10). In the presence of a dielectric, however, the energy related to that charge transfer process is distributed over the entire dielectric layer between the drop and the substrate. The energy gain per unit area is therefore by no means localized at the drop–substrate interface, if *interface* is something that is related to the border between two physical phases or two chemically distinct materials. Only if we decide to interpret the entire dielectric layer as part of the interface between the drop and the substrate in a more general manner we can speak of an *effective interface*. Given the typical separation of length scales between d and R and the proportionality between the electrostatic energy and the drop–substrate interfacial area A_{sl}, one may indeed formally choose to do so. Yet, one should be fully aware that the relation between such an effective interface and an ordinary phase boundary is much less direct than in the case of a metal–electrolyte interface. To stress this difference, it is useful to denote the resulting voltage-dependent interfacial energy in the presence of a dielectric as an *effective* interfacial tension in our definition (Eq. (5.5)).

5.2.3 Limitations of the Standard Model: Nonlinearities and Contact Angle Saturation

The entire analysis in the preceding sections was focused on a quantitative description of the EW in the YL regime. Yet, as already stated above, the appearance of the saturation regime in which the contact angle becomes independent of the applied voltage is very universal and one of the most intriguing aspects of EW. We will come back to a variety of specific aspects of this phenomenon in detail in Chapter 7. For now, let us note a few simple observations and conclusions that we can draw based on our discussion so far. According to macroscopic numerical calculations such as the ones described in the preceding section, the YL equation holds at least down to the smallest apparent contact angles studied of approximately 5°. No indication of an onset of deviations was found. This is in amazing contrast to experimental observations that typically report deviations starting to appear for angles of the order of 60–80°. While the details of the origin of contact angle saturation remain unclear at the time of the writing of this book, the discussion in Section 5.1 provides one important clue that is generally believed to be at the heart of the problem: The electric field diverges upon approaching the three-phase contact line. This fact was already pointed out by Berge and coworkers in their earliest report of the phenomenon in 1999 [9]. These authors discussed a series of remarkable phenomena that were correlated with the saturation of the contact angle: the appearance of spikes in the electrical current into the system, the appearance of optical luminescence of the (specifically chosen) ambient gas in the vicinity of the contact line [9], and the emission of small satellite droplets (Figure 5.11a). These phenomena can be related to a nonlinear response of the materials related to the diverging electric fields around the contact line. The emission of the luminescence light was attributed to a dielectric breakdown in the ambient air. Air is actually known to be a rather poor dielectric material. The dielectric strength of air, i.e. the maximum electric field that it can tolerate while still remaining nonconductive, is of the order of $3\,\mathrm{kV\ mm^{-1}}$. This value refers to dry air. In the vicinity of the drop surface, where we can expect the air to be saturated with water vapor, the value will be substantially lower.

The occurrence of spikes in the electrical current was attributed to some form of dielectric breakdown of the insulating layer. This effect can be captured by implementing the following very generic model of dielectric breakdown into a numerical simulation: A volume element of the material is considered to behave as a regular linear dielectric up to a certain critical value of the local electric field, the material's dielectric breakdown strength. If the local field exceeds this threshold value, however, the material within the specific volume element in question

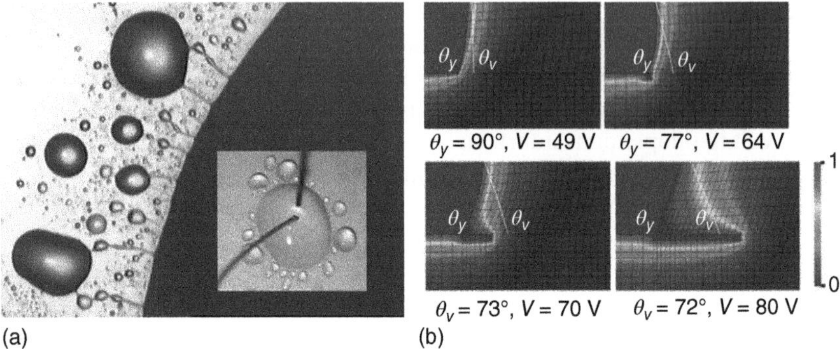

Figure 5.11 Field-induced microscopic phenomena that give rise to contact angle saturation. (a) Emission of highly charged satellite drops from the contact line of a water drop on a hydrophobized glass surface (see also [10]). (b) Numerical calculation of the electrostatic potential in the vicinity of the contact line for an insulator that turns locally conductive if the local electric field exceeds a certain threshold value. (See text for details.) Source: Reproduced with permission from [11].

becomes conductive, while the rest of the insulating layer keeps its dielectric properties. Upon implementing this model in a self-consistent numerical simulation in the spirit of the preceding section, Papathanasiou and coworkers [11] found that a horizontal *finger* develops on the top of the dielectric layer that turns conductive (Figure 5.11b). This now conductive part of the layer assumes the electrostatic potential of the drop, to which it is directly connected. As a consequence it screens the electric field from the contact line and therefore limits the further increase of the force acting on the contact line and hence the further reduction of the contact angle. Such a *chemical* conversion of the material is also consistent with the frequent observation of more hydrophilic regions on the surface around the contact line after excursions to very high voltages well into the saturation regime.

While the dielectric breakdown of the air and the insulating layer is caused by a nonlinear material response, the emission of the satellite drops is caused by an instability of the liquid surface. Like the nonlinear material response, this instability is caused by the diverging electric field and the related divergence of the local charge density along the contact line. In essence, the process is similar to electrospraying as frequently used in mass spectrometry of biological substances. If the local electric field and charge density are too high, surface tension is no longer capable of stabilizing the liquid surface. As a consequence, the drop breaks up and begins to emit (highly charged) satellite droplets. In the case of a cylindrically symmetric geometry as in electrospraying, analytical solutions for the drop shape (Taylor cone) are available for certain electrode configurations. For EW, the situation is qualitatively similar, yet, due to the lack of symmetry, no analytical solution is available. The absence of the instability in the two-dimensional and the axisymmetric simulations discussed above suggests that the instability is in fact caused by a transverse instability with an unstable mode along the contact line. While this scenario has been sketched already by Berge and coworkers, a quantitative fluid dynamic analysis has been hampered by the complexity of the geometry (see also Chapter 7). Contact angle saturation is believed to arise from the fact that the ejected satellite drops are charged and therefore screen the electric field, qualitatively similar to the *conductive finger* discussed above in the context of the local dielectric breakdown model.

Interestingly, a related effect has been observed on the atomic scale in molecular dynamics (MD) simulations. Such simulations are rather difficult to perform because of the different ranges of molecular interactions that are usually cut off after a few molecular diameters in MD simulations and the long range of electrostatic forces. The challenge is therefore to devise a scheme that allows for a consistent treatment of short-range molecular and long-range electrostatic forces at affordable computational costs. Carefully carried out simulations indeed reproduce all the macroscopic observations described above, provided that the applied potentials are not too high [12]. In particular, it turns out that the apparent contact angle in such simulations indeed follows the same YL equation as in the macroscopic theory, while the local contact angle remains essentially constant. In contrast to the macroscopic theory, however, MD simulations provide insight into another microscopic mechanism of contact angle saturation: At very high voltage, the force acting on individual ions in the drop becomes stronger than their cohesion with the solvent. As a consequence, the counterions that are supposed to screen the charge at the drop surface are ripped out of the liquid and deposited on the solid surface next to the drop. There, they act as screening charges that limit a further increase of the force acting on the drop, very much like emitted and evaporated satellite drops.

5.3 DC Versus AC Electrowetting

5.3.1 General Principles

The standard EW theory as described above does not depend on the sign nor the frequency of the applied voltage. If we modulate the applied voltage at AC frequencies, the observed response depends on the frequency range. Basically, there are two different types of characteristic frequencies in EW systems – first the mechanical (hydrodynamic) resonance frequencies ω_m of the droplet involved and second the electrical transition frequency ω^c_{el} from the perfect conductor regime to the dielectric regime – in which the electric field penetrates the fluid. For typical drop sizes in EW, hydrodynamic resonances are in the range of tens to hundreds of Hz. The electrical transition from conductive to dielectric behavior is governed by the conductivity, the dielectric constant, and the geometry of the drop, as we discussed in Section 2.2 in the context of the leaky dielectric model. For the conditions of typical EW experiments, the critical frequency is in the range of several tens to hundreds of kHz. Hence, we can distinguish three regimes of AC EW: (i) $\omega < \omega_m$, (ii) $\omega_m < \omega < \omega^c_{el}$, and (iii) $\omega > \omega^c_{el}$. In regime (i) with typical frequencies of a few Hz, the contact angle and the shape of the liquid follow the predicted equilibrium behavior quasi-statically. At somewhat higher frequencies, at the transition between regime (i) and (ii), mechanical eigenmodes of the drop are excited resonantly. This leads to a strongly frequency-dependent response with very large oscillation amplitudes whenever the excitation frequency precisely matches with one of the resonance frequencies. The exact shape of these hydrodynamic modes can be complex, depending on the geometric boundary conditions. In Chapter 6 we will illustrate the calculation for the example of the classical eigenmodes of a freely oscillating drop. At the intermediate frequencies of regime (ii), inertia prevents the drop from following the AC excitation. At the same time, the drop still behaves as a perfect conductor because we are still in the regime $\omega < \omega^c_{el}$, where the resistance of the drop is low enough to guarantee perfect screening. For such frequencies, the drop only responds to the time average of the applied voltage. For a sinusoidal voltage $U(t) = U_0 \sin \omega t$ with $\omega = 2\pi/\tau$, the fact that $f_{el}(t) = c\, U(t)^2/2$ is quadratic in U results in a time-averaged electrostatic force per unit:

$$f_{el}^{AC} = \frac{1}{\tau}\int_0^{\tau} f_{el}(t)\, dt = \frac{c}{2} U_{RMS}^2 \tag{5.13}$$

This expression is exactly the same as Eq. (5.6), except for the fact that we replaced the static voltage U by its root-mean-square (RMS) value. (We also omitted the minus sign, implicitly understanding that f_{el} is always oriented along the outward normal of the fluid.) Inserting this force into the force balance at the contact line, we obtain again at the standard YL expression for the reduction of the time-averaged apparent contact angle:

$$\cos\theta_{RMS}(U_{RMS}) = \cos\theta_Y + \frac{c_d}{2\gamma} U_{RMS}^2 \tag{5.14}$$

which again is equivalent to the YL equation, Eq. (5.4), except for the fact that U_{RMS} replaces the static voltage. Not too surprisingly, AC-EW is hence perfectly equivalent to EW with DC voltage, as long as we insert the RMS value of the applied voltage. Given the equivalence of the response, the explicit reference to the RMS values is often skipped in the literature.

At even higher excitation frequencies, for $\omega \approx \omega^c_{el}$, the electric field begins to penetrate the drop phase. As a consequence, the Maxwell stress at the interface and the force acting on the contact line decrease. Hence the EW gradually disappears, and the contact angle at fixed voltage increases as the AC frequency is increased from regime (ii) into regime (iii). The field penetration and the gradual reduction of the EW response in this frequency range can be seen directly in Figure 5.12.

To calculate the contact angle response at such high frequencies, we need to extend our treatment of the preceding section and describe the response of the drop by the leaky dielectric model that we introduced in Section 2.2. For the leaky dielectric, we need to calculate the potential distribution by solving the Laplace equation inside the drop, i.e. in domain *1* of Figure 5.5b, as well. In addition, we need to apply appropriate boundary conditions as the interface between the drop and the ambient dielectric. Following our analysis in Section 2.2, the actual charge at the interface is composed of a combination of polarization charge and free charge. Charge conservation at the interface leads to the boundary condition

$$\vec{j}_1\vec{n} = (\lambda_1\vec{E}_1 + \epsilon_0\epsilon_1\partial_t\vec{E}_1)\,\vec{n} = (\epsilon_0\epsilon_2\partial_t\vec{E}_2)\,\vec{n} = \vec{j}_2\vec{n} \tag{5.15}$$

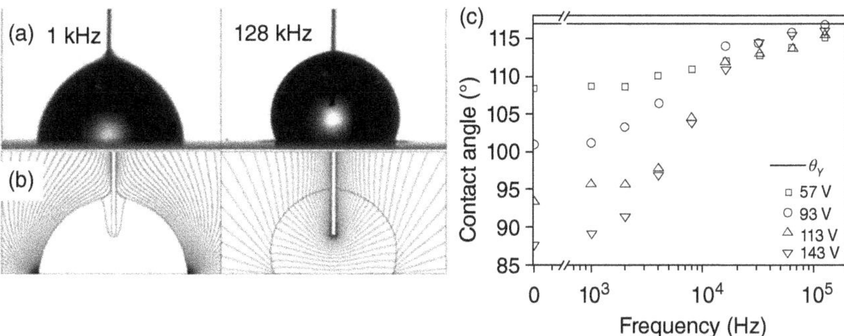

Figure 5.12 Frequency-dependent contact angle response for a typical millimeter-sized sessile drop with a conductivity of 5×10^{-4} S m^{-1} at low at high AC frequency. (a) Optical view graphs. (b) Equipotential lines from a numerical simulation. (c) Experimental contact angle versus frequency for a series of applied RMS voltages as indicated. Source: Hong et al. 2008 [13]. Adapted with permission of Springer Nature.

where $\vec{n}$ is the local surface normal of the drop surface. Solving the appropriate equations, it turns out that the potential within the drop is indeed constant at low frequencies, as expected for a perfect conductor with perfect screening. At high frequencies, however, the electric field penetrates into the drop, as shown by the distribution of equipotential lines within the drop in Figure 5.12b. The decrease in density of equipotential lines at higher frequencies implies a reduction of the local electric field and hence a reduction of the EW effect (Figure 5.12c), as anticipated. Knowing the potential distribution, we can calculate the Maxwell stress tensor and subsequently all details of the stress distribution and the equilibrium configuration of the interface using the standard formulas discussed in Chapter 2, Eqs. (2.61)–(2.64). The reduction of the peak electric field also has positive aspects: It reduces the electrical load on the materials and therefore helps to reduce the threat of field-induced damage of the dielectric material. This effect may contribute to the observation that EW devices operated by AC voltage are frequently more robust and have a longer lifetime.

While the treatment of leaky dielectrics under AC voltage given here reproduces the reduction of the EW effect at high frequencies, the complete response of leaky dielectrics at high frequencies is actually even more complex. Due to the finite ohmic resistance of the drop, the electrical currents within the drop give rise to temperature gradients. The temperature gradients generate gradients in electrical conductivity, dielectric constant, and density. The gradients give rise to additional body forces as discussed in Chapter 2 (see, e.g. Eq. (2.61)) that drive flows within the drop. The calculation of these *electrothermal flows* is beyond the scope of this book. The interested reader is referred to the original literature [14, 15]. A few practical examples will be discussed qualitatively in Chapter 8.

5.3.2 Application Example: Parallel Plate Geometry

For a three-dimensional drop, the calculation of the frequency dependence of the EW response in the leaky dielectric regime requires a full numerical solution of the governing equations, as described above. However, the same transition can also be observed and analyzed by the simple analytical formulas of the leaky dielectric model in a simpler geometry of a liquid sandwiched between two parallel electrodes, as shown in Figures 5.13 and 5.14. Figure 5.13a–c shows the response of the contact angle for three different conductivities of such sandwiched drops in a frequency range between DC voltage and 100 kHz. Clearly, the response of the contact angle strongly decreases with increasing frequency for the lowest salt concentration, but remains frequency independent at the highest. From a practical perspective this result directly leads to the comforting conclusion that adding even small amounts of salt (the highest conductivity in Figure 5.13 corresponds to concentration of a few mM) is sufficient to suppress finite conductivity effects for typical EW applications.

The advantage of the simple quasi-one-dimensional geometry is that the electric fields and currents can be regarded as being oriented normal to the surfaces. In this case, the electrical conductivities λ and the dielectric constants

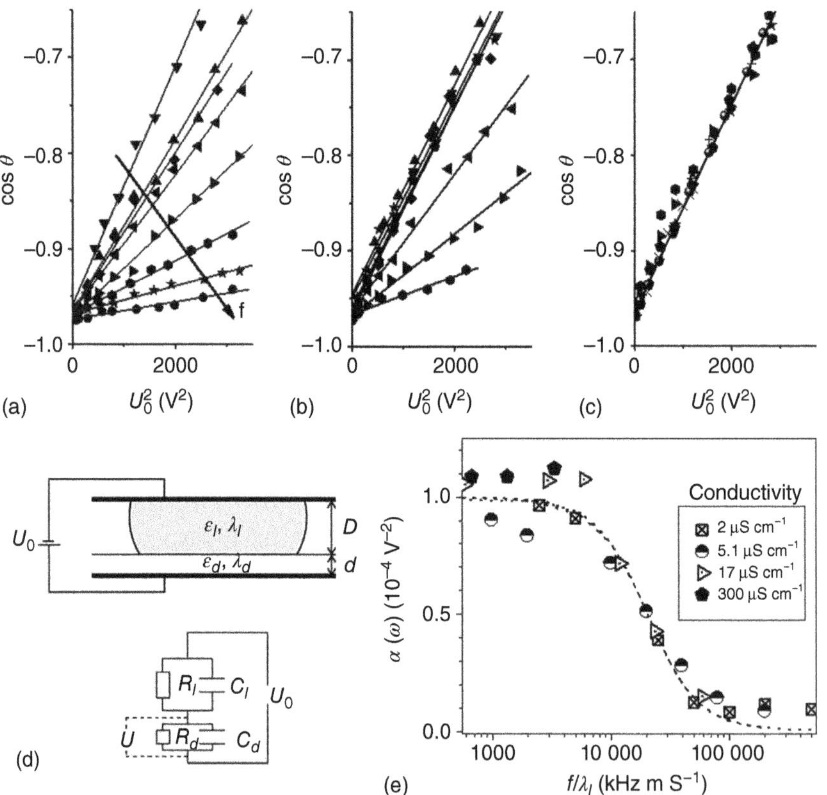

Figure 5.13 Frequency dependence of the EW response at variable conductivity. (a)–(c) $\cos\theta$ versus U_0^2 for conductivities of 5 (a), 17 (b), and 300 (c) $\times 10^{-4}$ S m^{-1} at frequencies of DC, 0.5, 1, 5, 10, 20, 40, 100 kHz increasing top to bottom. (d) Experimental sandwich geometry and corresponding equivalent circuit. (e) Master curve of the frequency cutoff for variable conductivity. Source: Data taken from [16].

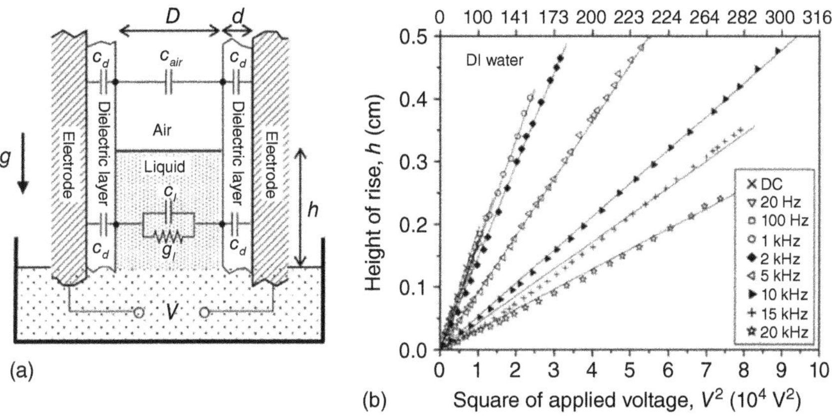

Figure 5.14 (a) Capillary rise experiment with EW-functionalized capillary walls and illustration of the capacitive and resistive elements of the equivalent circuit diagram. (b) Resulting coefficient *K* of the linear relation $h(\omega) = K(\omega)\, U^2$ (cf. Eq. 5.18). Source: Jones et al. 2004 [17]. Reproduced with permission of ACS.

ϵ of the materials, the fluid and the dielectric layer, are easily combined with the corresponding layer thicknesses to effective capacitors and resistors in a simple equivalent circuit model (Figure 5.13d). If we treat the dielectric layer as perfect dielectric with infinite resistivity, we can neglect the resistance R_d in Figure 5.13d. Then the system is characterized by the ohmic resistance and the capacitance of the liquid $R_l = D/A\lambda_l$ and $C_l = \epsilon_l\epsilon_0\, A/D$ and the capacitance $C_d = \epsilon_d\epsilon_0 A/d$ of the dielectric layer, where A is the footprint area of the drop. Using standard complex notation, we find that the effective impedance of the entire circuit is given by $Z_{eff} = i\omega C_l/(1 + i\omega R_l C_d)$. The relevant voltage for the EW response is given by the potential drop U_d across the dielectric layer. If we write the applied voltage as $U(t) = Re\,\{U_0 e^{i\omega t}\}$,we find after a few lines of algebra

$$U_d = Re\left\{\frac{1 + i\omega R_l C_l}{1 + i\omega R_l(C_l + C_d)}\, U_0\right\} = \frac{1 + \omega^2 R_l^2 C_l(C_l + C_d)}{1 + \omega^2 R_l^2(C_l + C_d)^2} U_0 \tag{5.16}$$

As discussed in Section 2.2, this expression is governed by two dimensionless parameters, the characteristic time $\tau_0 = R_l C_d = D\epsilon_d\epsilon_0/\lambda d$ and the capacitance ratio $\alpha = C_l/C_d$. $\tau_0^{-1} \approx \omega_{el}^c$ is the characteristic transition frequency from the perfectly conductive to the dielectric response regime. Note that the relaxation time τ_0 depends on the geometric parameters D and d of the system. It is usually much longer than the intrinsic charge relaxation time $\tau_c = \epsilon_l\epsilon_0/\lambda$ of the liquid in the bulk. This dependence on the details of the geometry is characteristic for the regime of dielectrophoretic forces, in which the electric field is less localized and frequently extends over the entire system.

Historically, the first systematic experiments to explore the dependence of EW on the AC frequency were carried out by T.B. Jones and coworkers in the early 2000s. His experiments were very important in establishing the understanding of the relation between EW and dielectrophoresis, despite the fact that they did not measure variations of the contact angle but rather focused on the ponderomotive force exerted by the electric field in EW-induced capillary rise experiments (see Figure 5.14). EW-functionalized plates were immersed in a bath of conductive fluid, and the height of rise of the meniscus was monitored as a function of the amplitude U and – in particular – of the frequency of the applied AC voltage. Force balance between gravity, capillary forces, and the EW force per unit length directly shows that the height of rise is given by

$$h(U) = h_0 + K(\omega)U^2 \tag{5.17}$$

where $h_0 = 2\gamma\cos\theta_Y/\rho gD$ is the height of rise due to capillarity alone with D being the distance between the plates and ρ being the density of the liquid:

$$K(\omega) = \frac{\epsilon_0\epsilon_d^2\left(\epsilon_0^2\omega^2(\epsilon_l - 1)\left(\epsilon_l d + \frac{\epsilon_d D}{2}\right) + \lambda_l^2 d\right)}{4\rho g(2d + \epsilon_d D)\left(\left(\epsilon_0^2\omega^2\left(\epsilon_l d + \frac{\epsilon_d D}{2}\right)^2 + \lambda_l^2 d^2\right)\right)} \tag{5.18}$$

is a frequency-dependent coefficient that arises from the analysis of the equivalent circuit model along the same lines as discussed above (see Figure 5.14a). The reader is invited to derive the expression for $K(\omega)$ and analyze its low and high frequency limits (see Problem 5.4).

Problems

5.1 Calculate the secular equation for the electric field distribution as sketched in Figure 5.8b. See also Ref. [4].

5.2 Show that linearizing the sinh function in Eq. (5.11) indeed leads to an upper estimate of ϕ_s.
Hint: Use a graphic representation of the equation and rationalize the solution in terms of properties of the sinh function.

5.3 Calculate the surface potential using Eq. (5.12) for a typical dielectric layer ($\epsilon_d = 2$) with thicknesses of $d = 0.01, 0.1, 1$ µm and for aqueous solutions of a monovalent salt with bulk concentrations of $c = 0.1, 1, 10, 100$ mM. Indicate the conditions for which neglecting the electric double layer becomes questionable.

5.4 (a) Derive the expression Eq. (5.18) using the equivalent circuit model sketched in Figure 5.14a. (Note the specific definition of D and d in the figure.) (b) Repeat the derivation of the same expression using the Maxwell stress tensor formalism. Choose a suitable control volume. (c) Show that ratio of the height of rise in the low frequency limit to the high frequency limit is given by $K(0)/K(\infty) = \epsilon_d\ D/((\epsilon_l - 1)d)$. Interpret this ratio in terms of the contact line force in EW and the expected dielectrophoretic force.

References

1 Paneru, M., Priest, C., Sedev, R., and Ralston, J. (2010). Static and dynamic electrowetting of an ionic liquid in a solid/liquid/liquid system. *J. Am. Chem. Soc.* 132: 8301–8308.

2 Banpurkar, A.G., Nichols, K.P., and Mugele, F. (2008). Electrowetting-based microdrop tensiometer. *Langmuir* 24: 10549–10551.

3 Buehrle, J., Herminghaus, S., and Mugele, F. (2003). Interface profiles near three-phase contact lines in electric fields. *Phys. Rev. Lett.* 91 (8): 086101.

4 Mugele, F. and Buehrle, J. (2007). Equilibrium drop surface profiles in electric fields. *J. Phys. Condens. Matter* 19 (37): 375112.

5 Papathanasiou, A.G. and Boudouvis, A.G. (2005). Manifestation of the connection between dielectric breakdown strength and contact angle saturation in electrowetting. *Appl. Phys. Lett.* 86 (16): 164102.

6 Berry, S., Kedzierski, J., and Abedian, B. (2006). Low voltage electrowetting using thin fluoroploymer films. *J. Colloid Interface Sci.* 303 (2): 517–524.

7 Kang, K.H. (2002). How electrostatic fields change contact angle in electrowetting. *Langmuir* 18 (26): 10318–10322.

8 Chatterjee, D., Hetayothin, B., Wheeler, A.R. et al. (2006). Droplet-based microfluidics with nonaqueous solvents and solutions. *Lab Chip* 6 (2): 199–206.

9 Vallet, M., Vallade, M., and Berge, B. (1999). Limiting phenomena for the spreading of water on polymer films by electrowetting. *Eur. Phys. J. B* 11 (4): 583–591.

10 Mugele, F., Klingner, A., Buehrle, J. et al. (2005). Electrowetting: a convenient way to switchable wettability patterns. *J. Phys. Condens. Matter* 17 (9): S559–S576.

11 Drygiannakis, A.I., Papathanasiou, A.G., and Boudouvis, A.G. (2009). On the connection between dielectric breakdown strength, trapping of charge, and contact angle saturation in electrowetting. *Langmuir* 25 (1): 147–152.

12 Liu, J., Wang, M., Chen, S., and Robbins, M.O. (2012). Uncovering molecular mechanisms of electrowetting and saturation with simulations. *Phys. Rev. Lett.* 108 (21): 216101.

13 Hong, J.S., Ko, S.H., Kang, K.H., and Kang, I.S. (2008). A numerical investigation on AC electrowetting of a droplet. *Microfluid. Nanofluid.* 5 (2): 263–271.

14 Garcia-Sanchez, P., Ramos, A., and Mugele, F. (2010). Electrothermally driven flows in AC electrowetting. *Phys. Rev. E* 81: 015303.

15 Lee, H., Yun, S., Ko, S.H., and Kang, K.H. (2009). An electrohydrodynamic flow in ac electrowetting. *Biomicrofluidics* 3 (4): 044113.

16 Kumar, A., Pluntke, M., Cross, B. et al. (2005). Charged droplet generation and finite conductivity effects in AC electrowetting, Materials Research Society Fall Meeting, Boston.

17 Jones, T.B., Wang, K.L., and Yao, D.J. (2004). Frequency-dependent electromechanics of aqueous liquids: electrowetting and dielectrophoresis. *Langmuir* 20 (7): 2813–2818.

6
Elements of Fluid Dynamics

The dynamics of the response of the fluid are an important element in almost every application of electrowetting (EW). In the vast majority of applications, the electrodynamic aspects of the problem are rather simple. The activation of the electrodes and the buildup of the electric double layers discussed in the preceding chapters occur on time scales that are much faster than the hydrodynamic response time of the drop. Moreover, we are in most cases only interested in the dynamics of the drop on a global scale, i.e. we do not worry about most of the microscopic deformations of the surface in the vicinity of the contact line that were discussed in Chapter 5. Under such conditions, activating or deactivating an electrode corresponds to an abrupt change of the local contact angle on top of the activated electrode. The subsequent dynamics of the drop are governed by the conventional hydrodynamic response of the drop to this change of a boundary condition. As a consequence, the dynamic response of EW systems is governed by the same general Navier–Stokes equations as conventional fluid dynamics. Understanding these equations along with a few generic examples is the topic of this chapter. In Section 6.1, we review the basic governing equations and discuss the general simplifications that typically arise in microfluidic systems. Section 6.2 is devoted to two flow geometries with a separation of length scales, namely, first the Hele–Shaw approximation of flows in quasi-two-dimensional flows, such as the typical sandwich geometry of lab-on-a-chip systems in EW, and second the lubrication approximation, in which the thin flowing layer is bounded by a free liquid surface. In both cases, the liquid films are much thinner than the characteristic lateral extension of the system. In Section 6.3, we apply the lubrication approximation to discuss basic elements of contact line dynamics, a hot topic in microscale fluid dynamics that is crucial for the dynamics of many EW systems. In Section 6.4, we discuss the spectrum of surface waves and the eigenmodes of drops that result from a balance of capillary and inertial forces. The corresponding time scales provide a limit to the fastest response time of EW systems.

6.1 Navier–Stokes Equations

Fluid mechanics may look scary at first glance, and in fact solving fluid dynamic problems can be rather tedious because it typically involves solving partial differential equations. Nevertheless, for a conceptual point, the governing equations of

Electrowetting: Fundamental Principles and Practical Applications,
First Edition. Frieder Mugele and Jason Heikenfeld.

fluid mechanics are not more than an extension of Newton's laws of the mechanics of point masses to continuous media. In this section, we discuss the basic principles of this step, which leads to the derivation of the Navier–Stokes equations of fluid dynamics. Along the way, we will introduce the continuity equation of matter and the viscous stress tensor.

6.1.1 General Principles: from Newton to Navier–Stokes

The starting point of our consideration is Newton's second law

$$m\vec{a} = m\frac{d\vec{v}}{dt} = \vec{F} \tag{6.1}$$

which describes the acceleration of a point mass m under the influence of an external force $\vec{F}$. In discrete approaches to the dynamics of fluids such as molecular dynamics simulations, we write down Eq. (6.1) for each particle. In this case, the force $\vec{F}$ typically depends on the position of all (or at least many) of the other particles resulting in complex sets of hundreds of thousands of coupled equations of motion. Collective behavior is then obtained by averaging, e.g. the local velocities over certain finite volumes in space. In continuum mechanics, the molecular character of matter is ignored and point mass does not exist. Instead we consider the fluid as a continuous medium that is characterized by an in general position- and time-dependent mass density $\rho(\vec{r}, t)$ and by a velocity field $\vec{v}(\vec{r}, t)$. Additional material properties include the viscosity μ, the dielectric constant ϵ, possibly an electric charge density ρ_{el}, etc. To describe the dynamics of the continuous medium, we consider volume element dV (Figure 6.1), as we already did in the context of static equilibria in the Appendix of Chapter 1 and in Chapter 2, Section 2.4.2, but now in the presence of an arbitrary flow field $\vec{v}(\vec{r}, t)$. Because we now consider dynamic situations, we have to worry about the flux of matter into and out of our control volume. Matter is a conserved quantity. Hence, any change in mass must be caused by a mass flux into or out of dV through its surface. The mass flux density is given by $\vec{j} = \rho\vec{v}$, and hence the conservation of mass reads

$$\frac{d}{dt}M = \partial_t \int \rho \, dV = -\oint \vec{j} d\vec{A} = -\int div\, \rho\vec{v} dV \tag{6.2}$$

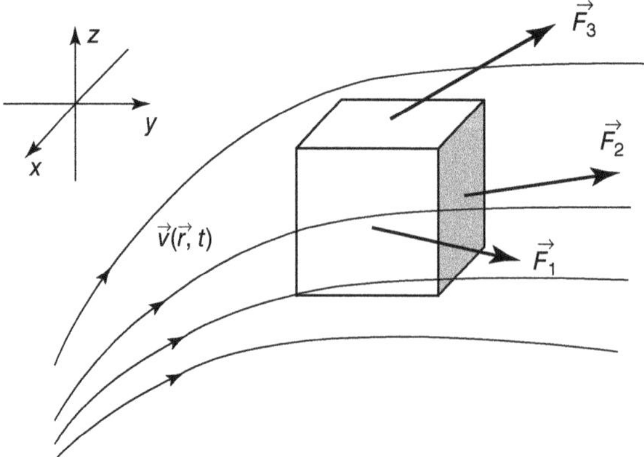

Figure 6.1 Control volume of fluid in an arbitrary flow field. Forces exerted by the ambient fluid and/or body forces acting on the visible surface elements of the control volume are illustrated. Fluxes $\vec{j}$ cross the surfaces of the volume element in analog manner.

Because this conservation law must hold universally for any control volume, we can rewrite the expression as a local conservation equation, the continuity equation, in differential form:

$$\partial_t \rho + div\ \rho \vec{v} = 0 \tag{6.3}$$

For the purpose of the present book, fluids can be considered as incompressible and homogeneous in density, i.e. $\rho(\vec{r}, t) \equiv \rho$, where ρ is a fixed mass density. In fact this approximation is almost universally applicable in microfluidics, except for applications involving ultrasound and sometimes for gas flows. For conventional incompressible flows, the continuity equation takes the simple form

$$div\ \vec{v} = \partial_x v_x + \partial_y v_y + \partial_z v_z = 0 \tag{6.4}$$

The continuity equation thus establish a relation between the gradients in flows in different directions. Such relations are frequently useful to simplify the solution of fluid mechanical problems, as we will see, e.g. in the context of thin film flows in Section 6.2.

Next we consider the acceleration of our volume element dV on the left-hand side of Eq. (6.1). Because $\vec{v} = \vec{v}(\vec{r}(t), t)$, the total time derivative in Eq. (6.1) formally involves partial derivatives with respect to both time and space, where the latter is followed by an inner time derivative of $\partial_t \vec{r} = \vec{v}$:

$$\frac{d\vec{v}(\vec{r}, t)}{dt} = \partial_t \vec{v} + (\vec{v}\nabla)\vec{v} = \partial_t \vec{v} + (v_x \partial_x + v_y \partial_y + v_z \partial_z)\vec{v} \tag{6.5}$$

Physically, this implies that the velocity of our volume element changes for two reasons: first of all because of an increase of the velocity at the position $\vec{r}$ and second of all because the volume element dV itself is transported (*advected*) along with the flow to another position where the local velocity is different. This advective term gives rise to a nonlinear dependence of the equation of motion, which is at the origin of many complex phenomena in fluid dynamics, including in particular turbulence. As we will see below, it is usually negligible in microfluidic applications.

Having translated the inertial term of Newton's equation into a continuum mechanics language, we now need to analyze the right-hand side of the equation, i.e. the forces. From our considerations of gravitational and electrostatic forces in Chapters 1 and 2, we already know that the matter in our volume element dV can experience body forces such as $\vec{f}_g = -\rho g \vec{e}_z$ and $\vec{f}_{el} = \rho_{el}\vec{E} - \epsilon_0 E^2 \nabla\epsilon/2$. Moreover, we already considered the forces acting as stresses on the surface of dV. The total force experienced by dV is therefore

$$\vec{F}_i = \oint \sum_j T_{ij}\hat{n}_j\, dA + \int \vec{f}_{b,i} dV \tag{6.6}$$

where $\vec{f}_b$ is the sum of all body forces and T_{ij} is the total stress tensor.

In the absence of flow, the stress tensor is diagonal and isotropic. Its absolute value is given by the local hydrostatic pressure p, i.e. we have $T_{ij} = -p\ \delta_{ij}$, implying that all stresses are normal to the surface elements of dV. In the presence of fluid flow, however, this is no longer the case. Viscous forces give rise to shear stresses that appear as off-diagonal elements in the stress tensor. This will give

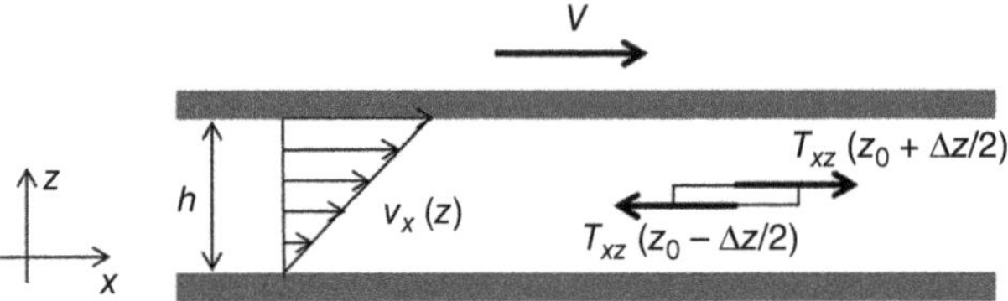

Figure 6.2 Shear flow between two parallel plates and force balance on a control volume of height Δz placed at an arbitrary vertical position z_0.

rise to deformations of our volume element during flow. These internal deformations, which are absent by definition in a system consisting of point masses, give rise to internal *viscous* dissipation. To understand the origin and the nature of viscous stresses, we go back to the original observation by Newton who studied the force required to slide two parallel plates separated by a liquid layer of thickness h with respect to each other at some constant speed V; see Figure 6.2. He noted that the force is given by

$$F = \mu \frac{V}{h} A \tag{6.7}$$

where A is the area of the plate and μ is a material property of the liquid, namely its shear viscosity. Since then, we denote liquids that display this kind of mechanical response to shear stresses as *Newtonian liquids*. Many fluids including water, oil, and common solvents that we typically encounter in microfluidic experiments display such *Newtonian* behavior. Exceptions are so-called complex fluids, typically solutions and suspensions of molecules and particles at sufficiently high concentrations or pure liquids consisting of molecules with intrinsically complex architecture and interactions such as polymer melts and liquid crystals. Such materials can be anisotropic or display a nonlinear response as a function of the applied shear force, in which case the viscosity either increases (shear thickening) or decreases (shear thinning) with increasing shear rate. Complex fluids can also display a finite shear stress, and their *viscoelastic* response depends in general on the frequency of the applied external forcing. For the rest of this book, we limit our considerations to simple Newtonian fluids with a fixed shear viscosity μ.

To proceed, we need to break down the macroscopic behavior described in Eq. (6.7) and explore its consequences for a microscopic control volume inside the fluid. First of all, we can rewrite Eq. (6.7) as a stress per area of the plate $F/A = \mu V/h$. In a stationary state, this stress that we apply to the top plate is balanced by an equal and oppositely oriented stress that a fluid element just next to the plate exerts on the solid. Our fluid element at the solid–liquid interface experiences the same stress F/A toward the right from the top surface. To remain in steady state, it must also experience an oppositely oriented stress of the same magnitude from the next fluid layer underneath. Continuing this reasoning, we realize that there is a constant stress $T_{xz}^{visc} = F/A$ present everywhere in the fluid between the two plates. For symmetry reasons the velocity in the parallel plate geometry varies linearly from zero to V across the gap, i.e. $v_x(z) = Vz/h$. We find thus that the stress in the sheared fluid is related to the velocity gradient by $T_{xz}^{visc} = \mu \partial_z v_x$. This linear relation between the shear stress and the velocity gradient, which is equal to the deformation (strain) rate, is called the constitutive equation of a Newtonian fluid. Generalizing to arbitrary spatial coordinates and symmetrizing, the final expression for the constitutive equation relating the viscous stress tensor to the deformation rates $\partial_i v_j$ reads

$$T_{ij}^{visc} = \frac{\mu}{2}\,(\partial_j v_i + \partial_i v_j) \tag{6.8}$$

If we know that a liquid behaves Newtonian, we can use Eq. (6.8) to calculate the stresses acting on a control volume inside the fluid in the presence of an external flow field. For our control volume on the right-hand side of Figure 6.2 placed at an arbitrary height z_0, we can write $F_x/A = T_{xz}^{visc}(z_0 + \Delta z/2) - T_{xz}^{visc}(z_0 - \Delta z/2) = \partial_z T_{xz}^{visc}(z_0)\Delta z$. Inserting the constitutive equation, this yields a shear stress $F_x/A = \mu\partial_{zz}v_x\ \Delta z$ and a total viscous force per volume element $f^{visc} = \mu\partial_{zz}v_x$, which is homogeneous across the entire system for this simple flow geometry.

Generalizing the preceding considerations to a volume element in an arbitrary flow field $\vec{v}(\vec{r}, t)$, as shown in Figure 6.1, the net force in the ith direction is given by

$$F_i^{visc} = \sum_j \oint T_{ij}^{visc}\,\hat{n}_j dA = \sum_j \partial_j T_{ij}^{visc}\,\Delta V \tag{6.9}$$

Inserting again the constitutive equation for a Newtonian fluid, we obtain a net viscous force per unit volume of

$$f_i^{visc} = \mu \sum_j \partial_{jj} v_i = \mu\,\nabla^2 v_i \tag{6.10}$$

Having collected all these terms, we are now in the position to rewrite Eq. (6.1) complete for a liquid, i.e. we can write down the Navier–Stokes equations:

$$\rho(\partial_t \vec{v} + (\vec{v}\cdot\nabla)\vec{v}) = -\nabla P + \mu\nabla^2\vec{v} + \vec{f}_b \tag{6.11}$$

In this version, we imply that the isotropic pressure P includes contributions to the stress tensor caused by the body forces $\vec{f}_b$. This becomes more clear if we express the first two terms on the right-hand side using the divergence of the total stress tensor and write in tensor notation

$$\rho(\partial_t v_i + (v_j\partial_j)v_i) = \partial_j T_{ij} + f_{b,i} \tag{6.12}$$

The Navier–Stokes equations are thus a set of coupled nonlinear partial differential equations for the flow field $\vec{v}(\vec{r}, t)$. In addition to the flow field, there is the (scalar) pressure field $p(\vec{r}, t)$ that needs to be calculated along with the solution for $\vec{v}(\vec{r}, t)$.

6.1.2 Boundary Conditions

As for any other differential equation, solving the Navier–Stokes equations requires boundary conditions regarding both time and the spatial coordinates. While the temporal boundary conditions are usually determined by the specific situation of an experiment – or sometimes not interesting at all if we search for a steady state or a periodic solution – the spatial boundary conditions are determined by the specific behavior of the velocity field at the interfaces with the geometric boundaries that constrain the flow. Typically, these boundaries consist of the surfaces of the solid containers or tubing that surrounds the liquid. In addition, we frequently encounter free interfaces between two immiscible liquid phases such as water and oil and water and air in applications of EW.

If we consider the velocity field at an interface, we can distinguish between the component tangential to the interface and the component normal to the interface. For the normal component, the consideration is very simple: Fluid cannot penetrate the adjacent solid phase, and there is also no fluid coming out of the solid. As a consequence, the normal component of the velocity field of the liquid $\vec{v}_{n,liq}$ must be the same as the normal component of velocity of the interface $\vec{v}_{n,int}$ itself. The same reasoning holds equally for the interface between two immiscible fluids:

$$\vec{v}_{n,liq} = \vec{v}_{n,int} \tag{6.13}$$

Formally, we can derive this so-called kinematic boundary condition by considering the continuity equation, Eq. (6.3), for a control volume that contains the interfaces, as sketched in Figure 6.3a. For interfaces between two fluid phases, we need to consider in addition the discontinuity of the pressure field caused by the interfacial tension. This discontinuity is given by Laplace's equation, Eq. (1.6), as we discussed in great detail in Chapter 1. If we place ourselves into a reference frame that moves along with the interface, we can see that our reasoning discussed in Figure 1.4 for a static situation still holds for the more general case of a dynamic system that involves flow. That is, we still have

$$\Delta p = p_1 - p_2 = \Delta P_L = \frac{\gamma}{2}\kappa \tag{6.14}$$

along with all the necessary caveats regarding the sign of the mean curvature κ that we discussed in Chapter 1. (The subscripts 1 and 2 denote the two adjacent fluids.)

For the tangential component of the velocity field, there is no unique equivalent of Eq. (6.13). For solid–liquid interfaces, it is generally assumed that molecular interaction forces attach the liquid molecules locally very strongly to the solid

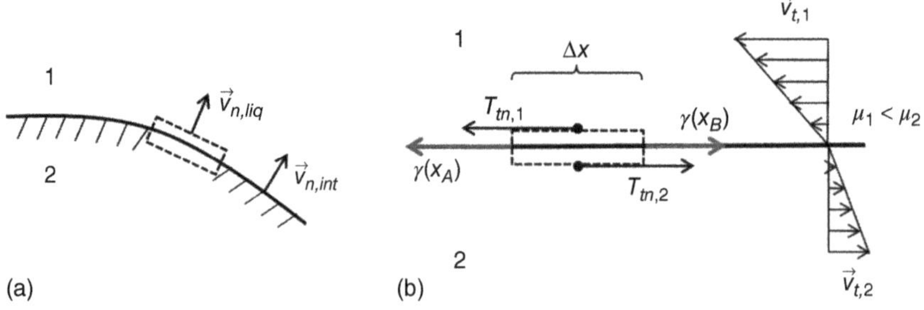

Figure 6.3 Boundary conditions for the velocity field at interfaces. (a) Normal component. (b) Tangential component. Very right: tangential velocity profiles in co-moving reference frame.

surface. As a consequence, the liquid moves along with the solid and assumes the same tangential velocity $\vec{v}_t$ as the solid:

$$\vec{v}_{t,liq} = \vec{v}_{t,sol} \tag{6.15}$$

This is known as the no-slip boundary condition. While not justified any fundamental consideration such as the kinematic boundary condition for the normal component of the velocity, it has proven to hold in most conditions ever since its introduction in the nineteenth century. Only very recently it was found in the context of specific liquid-repellent surfaces that there might be a tiny amount of slip involving a small tangential *slip* velocity at the interface. However, this is a nanoscale effect that hardly matters on the scale of micrometers to millimeters..

For fluid–fluid interfaces, the situation is clearer. If we consider again a control volume that contains the interface (Figure 6.3b), the fluid inside this control volume experiences tangential viscous stresses from both sides of the interface. In addition, there are tangential stresses due to the interfacial tension γ. In a reference frame, in which the control volume moves along with the interface, the balance of tangential forces reads

$$\gamma(x_A) + \mu_1 \partial_z \vec{v}_{1,t}\, \Delta x = \mu_2 \partial_z \vec{v}_{2,t}\, \Delta x + \gamma(x_B) \tag{6.16}$$

where $x_{A/B} = x_0 \pm \Delta x/2$ are the left and the right ends of the control volume.

In most situations, γ does not depend on the position along the interface. Then Eq. (6.16) reduces to

$$\mu_1 \partial_z \vec{v}_{1,t} = \mu_2 \partial_z \vec{v}_{2,t} \tag{6.17}$$

While the tangential component of the velocity itself is continuous, its gradient thus displays a jump that depends on the ratio between the viscosities μ_1 and μ_2 of the two fluids. It is worthwhile to note the consequences of Eq. (6.17) for situations with a very large ratio of the viscosities of the adjacent fluids, as commonly encountered for liquid–gas interfaces: If one of the two viscosities is negligibly small, this implies that the velocity gradient in the more viscous phase vanishes upon approaching the interface, i.e.

$$\partial_z \vec{v}_{t,liq} = 0 \tag{6.18}$$

In the presence of gradients of temperature or concentration of surfactants, however, γ does depend on these local properties and can therefore vary along the interface. Expanding the interfacial tension forces in Eq. (6.16) around x_0, we find a more general version of Eq. (6.17), namely,

$$\partial_{\|} \gamma = \mu_2 \partial_z \vec{v}_{2t} - \mu_1 \partial_z \vec{v}_{1t} \tag{6.19}$$

where $\partial_{\|} \gamma$ denotes the gradient of γ tangential to the interface. Flows resulting from such tangential gradients in interfacial tension are generally known as Marangoni flows. They can be particularly strong in microfluidic situations. In EW applications, Marangoni flows can be encountered upon merging drops that contain different surface-active solutes or different concentrations of the same surface-active solute.

If the interface is charged, as, for instance, in the case of an electric double layer, tangential components of the Maxwell stress tensor give rise to additional tangential components in the force balance in Eq. (6.15). These stresses are used extensively in microfluidics to drive so-called electroosmotic flows.

6.1.3 Nondimensional Navier–Stokes Equation: The Reynolds Number

In many situations, we can specify characteristic dimensions and flow speeds in a given fluidic problem. This is very useful to estimate the relative importance of the different terms in the Navier–Stokes equations. The spirit is similar to the comparison of the effects of surface tension and gravity in Chapter 1, which led to the definition of the Bond number and the capillary length (see Section 1.4). A flow problem is typically characterized by some characteristic flow velocity U and some geometric dimension L. In specific situations, there may also be more characteristic length scales, as we will encounter in Section 6.2. For the time being, let us assume that one length scale is enough. Based on these two quantities, we can define a characteristic time scale $\tau = L/U$. This is the time on which a given element of fluid is transported across the system. It is thus characteristic for the nonlinear advective term in the Navier–Stokes equation. Hence, we can rewrite $(\vec{v}\nabla)\vec{v} = U^2/L\ (\hat{v}\hat{\nabla})\hat{v}$, where symbols with hats are nondimensional quantities measured in units of U and L. Because U and L are the characteristic dimensions of the problem, the resulting nondimensional quantities are of order unity. In addition to τ, there may also be an additional external time scale T that gives rise to an explicit time dependence of the term $\partial_t\vec{v}$. An example would be a time-dependent boundary condition such as a periodic variation of the contact angle in an EW experiment involving AC voltage. In such a situation, the explicitly time-dependent term is best written as $\partial_t\vec{v} = U/T\ \partial_{\hat{t}}\hat{v}$. Using the characteristic units U and L, we also rewrite the viscous term as $\mu\nabla^2\vec{v} = \frac{\mu U}{L^2}\hat{\nabla}^2\hat{v}$. At the same time, we can identify a characteristic pressure gradient $\Delta P/L = \mu U/L$ due to viscous stresses involved in the flow. Rewriting Eq. (6.11) in units of these characteristic dimensions, we find

$$Re\left(\frac{\tau}{T}\,\partial_{\hat{t}}\hat{v} + (\hat{v}\hat{\nabla})\hat{v}\right) = -\hat{\nabla}\hat{P} + \hat{\nabla}^2\hat{v} \tag{6.20}$$

Here, we introduced the Reynolds number $Re = \rho UL/\mu$, which characterizes the relative importance of inertial and viscous effects. The Reynolds number thus describes the relative importance of the nonlinear term in the Navier–Stokes equations to the viscous one. It is known that strong nonlinear effects such as the transition to turbulence typically occur for $Re = O(10^3–10^4)$. Such high Reynolds numbers are usually not achieved in microfluidics, because of the very fact that the characteristic linear dimension L appearing in the expression for Re is by definition small in small systems. Therefore, the nonlinear $(v\nabla)v$ term in Eq. (6.20) is usually negligible. It should be noted, though, that inertial effects nevertheless play a role in many microfluidic applications. In particular in EW, drops are routinely exposed to abruptly changing boundary conditions. Figure 6.4 shows an illustration of such a process, in which the equilibrium contact angle is suddenly reduced by applying a voltage to interdigitated EW

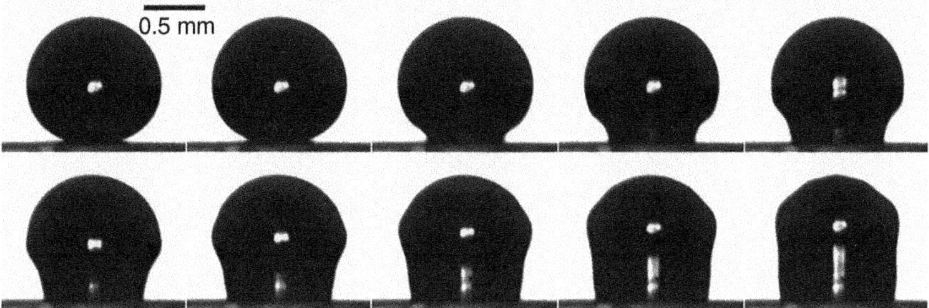

Figure 6.4 Time sequence of a drop on an electrowetting substrate with embedded interdigitated electrodes. A voltage is suddenly applied at $t = 0$ (top row, left image), leading to an abrupt change of the contact angle. The resulting perturbation of the drop subsequently propagates along the liquid surface and reaches the apex of the drop after approximately 2 ms. (Time between images: 0.2 ms.)

electrodes submerged under a dielectric coating on the substrate. θ is reduced by approximately 80° on a time scale T that is too fast for the inertial mass of the drop to follow instantaneously. Because $T \ll \tau$, the response of the liquid under such conditions is governed by the first, explicitly time-dependent term on the left-hand side of Eq. (6.20). In such situations, the Navier–Stokes equations reduce to the so-called time-dependent Stokes equations. Returning to dimensional units, these equations read

$$\rho \partial_t \vec{v}(\vec{r}, t) = -\nabla P + \mu \nabla^2 \vec{v} + \vec{f}_b \tag{6.21}$$

These three equations, one for each spatial coordinate, have the important property of being linear in the velocity field. As a consequence, they are amenable to decomposition into Fourier modes and other eigenmodes, and the superposition principle holds. While nonlinearities due to the boundary conditions may still arise (e.g. due to the capillary equation), the time-dependent Stokes equations nevertheless provide the proper framework for most problems in dynamic EW. In the following sections of this chapter, we will discuss several examples of EW-related flow problems and solution techniques such as the eigenmode decomposition of oscillating drops.

6.1.4 Example: Pressure-Driven Flow Between Two Parallel Plates

The arguably most generic flow profile next to the linear Couette flow profile discussed above in the context of the definition of the shear viscosity (Figure 6.2) is the so-called Poiseuille flow of a pressure-driven flow between two parallel walls; see Figure 6.5. We look for a steady-state solution at low Re numbers. The governing equations, the time-independent equivalent of Eq. (6.21), are known as Stokes equations:

$$\nabla P = \mu \nabla^2 \vec{v} + \vec{f}_b \tag{6.22}$$

We assume a simple pressure-driven pipe flow with some inlet pressure P_{in} larger than the outlet pressure P_{out} in the absence of any body force, i.e. we take

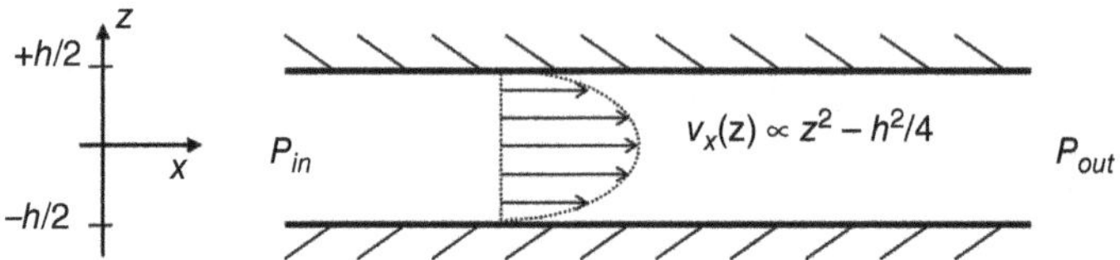

Figure 6.5 Pressure-driven flow between two parallel walls. Illustration of the parabolic velocity profile.

$\vec{f}_b = 0$. For symmetry reasons, flow and pressure gradient are directed exclusively in the x-direction along the channel. As a consequence, the only nontrivial equation left is

$$\mu \partial_{zz} v_x(z) = \partial_x P \tag{6.23}$$

where $\partial_x P = (P_{out} - P_{in})/L\ (< 0)$ is the homogeneous pressure gradient along the channel of length L. The no-slip boundary condition at the walls positioned at $z = \pm h/2$ implies that $v_x(\pm h/2) = 0$. Moreover, $\partial_z v_x(0) = 0$ for symmetry reasons. Integrating Eq. (6.23) twice respecting these boundary conditions yields a parabolic, so-called *Poiseuille* flow profile:

$$v_x(z) = -\frac{\partial_x P}{2\mu}((h/2)^2 - z^2) \tag{6.24}$$

The minus sign in the equation ensures that the flow is directed opposite to the pressure gradient, as it should be.

The maximum flow rate in the middle of the channel is thus $v_{max} = v_x(0) = -\partial_x P\, h^2/8\mu$. The corresponding total flux rate of the fluid is given by

$$Q = \int_{-h/2}^{+h/2} v_x(z) dz = -\frac{h^3}{6\mu} \partial_x P \tag{6.25}$$

Because the total flux is given by the product of the channel width and the average flow velocity V, we can read from this expression the average flow velocity $V = Q/h = h^2 \Delta P/6\mu L$, with the (positive) pressure difference $\Delta P = P_{in} - P_{out}$. Equation (6.25) also provides a linear relation between Q and ΔP, namely, $Q = (h^3/6\mu L)\, \Delta P$. This expression is analog to Ohm's law $I = U/R$ of electricity relating the electrical current I to the applied voltage U in a conductor. This analogy between Eq. (6.25) and Ohm's law along with the analogy between the continuity equation and Kirchhoff's law forms the basis of microfluidic *plumbing rules* that allow to map the distribution of pressures and fluxes in networks of microfluidic pipes and tubes to the distribution of potentials and currents in resistive electrical circuits. This approach will be addressed in Problem 6.7. Using the analogy, we identify

$$R_H = \frac{6\mu L}{h^3} \tag{6.26}$$

as the hydraulic resistance of a slit pore. As expected, the resistance scales linearly with the viscosity of the fluid and with the length of the tube. Note that R_H also scales as $1/h^3$. The same pressure gradient in a ten times thinner channel thus produces a 1000 times lower flux. Physically, this strong dependence arises from the no-slip boundary condition: Due to this boundary condition, the fluid is forced to stick the two walls at $\pm h/2$. Being held back on two sides at a close distance, it is very difficult to force the fluid in between. For cylindrical tubes the situation is even worse because the liquid is surrounded by solid in all directions. Hence, the analog calculation in cylindrical coordinates yields an even stronger dependence on the tube radius r, $R_H \propto 1/r^4$. This law is widely known as Hagen–Poiseuille law. This dramatic size dependence of R_H should always be kept in mind whenever designing micro- and nanofluidic devices.

The flow within a liquid drop sandwiched between two parallel plates, as it is encountered in many EW-based lab-on-a-chip systems, is also pressure-driven flow. Upon activating the electrodes to move the drop, the contact angle decreases by $\Delta\theta$ on one side of the drop, while it remains equal to Young's angle on the other side. For a simple two-dimensional system with identical contact angles on the top and bottom surface as sketched in Figure 6.6a, this generates a pressure drop $\Delta P = 2\gamma\Delta\cos(\theta)/h$ along the drop, which sets the liquid in motion. In steady state, the drop moves at a constant velocity V. In the central region of the drop, a parabolic pressure profile develops, as sketched in Figure 6.6a. Close to the edges, the flow profile must be different to conserve the shape of the menisci during the steady drop motion. It becomes particularly clear if we consider the flow field in a moving reference frame, in which the drop is stationary (see Figure 6.6b). In that frame, the solid walls and hence the fluid at the interface move toward the left with a velocity $-V$. In the center of the channel, the liquid flows in the positive x-direction. Close to the front of the drop, liquid is moving forward along the center is deflected toward the walls and is subsequently recirculated toward the back. This results in a net circulation

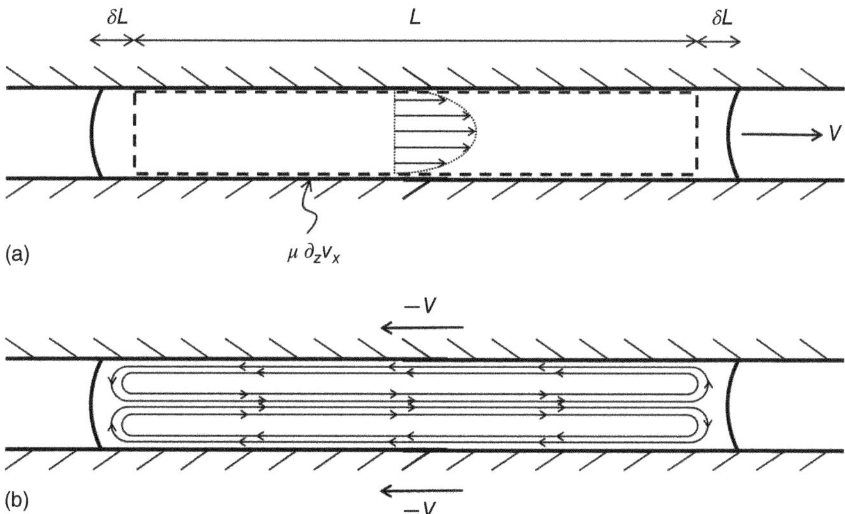

Figure 6.6 Drop confined between two parallel plates moving due to a difference in contact angle. (a) Flow field as seen in the laboratory reference frame. (b) Recirculating flow as seen in a moving reference frame.

within the drop that plays an important role of mixing the content of the drop (see Chapter 8). The local recirculation from the center toward the walls (and vice versa) is found to take place with a region $\delta L \approx h$.

Moving back to the laboratory frame and ignoring the effect of the local recirculations for the moment, we can estimate the expected steady-state flow velocity of the drop by balancing the forces acting on the dashed control volume in Figure 6.6a. The pressure difference gives rise to driving capillary force $F_c \approx \Delta P\, h = 2\gamma \Delta\cos(\theta)$, where we neglect the finite pressure drop within the regions δL on both ends of the drop. This force is opposed by the viscous drag at the solid–liquid interface. Integrating the shear stress $\mu\, \partial_z v|_{h/2}$ along the two interfaces, we find $F_{visc} = 2L\mu\, \partial_z v|_{h/2} = 6L\mu V/h$, where we expressed the velocity gradient at the interface in terms of V. Balancing the two forces, we find $V = \gamma h\, \Delta\cos(\theta)/3\mu L$. For the same $\Delta\cos(\theta)$ reducing the separation between the plates thus results in a decrease of the expected flow velocity because the increase in hydraulic resistance dominates over the increase in Laplace pressure difference. This is indeed confirmed by experiments. Yet, as we will see later, the actual drop velocities in typical EW experiments are substantially lower than predicted here. This is caused by additional dissipation due to the local flow fields in the region δL, including in particular the motion of the contact lines that we will discuss explicitly in Section 6.3.

6.2 Lubrication Flows

6.2.1 General Lubrication Flows

At the end of the preceding section, we invoked certain symmetry relations and assumptions of small velocities to calculate the flow in a drop between two strictly parallel walls. The simplifications leading to the parabolic flow profiles discussed above hold for a much wider and more general class of flow problems that are characterized by a quasi-two-dimensional geometry. They can be derived systematically whenever liquid layers are confined between two interfaces at some distance h that is small compared with the typical longitudinal dimension λ of the system. The two interfaces can either be two solid surfaces, as, for instance, in thin lubrication layer in bearings or in typical EW-based lab-on-a-chip devices, or one of the two interfaces can also be a free liquid–liquid or liquid–vapor interface, as, for instance, in the vicinity of a moving contact line and for thin liquid films covering a solid surface (see Figure 6.7). Moreover, the two interfaces do not need to be strictly parallel. The resulting flows are known as lubrication flows, and they are very common. Here, we discuss some of their most general physical properties. For further details, the interested reader is referred to the excellent review by [1].

The separation of length scales $h \ll \lambda$ implies that the flow velocities in the z-direction perpendicular to the confining surfaces are small compared with the transverse velocities in the x–y-plane. This is natural because otherwise, the thickness of a free liquid film would either very quickly vanish or become *large*, implying that the flow would no longer qualify as a lubrication flow. Formally,

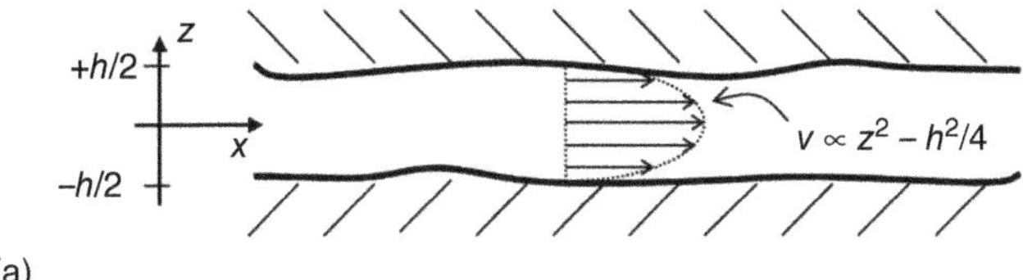

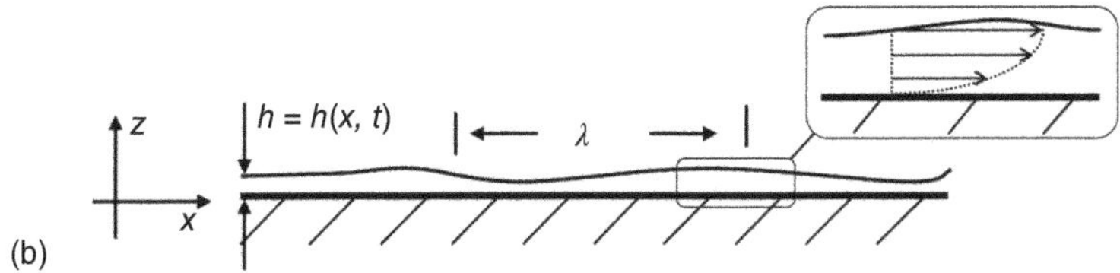

Figure 6.7 Lubrication flow in thin liquid films. (a) Confined layer between two solid surfaces with parabolic Poiseuille flow profile. (b) Liquid film with free surface and Poiseuille flow profile with shear-free boundary condition at the liquid surface.

we can derive the relative strength of the flow rates by considering the continuity equation $\nabla\vec{v} = \partial_x v_x + \partial_z v_z = 0$. Considering that $\partial_x \sim 1/\lambda$ and $\partial_z \sim 1/h$, we find

$$v_z \approx \frac{h}{\lambda} v_x \ll v_x \tag{6.27}$$

The same argument then leads to the conclusion that the dissipation in the flow is dominated by variations of the viscous shear stresses in the z-direction:

$$\partial_{xx} v_x \approx \left(\frac{h}{\lambda}\right)^2 \partial_{zz} v_x \ll \partial_{zz} v_x \tag{6.28}$$

An analog equation holds in the y-direction. (To simplify the notation, we will drop the y-dependence in the following.) With these simplifications in place, we can also see that the nonlinear term in the Navier–Stokes equations is typically negligible in lubrication flows because $h/\lambda \ll 1$. The Reynolds number for these flows reads

$$Re = \frac{\rho v_x^2/\lambda}{\mu v_x/h^2} = \frac{\rho v_x h}{\mu} \frac{h}{\lambda} \ll 1 \tag{6.29}$$

As a result, the velocity is only oriented in the x-direction and the Navier–Stokes equations reduce to the Stokes equation

$$\mu \partial_{zz} v_x = \partial_x p \tag{6.30}$$

where $p = p(x)$ is the in-plane pressure gradient that now depends on the lateral coordinate x because of possible lateral variations of the film thickness. Note that pressure gradients in z-direction are negligible to first order in h/λ because $\partial_z p \sim \partial_{zz} v_z \sim \frac{h}{\lambda} \partial_x p$. Because p does not depend on z, Eq. (6.34) is readily integrated along the z-direction. This separates the dependences on the directions along and transverse to the flow. Thus, we end up with the same governing equation as in the preceding section. For the case of a liquid film sandwiched between two solid surfaces (Figure 6.7a) with no-slip boundary conditions, we obtain again a parabolic Poiseuille flow profile,

$$v_x(z) = -\frac{\partial_x p(x)}{2\mu}\left(z^2 - \left(\frac{h(x)}{2}\right)^2\right) \tag{6.31}$$

but now with a position-dependent pressure gradient and local film thickness $h(x)$. Because such variations only take place on length scales $\lambda \gg h$ by virtue of our initial assumptions, the lubrication approximation implies that the two interfaces are locally sufficiently parallel to guarantee a parabolic flow profile based on the local thickness and the local pressure gradient. The parabolic flow profile thus holds for more general conditions than in our initial derivation, and we see that it depends only on the separation of transverse and longitudinal length scales rather than the more restrictive symmetry requirement discussed above.

For a free film on a solid surface (Figure 6.7b), the same general conclusions hold. The only difference is that the flow is slightly less restricted. The governing equation is the same, Eq. (6.30), and so is the parabolic nature of the flow profile. Yet, in this case the liquid is held back by a no-slip boundary condition only on one side at $z = 0$. On the other side, at the free surface, the no-stress boundary condition $\partial_z v_x(h) = 0$, Eq. (6.18), applies. As a consequence, the apex of the parabolic flow profile is no longer located in the middle of liquid layer but shifts to the free surface at $z = h$. The resulting flow profile then reads

$$v_x(z) = \frac{\partial_x p}{2\mu} z(z - 2h) \tag{6.32}$$

The corresponding flux still scales as $1/h^3$, but with a different numerical prefactor:

$$Q = -\frac{\partial_x p}{3\mu} h^3 \tag{6.33}$$

Let us return briefly to a vector notation with two transverse directions x and y to illustrate some generic properties of lubrication flows. We can rewrite Eq. (6.30) as

$$\mu\, \partial_{zz}\, \boldsymbol{v}_{\|} = -\boldsymbol{\nabla}_{\|} p \tag{6.34}$$

where the subscript $\|$ denotes the transverse x, y-directions. Because p does not depend on z as discussed above, we can make an ansatz to separate the x-, y-, and the z-dependence of the flow and write $\boldsymbol{v}_{\|}(x, y, z) = \boldsymbol{v}_{\|}(x, y, 0) f(z)$. Here $f(z) = 1 - (2z/h)^2$ for the case of a liquid film between two solid surfaces. The qualitative shape of the flow field thus does not depend on z. The streamlines look the same in every plane, and only the absolute value of the flow velocity varies with z following the parabolic Poiseuille profile. As a result, Eq. (6.34) reduces to

$$v_{\|}(x, y, 0) = -\frac{h^2}{2\mu}\, \boldsymbol{\nabla}_{\|} p(x, y) \tag{6.35}$$

Hence, the flow field $\boldsymbol{v}_{\|}(x, y, 0)$ can be written as a gradient of the in-plane pressure distribution. The characteristics of two-dimensional lubrication flows are thus very similar to three-dimensional inviscid flows, for which the velocity field can also be written as the gradient of a velocity potential (see Section 6.1). Like for three-dimensional potential flows, two-dimensional lubrication flows are thus irrotational, i.e. $\nabla_{\parallel} \times v_{\parallel}(x, y) = 0$. This conclusion may seem counterintuitive at first glance given the prominence of viscous dissipation in flows in thin channels. Remember, however, that the effect of dissipation is taken into account completely by the parabolic Poiseuille flow profile along the z-direction that we integrated out upon reducing Eq. (6.34) to Eq. (6.35). The remaining flow in the x–y-plane is indeed inviscid within lubrication approximation.

6.2.2 Lubrication Flows with a Free Liquid Surface

Flow problems involving free liquid surfaces are in general rather complex because the shape of the interface adapts itself according to the flow. A full fluid mechanic solution therefore requires a self-consistent solution for the flow field and the shape. In the case of lubrication flows, the situation is somewhat simplified because we know that the flow field is parallel to the interface and displays a parabolic dependence on z. In many cases, this information about the flow field is enough, and the primary question of interest is the evolution of the film thickness. That is, the problem is solved once we know $h(x, t)$. (We omit again the y-dependence for simplicity.) The evolution of the film thickness can be easily related to the transverse flux in the film by taking into account material conservation. If the influx $Q(x)$ into a small area element dx on the surface exceeds the outflux $Q(x + dx)$, the local film thickness increases and vice versa (see Figure 6.8). That is,

$$\partial_t h(x) = -\partial_x Q(x) \tag{6.36}$$

Inserting the expression for the flux derived in the preceding section, Eq. (6.33), we find the well-known thin film equation

$$\partial_t h(x, t) = \frac{1}{3\mu} \partial_x (h^3 \partial_x p) \tag{6.37}$$

This is a highly nonlinear partial differential equation that couples the evolution of the film thickness to the evolution of the pressure field. To close the problem, we need to express the pressure field in terms of the film height.

In general, the coupling can depend on many parameters and external fields that enter the boundary condition at the liquid surface. A contribution that is always present is obviously the capillary pressure $\Delta p_L = \gamma\kappa$. Because the thin film equation is derived under the assumption $h \ll \lambda$, it is obvious that the gradients of the liquid surface are always small. As a consequence, we can use the linearized expression for the curvature $\kappa = -\partial_{xx} h$. We introduce here the minus sign to take into account the fact that a positive curvature of the function $h(x)$ in Figure 6.8 implies a negative Laplace pressure in the fluid and vice versa. Hence

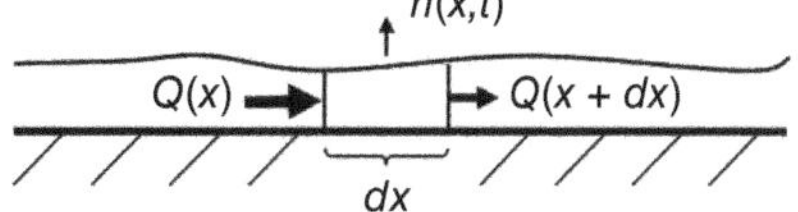

Figure 6.8 Mass conservation and evolution of film thickness. The liquid volume of the control area between x and $x + dx$ changes at a rate $dx\, \partial_t h(x) = Q_{in}(x) - Q_{out}(x + dx) = -\partial_x Q\, dx$.

we have reduced the problem to a partial differential equation for the desired function $h(x, t)$:

$$\partial_t h(x,t) = -\frac{\gamma}{3\mu}\partial_x(h^3\partial_{xxx}h) \quad (6.38)$$

Being a fourth-order highly nonlinear differential equation, calculating the evolution of such a thin liquid film is in general a challenging problem. The thin film equation has many applications. In the following we will discuss two of them that are frequently encountered in EW.

6.2.3 Application I: Linear Stability Analysis of a Thin Liquid Film

Frequently, it is of interest to assess the stability of thin liquid films against a small perturbation $\delta h(x, t)$ around the average film thickness h_0. We calculate whether such a random perturbation will decay with time and restore the original unperturbed film or whether it will grow and thereby destroy the original configuration. To enable a simple analytical analysis, we first linearize the thin film equation. For the perturbed film profile, we write $h(x, t) = h_0 + \delta h(x, t)$ with $\delta h \ll h_0$. Associated with this thickness perturbation, there will be a pressure perturbation. This pressure perturbation is also small such that we can write $p(x, t) = p_0 + \delta p(x, t)$ with $\delta p \ll p_0$. Inserting these two expressions into Eq. (6.37) and keeping only first-order terms in the small quantities, we find

$$\partial_t h(x,t) = \partial_t \delta h(x,t) = \frac{1}{3\mu}\partial_x((h_0 + \delta h)^3 \partial_x \delta p) \approx \frac{h_0^3}{3\mu}\partial_{xx}\delta p \quad (6.39)$$

This equation is linear in the perturbations. Hence it is now possible to describe any arbitrary perturbation as a superposition of Fourier modes, and it is sufficient to analyze each Fourier mode individually (see Figure 6.9a). Therefore, we write $\delta h(x, t) = \alpha(t)\cos(qx)$, where $q = 2\pi/\lambda$ is the wavenumber of the perturbation. Subsequently, it is necessary to establish a relation between δh and the pressure perturbation $\delta p = \delta p_0(t) cos(qx)$. A classical example of a linear thin film instability is the so-called Rayleigh–Taylor instability of a hanging liquid film, such as a layer of paint put on the ceiling of the room. If the solvent of the paint does not dry sufficiently fast (or if the paint layer is too thick), the film becomes unstable and decomposed into drops that may eventually fall from the ceiling. In the case of the Rayleigh–Taylor instability, there are two contributions to the pressure, the hydrostatic pressure $\delta p_h = -\rho g \delta h$ and the capillary pressure $\delta p_L = -\gamma \partial_{xx}\delta h$.

Inserting the Fourier mode, we find $\delta p = \rho g(1 - \lambda_c^2 q^2)\delta h$. Inserting our ansatz $\delta h = \alpha(t)\cos(qx)$ into Eq. (6.39), we find

$$\partial_t \alpha(t) = -\frac{h_0^3}{3\mu}\rho g\, q^2(-1 + \lambda_c^2 q^2)\alpha(t) \equiv -\tau(q)\alpha(t) \quad (6.40)$$

The solution to this equation is an exponential function with a time constant $\tau(q)$ that either grows or decays, depending on the sign of τ. For short wavelengths, $\lambda_c^2 q^2 > 1$, surface tension dominates over gravity and restores a flat configuration of the film. Conversely, for long wavelengths, $\lambda_c^2 q^2 < 1$, gravity

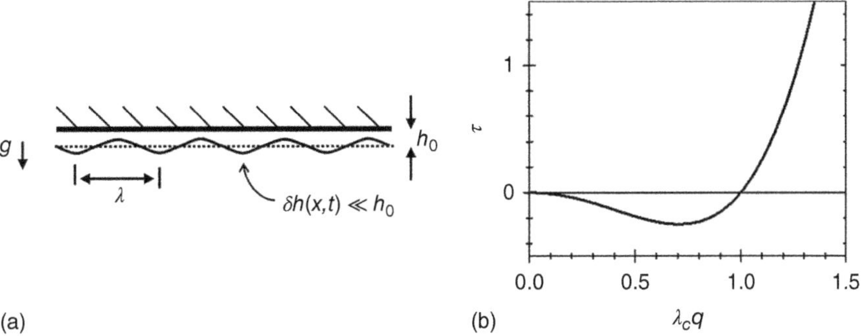

Figure 6.9 Rayleigh–Taylor instability. (a) Schematic drawing of perturbed liquid film. (b) Growth rate $\tau(q)$.

dominates. Fourier modes with such wavelengths are unstable, and the corresponding perturbations grow with time. In practice, the dominant mode that is observed experimentally is typically the fastest-growing mode, corresponding to the minimum of $\tau(q)$ in Figure 6.9b with $q^* = 1/\sqrt{2}\lambda_c$.

While surface tension always tends to restore the flat configuration of the film, several other forces can take over the role of the destabilizing gravitational force in the Rayleigh–Taylor instability. For instance, if a very thin film of a partially wetting fluid is deposited on a surface, e.g. by a spin-coating or dip-coating process, such a film can be subject to destabilizing molecular interaction forces that lead to an attractive disjoining pressure. Such a situation will be discussed in Problem 6.8.

In EW, it is rather common that thin dielectric oil films become entrapped between a solid surface and the conductive (frequently aqueous) droplet phase. In this case, the Maxwell stress $\pi_{el} = \epsilon_0 \epsilon_{oil} E_{oil}^2/2$ exerts a pressure on the oil–water interface that destabilizes the oil film. The occurrence of this instability affects the performance of many EW-driven devices, including in particular displays. The geometry is sketched in Figure 6.10. If we apply a voltage U between the electrode at the bottom and the aqueous phase on the top, the oil film and the solid dielectric layer act as two capacitors in series. For a dielectric layer of thickness d and the dielectric constants of oil and dielectric as ϵ_{oil} and ϵ_d, we can write the electric field at the unperturbed oil–water interface as $E_o = U/H$, where $H = h + \frac{\epsilon_{oil}}{\epsilon_d} d$ is the effective dielectric thickness of the stacked oil and dielectric layer. In the presence of a perturbation δh such that $h = h_0 + \delta h$, the electric field and hence the Maxwell stress at the oil–water interface increase at locations of reduced film thickness and vice versa. The resulting pressure perturbation reads

$$\delta p_{el} = \frac{\gamma}{H} \frac{\eta}{\left(1 + \frac{\delta h}{H}\right)^2} \cong \frac{\gamma \eta}{H} \left(1 - 2\frac{\delta h}{H}\right) \tag{6.41}$$

where we introduced the EW number $\eta = \epsilon_0 \epsilon_{oil} U^2/\Upsilon H$ based on the effective dielectric thickness H. Combining this expression with the standard contribution for the capillary pressure, $\delta p_L = -\gamma \partial_{xx} \delta h$ with $\delta h = \alpha(t)\cos(qx)$, we find an ana-

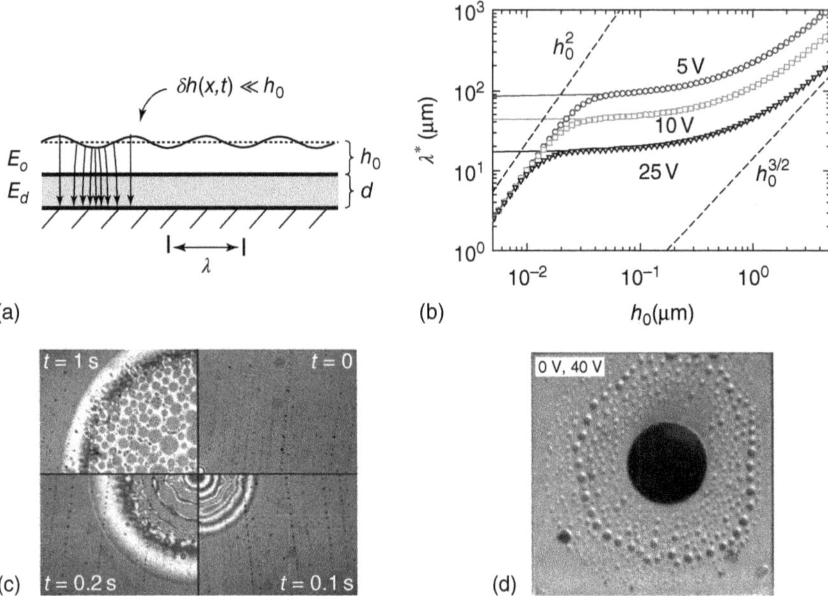

Figure 6.10 Electrically driven breakup of a thin oil film under the influence of a Maxwell stress. (a) Schematic of dielectric layer, oil layer, and electric field distribution; gray: dielectric layer. (b) Fastest-growing wavelength versus initial oil film thickness. The transition from the plateau to the quadratic regime for small h_0 arises from another instability mechanism caused by van der Waals interaction (see Problem 6.8). (c) Time-lapse bottom view of a water droplet spreading in ambient oil upon increasing the voltage. An oil film is first entrapped and subsequently breaks up into droplets. (b, c) Source: Staicu and Mugele 2006 [2]. Reproduced with permission of APS. (d) Top view of a drop after spreading upon a sudden increase in voltage. Source: Sun and Heikenfeld 2008 [3]. Reproduced with permission of IOP Publishing.

log to Eq. (6.40):

$$\partial_t \alpha(t) = -\frac{h_0^3}{3\mu}\gamma\, q^2 \left(q^2 - \frac{2\eta}{H^2}\right) \alpha(t) \tag{6.42}$$

Hence, the Maxwell stress destabilizes the flat configuration of the interface, while the capillary pressure stabilizes it. As usual, δp_L always dominates at short length scales. From Eq. (6.42) we can read immediately that the critical wavevector beyond which perturbations are unstable is given by $q_c = \sqrt{2\eta}/H$, which increases linearly with U. Because $\eta \propto 1/H$, this expression also implies that the corresponding critical wavelength λ_c scales as $h_0^{3/2}$ for $h_0 \gg \epsilon_{oil} d/\epsilon_d$. For $h_0 \ll \epsilon_{oil} d/\epsilon_d$, it converges toward an h_0-independent but still voltage-dependent plateau value; see Figure 6.10b. (The transition to the quadratic regime in Figure 6.10b at very small h_0 is caused by van der Waals interaction, which is neglected in our present analysis; see [2].)

6.2.4 Application II: Entrainment of Liquid Films

Another important application of the thin film equation is the description of the entrainment of thin films of ambient fluid by moving bubbles or drops, as

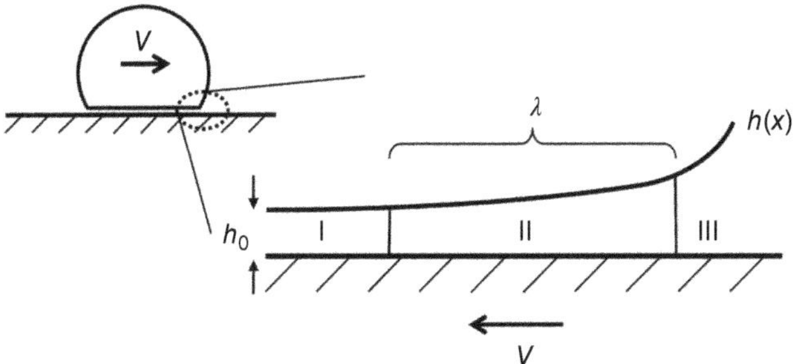

Figure 6.11 Liquid entrainment by a drop or bubble of macroscopic radius *R* moving with velocity *V* (top left). Zoom: in the rest frame of the drop, the surface profile *h*(*x*) is stationary and can be divided in the static zones I and III, connected by the dynamic meniscus II.

sketched in Figure 6.11. As the drop moves past the solid surface, it squeezes out ambient fluid close to its leading edge. The ambient fluid is held in place by its adhesion to the solid substrate. Depending on the drop speed *V*, there may or may not be enough time to squeeze out the ambient fluid down to its equilibrium thickness that is determined by the disjoining pressure of the system. Instead, a dynamically entrained film of thickness h_0 is entrapped with a thickness that can reach hundreds of nanometers, as the appearance of optical interference fringes shows. This problem is known in fluid dynamics as Bretherton problem or as Landau–Levich problem. Bretherton investigated the buoyancy-driven rise of gas bubbles in cylindrical liquid-filled tubes. Landau and Levich studied the closely related inverse problem of pulling a solid object at a fixed velocity *V* out of a stationary bath of a completely wetting fluid. Like in the Bretherton problem, the question of interest is to determine the thickness h_0 of the entrained liquid film as a function of the velocity *V* and the other relevant material parameters.

Because of its direct relevance for EW, we discuss here problem of a moving drop. We focus on the region close to the leading edge of the drop, and we place ourselves in a reference frame that moves along with the drop. This has the advantage that the profile $h(x)$ of the liquid–liquid interface that we are looking for is stationary. The key idea of the Landau–Levich and the Bretherton solution is to realize that the profile of the entrained film can be split in three regions. In region *I* (see Figure 6.11), far behind the trailing edge, the film thickness assumes the constant value h_0. This is the quantity that we want to determine. Sufficiently far on the right from the drop edge, in region *III*, the film profile merges into the macroscopic drop profile, which we assume to be a circle of radius *R* in our present two-dimensional analysis. In regions *I* and *III*, the profiles are static and not affected by the dynamic processes in the vicinity of the drop edge. For these asymptotic regions, we thus know the drop profile and the corresponding capillary pressure drop across the liquid interface, $\Delta p_{L,I} = 0$ and $\Delta p_{L,III} = \gamma/R$. The two asymptotic regions are merged together by a transition zone, region *II*, in which the curvature of the interface gradually changes and the fluid is funneled into the thin film that we are interested in. This transition region is similar to the region close to contact line in EW where the local electric fields distort the liquid profile, as discussed in Section 5.1. The difference is that in the present dynamic problem viscous stresses in the liquid film are responsible for the distortion of the surface profile rather than electrical ones. The width λ of region II is not known a priori and has to be determined as part of the solution.

Analyzing the problem in the reference frame of the moving drop, we note that the flux of the ambient fluid in the thin film is constant in steady conditions and equals $Q = -Vh_0$. Describing the shape of the dynamic meniscus in region II in lubrication approximation, the analog of Eq. (6.37) reads

$$-Vh - \frac{\partial_x p}{3\mu} h^3 = -Vh_0 \tag{6.43}$$

Writing $h(x) = h_0 + \delta h(x)$ and noting that the pressure gradient is determined by the capillary pressure $\delta p(x) = -\gamma \partial_{xx} \delta h$, this leads to an ordinary differential equation for the film profile:

$$\partial_{xxx} \delta h = \frac{3\mu V}{\gamma} \frac{\delta h}{h^3} = 3Ca \frac{\delta h}{h^3} \tag{6.44}$$

Here, we introduced the capillary number

$$Ca = \frac{\mu V}{\gamma} \tag{6.45}$$

which is the nondimensional number that characterizes the relative importance of surface tension and viscous forces. The exact solution of Eq. (6.44) is mathematically rather complex and requires a systematic procedure of *stitching* the solution of the film profile in region II to the two other solutions in regions I and III. The interested reader is referred to the more specialized literature for this so-called asymptotic matching method [4].

Here, we content ourselves with a qualitative scaling argument: The characteristic transverse length scale of our problem is λ, the width of the dynamic region. The characteristic vertical scale is $h \approx h_0$. To remain consistent with our lubrication approximation, we must request $h_0 \ll \lambda$. Inserting these characteristic scales in Eq. (6.44), we find $h_0 \sim Ca^{1/3}\, \lambda$. Consistency with the lubrication approximation thus limits the validity of our solution to situations with $Ca \ll 1$ or characteristic speeds $V \ll \gamma/\mu$. (This is typically not a severe limitation because the characteristic velocity is $V_{cap} = \frac{\gamma}{\mu} \approx 10\ \mathrm{m\,s^{-1}}$ for most common liquids.) To close the problem, we request that the curvature of the thin film on the right-hand edge of region II must match the macroscopic curvature in region III, i.e. we request $\partial_{xx} h \sim h_0/\lambda^2 \sim 1/R$. This matching condition thus yields $\lambda = \sqrt{h_0 R}$. Combining all terms we obtain the Bretherton scaling

$$h_0 \sim \beta\, R\, Ca^{2/3} \tag{6.46}$$

with a prefactor β of order unity. For a drop of 100 μm and $Ca = 10^{-3}$, this yields $h_0 = O(1\ \mu\mathrm{m})$, which is much larger than any possible equilibrium thickness due to a disjoining pressure.

The formation of such dynamically entrained liquid films is crucial for the low dissipation and fast motion of contact lines and drops in oil–water two-phase flow microfluidic devices driven by EW. The calculation of the thickness of the entrained film under EW conditions is slightly more evolved than the analysis presented above because it involves the Maxwell stress acting on the liquid–liquid interface in addition to the capillary pressures. The treatment follows the lines discussed in the preceding application example. For a more detailed discussion, see [2].

6.3 Contact Line Dynamics

Contact line dynamics has been one of the most challenging topics in wetting science since Huh and Scriven noticed in the 1970s that the force required to move a contact line diverges according to the standard laws of fluid dynamics. If true, this would mean that infinite forces would be required to move a drop, or, as Huh and Scriven put it, "not even Hercules could sink a solid (e.g. a ship)." The fact that drops can move and that ships can sink is a direct manifestation of the fact that continuum fluid dynamics cannot be extrapolated all the way down to atomic scales. The very fact that contact lines can move demonstrates the relevance of physical processes on the nanometer scale for the macroscopic dynamics of fluids. In this subsection, we discuss basic aspects of this multi-scale problem and provide a comparison to selected experiments of particular interest to EW. The reader interested in more details is referred to review articles such as [5].

6.3.1 Tanner's Law and the Spreading of Drops on Macroscopic Scales

The fact that contact line friction is indeed very strong becomes apparent when we consider the spreading of drops upon deposition onto a solid surface. If we gently bring a drop in contact with a solid surface, it initially touches the surface in a singular point with a momentary contact angle of 180°. As soon as solid–liquid contact is established, interfacial tension forces act and start to pull on the contact line. As a consequence, the drop begins to spread over the solid. Initially, this process is governed by the balance of the contact line forces and the inertia of the liquid that is accelerated. After such a short inertial regime, the drop assumes a spherical cap shape and continues to spread by gradually decreasing the transient *dynamic* contact angle θ_d toward its equilibrium value θ_Y (see Figure 6.12a). This gradual spreading process is amazingly slow. For completely wetting fluids (i.e. $\theta_Y = 0$), the in-plane radius r of the drop typically grows as slowly as $r \propto t^{1/10}$ (see Figure 6.12b). This famous result is known as Tanner's law. Equivalently, it can be written as $\theta_d(t) \propto t^{-3/10}$.

To understand the origin of Tanner's law, we consider the contact line motion on a macroscopic scale. First of all, it is clear that the spreading process is driven by the imbalance of the surface tension forces at the contact line, which manifests itself by the fact that $\theta_d > \theta_Y$. Balancing the forces acting on the highlighted wedge-shaped fluid volume close to the contact line in Figure 6.12c, we find that the driving forces per unit length of the contact line are given by

$$f_\gamma = \gamma(\cos\theta_Y - \cos\theta_d) \cong \frac{\gamma}{2}(\theta_d^2 - \theta_Y^2) \tag{6.47}$$

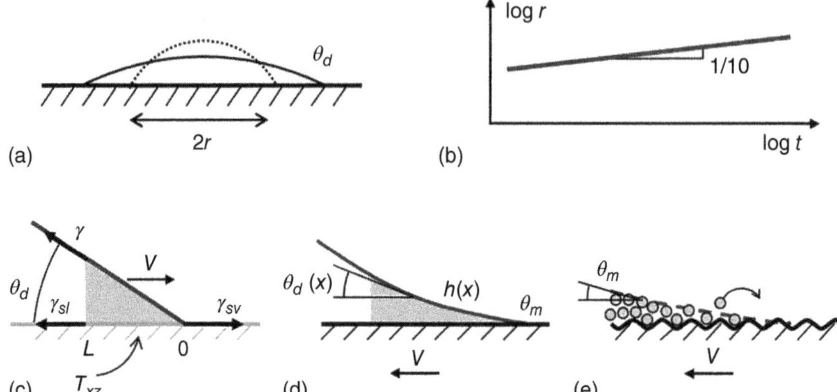

Figure 6.12 Drop spreading. (a) Schematic at different instances in time. (b) Tanner's law. (c) Macroscopic force balance. (d) Mesoscopic viscous contact line dissipation. (e) Microscopic thermally activated hopping dynamics.

The approximation on the right-hand side applies to small contact angles. Because the liquid is moving with respect to the solid, there is also a viscous stress acting along the drop–substrate interface. From dimensional arguments, it is clear that the resulting total viscous force per unit length of the contact line will be proportional to the viscosity of the fluid and the speed V of the moving contact line. Balancing this viscous force with the imbalanced surface tension force, Eq. (6.47), will result in a relation between the dynamic contact angle and the speed of the contact line, $\theta_d = \theta_d(V)$. Note that the origin of this dynamic contact angle is exclusively caused by viscous dissipation. It is not related to the (static) contact angle hysteresis that arises from surface heterogeneities that we discussed in Section 1.7.

To calculate the viscous force, we place ourselves in the reference frame of the moving contact line, as we did in the preceding section. We assume for simplicity that the contact angle and all slopes are small such that the lubrication approximation can be applied. Under these conditions, we find that the velocity profile in the drop is given by $v_x(z) = -V + \partial_x p\,(z^2 - 2hz)/2\mu$. Hence, the tangential stress at the solid–liquid interface is $T_{xz} = \mu\,\partial_z v_x|_0 = -\partial_x p\,h$. To determine the pressure gradient, we note that the total flux Q vanishes in the moving reference frame. Equating the left-hand side of Eq. (6.43) to 0, we find $\partial_x p = 3\mu V/h^2$. This leads to a total viscous force (per unit length of the contact line):

$$f_{visc} = \int_0^L \frac{3\mu V}{h} dx \tag{6.48}$$

On a macroscopic scale, we can approximate the surface profile by a wedge $h(x) = \tan\theta_d x \cong \theta_d\, x$. This implies that the integral diverges. This divergence is the essence of the problem noticed by Huh and Scriven. Luckily, however, the divergence is rather weak – logarithmic. Mathematically, we can solve the problem by introducing a cutoff length a of order *nanometers* that acknowledges the breakdown of the continuum fluid mechanical description on the atomic scale.

A variety of microscopic effects (molecular forces and disjoining pressure; local shear thinning due to diverging stresses; precursor film; diffuse interface) have been invoked to regularize the divergence. We will discuss any of these specific mechanisms here. Instead, we simply acknowledge the existence of some *zone of ignorance* in a region of order a around the contact line. Once we do that we can solve the rest of the hydrodynamic problem and calculate the viscous dissipation in the wedge as

$$f_{visc} \cong \frac{3\mu V}{\theta_d} \int_a^L \frac{1}{x} dx = \frac{3\mu V}{\theta_d} \ln \frac{L}{a} \tag{6.49}$$

Balancing this expression with the surface tension force, Eq. (6.47), yields the final result

$$\theta_d(\theta_d^2 - \theta_Y^2) = \frac{6\mu V}{\gamma} \ln \frac{L}{a} = 6\, Ca \ln \frac{L}{a} \tag{6.50}$$

which relates the dynamic contact angle θ_d to the speed of the contact line. For vanishing Young angle, this yields the result $\theta_d^3 \propto V$. Combining Eq. (6.50) and the realization that $\dot{r} = V$ with volume conservation and a few elementary geometrical relations for spherical caps, it is easy to derive Tanner's law (see Problem 6.9). The extreme sluggishness of the drop spreading process thus arises from the very strong viscous dissipation in the vicinity of the contact line.

6.3.2 Surface Profiles on the Mesoscopic Scale: The Cox–Voinov Law

The attentive reader may have noticed that the macroscopic scenario discussed in the preceding subsection is actually not fully consistent. On the one hand, we made use of the continuity equation to relate the pressure gradient $\partial_x p$ to the velocity V of the contact line. While it is clear that such a pressure gradient must be present to sustain the flow, we did not specify the physical origin of the pressure gradient. From our discussions in the preceding subsections, it is clear that this pressure gradient should arise from a gradient of the capillary pressure due to a surface profile with position-dependent curvature as sketched in Figure 6.12d. Yet, upon proceeding from Eqs. (6.48) to (6.49), we assumed the surface to be flat. This inconsistency can be resolved by going back to Eq. (6.43). As we noted above, the right-hand side of the equation should be set to 0 because there is no net flux in the presence of a three-phase contact line. For a consistent treatment of this problem, we should insert at this stage that any pressure gradient arises from the curvature of the interface by writing $\partial_x p = -\gamma \partial_{xxx} h$, as we did in the preceding section. As a result, we obtain an analog expression to Eq. (6.44) for vanishing net flux:

$$\partial_{xxx} h = \frac{3Ca}{h^2} \tag{6.51}$$

It is immediately clear that a straight wedge does not satisfy this equation. The exact solution is rather tedious to obtain and requires the matching and patching of solutions on different length scales similar to the solution of the Bretherton and Landau–Levich problem discussed above. We refer the interested reader to the literature for details (see [5]) and simply state the result, which is known as Cox–Voinov solution:

$$\theta_d^3(x) = \theta_m^3 + 9Ca \ln \frac{x}{a} \tag{6.52}$$

This result resembles in various aspects our previous macroscopic result. First of all, the microscopic cutoff a is still present and unavoidable. Hence, the solution still depends on a microscopic contact angle θ_m that arises from the atomic scale. This microscopic boundary condition is the same as θ_Y in Eq. (6.50). Second, the dependence of the third power of the dynamic contact angle on the capillary number and the logarithmic cutoff parameter is still the same. What is different, however, is the result that the local slope angle θ_d (remember $\tan \theta = \partial_x h$) depends on the position x in a logarithmic manner. That is, there is strictly speaking no well-defined dynamic contact angle. The value depends on the position of the measurement with respect to the contact line and hence on the resolution of the instrument. The existence of a nevertheless reasonably well measurable macroscopic dynamic contact angle is only saved by the separation of the length scales between the microscopic cutoff length $a = O$ (nm) and the macroscopic scale L of the measurement, which is typically larger than, say, 10 μm.

While we discussed the Cox–Voinov solution here in the context of the lubrication approximation, it is worth noting that the treatment can be extended to much larger contact angles as typically encountered in EW experiments and also to situations with two immiscible fluids. The resulting expressions for the dynamic contact angle are very similar to Eq. (6.52), and numerical deviations are of the order 1% for many conditions [5].

6.3.3 Dynamics of the Microscopic Contact Angle: The Molecular Kinetic Picture

To obtain a complete expression for the dynamic contact angle, we need to understand the speed dependence of the microscopic contact angle θ_m. To address this problem, Blake and Haynes proposed a rather generic model based on Eyring's microscopic model of viscosity (see [5] and references there). The general idea is that thermal fluctuations that we have disregarded in our macroscopic treatment so far must be important on small scales. On these molecular scales, thermally activated hopping processes should determine the response of the contact line to any externally applied driving force. The analysis starts by considering a contact line positioned somewhere in an energy landscape with local maxima and minima caused by some kind of heterogeneity on the substrate. Ultimately, we can think of individual atoms adsorbing and desorbing to and from atomic binding sites on the surface, as sketched in Figure 6.12e. Yet, also somewhat larger random surface heterogeneities on the scale of a few nm act as transient pinning sites that can be overcome by thermal activation: If we consider a surface defect of size l with an excess surface energy of $0.1\,\gamma$, the binding energy $\delta E_0 = 0.1\gamma\, l^2$ is comparable with thermal energies for $l \approx O\,(1\text{ nm})$, implying that thermally activated motion of the contact line at finite rates is possible for defect sizes up to a few nanometers, as has indeed been seen in molecular dynamics simulations on surfaces with nanoscale roughness. In equilibrium, i.e. for $\theta_m = \theta_Y$, the position of the contact line fluctuates due to random hopping processes with a rate of

$$\nu_{\pm} = \nu_0 e^{-\frac{\delta E_0}{k_B T}} \tag{6.53}$$

Here, ν_0 is a microscopic attempt rate and the $\pm$ sign indicates that the hopping rate in the advancing and receding direction is equal. In the nonequilibrium situation of an advancing contact line, the local contact angle is larger than θ_Y. Hence, the energy landscape is biased such that the energy barrier in the advancing direction is slightly lower than in the receding direction, i.e. $\delta E_+ = \delta E_0 - \epsilon$. Similarly, $\delta E_- = \delta E_0 + \epsilon$. This results in a net advancing motion of the contact line due to a biased hopping process with a frequency $\nu = \nu_+ - \nu_- = \nu_0(\exp(-\delta E_+/k_B T) - \exp(-\delta E_-/k_B T))$. If we apply the macroscopic imbalanced surface tension force f_γ to the fluctuating microscopic probe volume of fluid, we can estimate $\epsilon = f_\gamma\, l^2/2$. (This amount of energy is dissipated in every microscopic hopping event.) Inserting these expressions and Eq. (6.47), we obtain an expression for the contact line velocity:

$$V = l\,\nu = 2l\,\nu_0 e^{-\frac{\delta E_0}{k_B T}} \sinh\left(\frac{\gamma\, l^2}{2k_B T}(\cos\theta_Y - \cos\theta_m)\right) \tag{6.54}$$

This corresponds to a microscopic contact line friction force $f_m = \gamma(\cos\theta_Y - \cos\theta_m) = 2k_B T/l^2 \text{arcsinh}(V/v_m)$ with a characteristic microscopic velocity $v_m = l\,\nu_0 \exp(-\delta E_0/k_B T)$. For $V \ll v_m$, we can linearize this expression and obtain $f_m \approx \frac{2k_B T}{l^2 v_m}\, V = \xi\, V$, where ξ is a contact line friction coefficient. For small contact angles, we can rearrange Eq. (6.54) to obtain

$$\theta_m^2 = \theta_Y^2 + \frac{2k_B T}{\gamma\, l^2} \text{arcsinh}\left(\frac{V}{2l\nu_0} e^{\frac{\delta E_0}{k_B T}}\right) \tag{6.55}$$

Hence, also the microscopic contact angle θ_m becomes speed dependent if the atomistic hopping processes cannot keep up with the macroscopic speed of the contact line. Interestingly, the characteristic microscopic velocity scale v_m depends only on equilibrium properties of the system. Therefore it provides – at least in principle – a link to the static properties of the surface such as the contact angle hysteresis, which is determined by the same surface heterogeneity. If the viscous effects are negligible, we can equate θ_m in Eq. (6.55) with θ_D.

6.3.4 Comparison to Experimental Results

Contact line dynamics has been a very active field in recent decades that saw vivid debates. The difficulty in testing the models described above arises from the fact that most measurements only resolve the dynamic contact angle θ_d on the macroscopic scale (Figure 6.12c). Neither the viscosity-induced deformations of the liquid profile nor the deviation of the microscopic contact angle is easily accessible. Nor are the microscopic parameters a, l, v_m, and δE_0. As a result, far-reaching conclusions were sometimes based on the rather indirect evidence, in particular on macroscopic dynamic contact angle data; see Figure 6.13. By now, it is rather well established that both viscous dissipation on mesoscopic

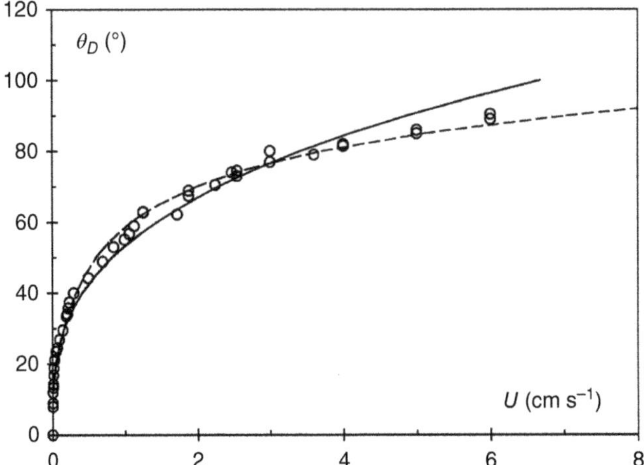

Figure 6.13 Dynamic contact angle versus contact line speed for an aqueous glycerol solution (viscosity 104 mPa s) on polyethylene-coated paper ($\theta_Y = 88^\circ$). Solid line: fit to Eq. (6.42); dashed line: fit to Eq. (6.55) with $\theta_m = \theta_D$. Source: Blake 2006 [6]. Reproduced with permission of Elsevier.

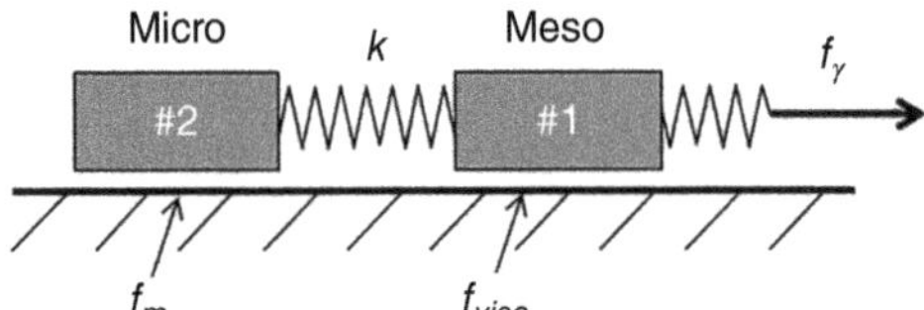

Figure 6.14 Qualitative sketch of the *series circuit* of contact line dissipation on microscopic and on mesoscopic scales.

scales and microscopic dissipation close to the fluctuating contact line matter. Viscous effects have been demonstrated in great detail for completely wetting viscous silicone oils (see [5] and references there). More volatile fluids, including in particular aqueous solutions of salts and many systems with large contact angles, tend to be better described by the molecular kinetic model, as illustrated in Figure 6.13.

Combined models that incorporate Eq. (6.55) as input for the Cox–Voinov relation, Eq. (6.52) have been also formulated. They give a consistent description of the dynamic contact angle as a function of the contact line speed for many systems over a wide range of speeds. Very qualitatively, we can picture the contact line as a train of two sleighs with different frictions as sketched in Figure 6.14. The driving force f_c due to the imbalanced surface tension forces is balanced by the combined friction of the mesoscopic sleigh #1 and of the microscopic sleigh #2. If the microscopic friction f_m is small, spring k remains unstretched, and we have $\theta_m \approx \theta_Y$. In this case the pulling force f_c is completely balanced by the viscous friction f_{visc} experienced by sleigh #1 on the mesoscopic scale. Vice versa, if $f_{visc} \ll f_m$, f_c is balanced on the microscopic scale (sleigh #2). Accordingly, k gets stretched and we have $\theta_m \approx \theta_d$. The degree of stretching of k thus measures the relative importance of viscous versus microscopic dissipation.

Contact line dynamics was also studied in detail in the presence of EW. In fact, the reduction of the dynamic contact angle was probably the first major

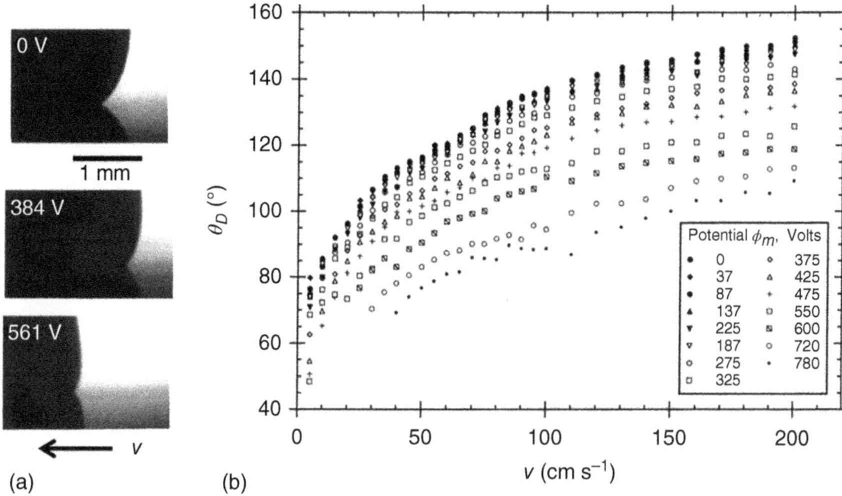

Figure 6.15 Electrowetting-induced reduction of the dynamic contact angle. (a) Snapshots of a liquid surface (liquid: black) next to a solid substrate that is moving at $v = 0.9\,\mathrm{m\,s^{-1}}$ from right to left at increasing voltage (top to bottom) as indicated, leading to a decreasing contact angle. (b) Dynamic contact angle versus speed for various voltages, increasing top to bottom as indicated. Source: Blake et al. 2000 [7]. Adapted with permission of ACS.

application of EW. In so-called curtain-coating technology, a lamella of liquid is falling down onto a solid substrate that is moving at a high speed of order meters per second. The liquid lamella is entrained by the solid surface and lays down as a liquid film that subsequently solidifies upon evaporation of the solvent. For decades, this used to be the standard fabrication technology for photographic films. The speed of this process was limited by the increase in contact angle with increasing speed, as illustrated in Figure 6.15b. Beyond a certain critical speed, an air film is entrained by the fast-moving solid underneath the liquid film, which destroys the liquid coating. As Figure 6.15 shows, the dynamic contact angle can be substantially reduced by applying a suitable voltage (or charge) to the moving solid carrier, thereby enabling higher deposition speeds. The voltage-dependent contact angle data can be fully described by Eq. (6.55), provided that the electrostatic force due to EW is taken into account, i.e. assuming that the equilibrium contact angle decreases following the Young–Lippmann equation, Eq. (5.1).

6.4 Surface Waves and Drop Oscillations

In this section, we discuss the dynamic response of liquid surfaces to periodic excitations with a fixed frequency ω. Such excitations give rise to surface waves with a continuous spectrum of wavelength λ for infinite surfaces and to discrete eigenmodes in the case of drops of finite size, as sketched in Figure 6.16. In both cases, our primary concern is to determine the relation between ω and the spatial index of the excitation, i.e. λ in case of the capillary waves and the mode index n in case of the discrete eigenmodes. Our analysis will provide the basic

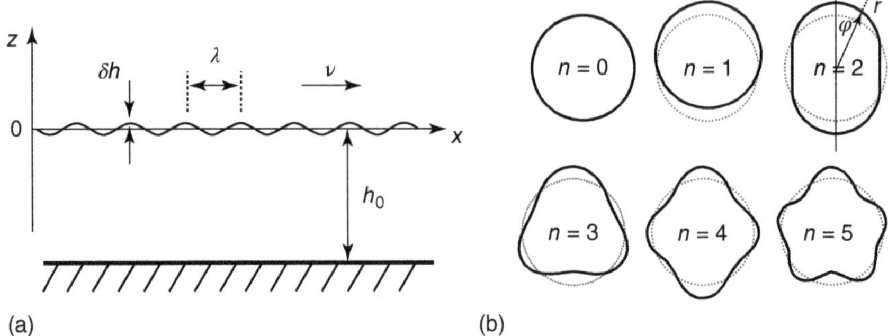

Figure 6.16 (a) Illustration of a propagating wave of wavelength λ on a liquid surface. (b) Discrete set of eigenmodes of an oscillating drop. n indicates the index of the corresponding eigenmode, described by the Legendre polynomial P_n.

principles to understand how drops respond to external forces as they arise in EW and in many other circumstances in science and technology. Before starting a formal treatment, let us analyze the situation based on qualitative physical considerations. First of all, we are looking for oscillatory solutions of the surface profile. Oscillatory behavior always arises in Newton's law from the balance of some restoring force and inertia. For a linear system, it results in a characteristic eigenfrequency $\omega = \sqrt{k/m}$, where k is the spring constant of the restoring force and m is the mass involved in the motion. In our present case, surface tension is the spring that provides the restoring force tending to reestablish a flat surface or a perfectly spherical drop shape. That is, we expect $k = \gamma$. For an oscillating drop, the relevant mass is simply the mass of the drop $m = \rho\frac{4}{3}\pi R_0^3$. Hence, we expect the eigenfrequency to scale as $\omega \propto \sqrt{\gamma/\rho R_0^3}$. The higher the surface tension, the higher the eigenfrequency, the larger the drop and the denser the lower eigenfrequency, as it should be. For a typical millimeter-sized drop of water, we find $\omega = 2\pi\ \nu \approx 100\ \text{rad s}^{-1}$. This estimate does not contain the mode index n. In fact, by taking the mass of the entire drop, we chose to limit ourselves to the lowest eigenmode. For the higher modes, the shape of the eigenmodes shown in (6.5) Figure 6.16b suggests that only a fraction of the drop is involved in the oscillatory motion. A detailed analysis should tell us how the relevant mass decreases with increasing mode number. For the surface waves, it is clear that the motion of the fluid will be localized close to the surface. Several wavelengths λ away from the surface, the fluid will remain at rest. Hence, we can choose $m \approx \rho\ \lambda^3$ as an estimate of the relevant mass. This results in an expected dispersion relation $\omega = \sqrt{\gamma/\rho\lambda^3}$ for wavelength dependence of capillary wave.

Having set out the general physical picture and our expectations, we can start to derive the formal solution for the two cases. Both derivations are classical results in fluid dynamics from the nineteenth century. While the important physical steps of the procedure are exactly the same, we will see that the technical calculation is a lot more cumbersome for the discrete eigenmodes of the drops than for the traveling capillary waves. Let us therefore start with the latter problem.

6.4.1 Surface Waves

The surface wave problem as sketched in Figure 6.16a has three length scales: the depth h_0 of the fluid layer, the wavelength λ, and the amplitude of the wave δh. We will assume that these three length scales are well separated, i.e. we assume $\delta h \ll \lambda \ll h_0$. For simplicity, we will take $h_0 = \infty$. Taking the limit of small amplitudes will allow for a number of simplifications in the modeling. First, we consider the full Navier–Stokes equations, Eq. (6.11), and estimate the relative importance of the explicitly time-dependent term and the nonlinear advective term. If the frequency of the wave is ω, the characteristic flow speed at the surface is of order $v = O(\omega\, \delta h)$. Hence, the order of magnitude of the explicitly time-dependent term is $\partial_t v = O(\omega^2\, \delta h)$. Taking the characteristic length scale to be λ, we find that the nonlinear term is given by $(v\nabla)v = O\left(\omega^2 \frac{\delta h^2}{\lambda}\right)$. Because $\delta h \ll \lambda$ by definition for waves of small amplitudes, we can indeed neglect this contribution. If we are interested primarily in the behavior of the purely oscillatory solution, we can also neglect the damping term. Taking into account the presence of gravity, our surface waves are thus governed by the equation

$$\rho \partial_t \vec{v} = -\nabla P - \rho g \vec{e}_z \tag{6.56}$$

The body force due to gravity points in the negative z-direction (see Figure 6.16a). Like the surface tension forces, it provides a restoring force that tends to restore the flat surface. Capillary pressures will enter the problem via the boundary conditions in the pressure term, Eq. (6.14). The pressure in the fluid is given by $P = P_0 - \rho g z + p(x, z, t)$, where P_0 is the ambient reference pressure at the surface in the absence of the perturbation $p(x, z, t)$ caused by the wave. We treat the problem in two dimensions, and we look for propagating wave solutions that can be written as

$$\begin{aligned} h(x,t) &= \delta h \exp(i(\omega t - qx)) \\ p(x,z,t) &= \delta p(z) \exp(i(\omega t - qx)) \\ \vec{v}(x,z,t) &= \delta \vec{v}(z) \exp(i(\omega t - qx)) \end{aligned} \tag{6.57}$$

Inserting this ansatz in Eq. (6.56), we find that the velocity field is determined by the gradient of the pressure perturbation, $\vec{v} = -\nabla p / i\omega\rho$. Together with the continuity equation, $\nabla \vec{v} = 0$, this implies that the pressure perturbation satisfies the Laplace equation, i.e. $\nabla^2 p = 0$. Hence, the amplitude of the pressure perturbation and the velocity field decays exponentially as a function of the distance from the surface (see Problem 6.4):

$$\begin{aligned} \delta p(z) &= \delta p_s \exp(qz) \\ \delta \vec{v}(z) &= \delta \vec{v}_s \exp(qz) \end{aligned} \tag{6.58}$$

Here, the subscript s denotes the amplitude at the liquid surface. (Note that this problem including the exponential decay is mathematically identical to the calculation of the electrostatic potential distribution next to the array of interdigitated electrodes with opposite charges that we considered in Chapter 2; see Figure 2.4.) To close the problem, we need to relate the amplitude of the pressure perturbation to the perturbation of the surface by making use of the boundary condition at the liquid surface. We determine δp_s by identifying

$$P = P_0 - \gamma \partial_{xx} \delta h = P_0 - \rho g\, \delta h + \delta p_s \tag{6.59}$$

Here, we made use again of the limit $\delta h/\lambda \ll 1$ to linearize the capillary pressure. Inserting our traveling wave ansatz, Eq. (6.57), we find the desired relation:

$$\delta p_s = (\gamma q^2 + \rho g)\, \delta h \tag{6.60}$$

To obtain a second relation between δp_s and δh, we use the kinematic boundary condition $v_s = \partial_t h = i\omega \delta h$ for the vertical component of the velocity at the surface. Writing $v_s = -\delta p_s / i\omega\rho$ (as obtained above) on the left-hand side of this expression, we find $\delta p_s = \omega^2 \rho\, \delta h$. Combining with Eq. (6.60), we can read the dispersion relation of surface waves:

$$\omega = \sqrt{gq(\lambda_c^2 q^2 + 1)} \tag{6.61}$$

This relation contains two limits, as illustrated in Figure 6.17. For wavelength long compared with the capillary length, i.e. $\lambda \gg \lambda_c$, the first term in the bracket under the square root is negligible. In this case, gravity is the dominant restoring force, and the frequency of the wave scales as $\omega \propto \sqrt{q}$. For shorter wavelengths, $\lambda < \lambda_c$, capillarity is the dominant restoring force. In this case, we find $\propto \sqrt{\gamma q^3/\rho}$, as expected based on our simple scaling analysis. Note that the boundary between the gravity-dominated regime and the capillarity-dominated regime is at $\lambda \approx \lambda_c$. This should not be a surprise, because the capillary length is *the* characteristic length scale for which gravitational and capillary force are of the same order of magnitude. We saw this already in Section 1.4 in the context of static problems – and the same conclusion continues to hold for dynamic problems such as surface waves.

6.4.2 Oscillating Drops

Let us now turn to oscillating drops with a discrete spectrum of eigenmodes. The shape of an oscillating liquid drop can be described as $r(\varphi, t) = r_0 + \delta r(\varphi, t)$, as

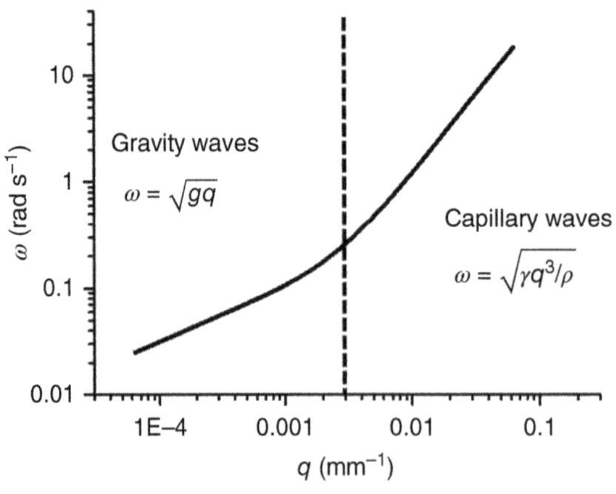

Figure 6.17 Dispersion relation of surface waves.

sketched in Figure 6.16b. Like in the case of the surface waves, we will assume that the perturbation is small, i.e. $\delta r \ll r_0$, and we will assume that the drop retains its cylindrical symmetry during the oscillation. In this case, we can decompose the shape of the drop into Legendre polynomial $P_n(\cos(\varphi))$ and write

$$r(\varphi, t) = r_0 + \sum_{n \geq 2} \widetilde{\delta r}_n(t) P_n(\cos(\varphi)) \tag{6.62}$$

Note that the decomposition for the perturbation starts with $n = 2$ as lowest index. The indices $n = 0$ and $n = 1$ do not describe oscillations but rather the equilibrium radius r_0 ($n = 0$) and the position of the center of mass ($n = 1$). The functional form of a few Legendre polynomials P_n is given in Table 6.1.

Like in the case of the surface waves, the full Navier–Stokes equations are simplified to Eq. (6.56) for small oscillation amplitudes. Because drops are generally not larger than the capillary length, we will neglect the gravity term as well.

Again, we are looking for periodic solutions, make an ansatz, and express the perturbation of shape, pressure, and velocity field in terms of the same Legendre polynomials P_n. We write

$$\begin{aligned} r(\varphi, t) &= r_0 + \sum_{n \geq 2} \delta r_n P_n(\cos\varphi)\, e^{i\omega_n t} \\ p(r, \varphi, t) &= p_0 + \sum_{n \geq 2} \delta p_n(r)\, P_n(\cos\varphi)\, e^{i\omega_n t} \\ v(r, \varphi, t) &= \sum_{n \geq 2} \delta v_n(r)\, P_n(\cos\varphi)\, e^{i\omega_n t} \end{aligned} \tag{6.63}$$

where the discrete index n replaces the continuous variable q in Eq. (6.57). Again, the pressure field satisfies the Laplace equation by virtue of Eq. (6.56) in combination with the continuity equation. In spherical coordinates with cylindrical symmetry, the Laplace equation reads

$$\nabla^2 p = \frac{1}{r^2}\partial_r(r^2\, \partial_r\, p) + \frac{1}{r^2 \sin\varphi}\partial_\varphi(\sin\varphi \partial_\varphi p) = 0 \tag{6.64}$$

Table 6.1 The first six Legendre polynomials $P_n(\cos(\varphi))$.

$P_0(\cos\varphi) = 1$
$P_1(\cos\varphi) = \cos\varphi$
$P_2(\cos\varphi) = \dfrac{3\cos^2\varphi - 1}{2}$
$P_3(\cos\varphi) = \dfrac{5\cos^3\varphi - 3\cos\varphi}{2}$
$P_4(\cos\varphi) = \dfrac{35\cos^4\varphi - 30\cos^2\varphi + 3}{8}$
$P_5(\cos\varphi) = \dfrac{63\cos^5\varphi - 70\cos^3\varphi + 15\cos\varphi}{8}$

The general solution can indeed be written in terms of Legendre polynomials including an algebraic dependence of the prefactor on r:

$$p(r,\varphi,t) = \sum_{n\geq 2} \delta p_{n,s} \left(\frac{r}{r_0}\right)^n P_n(\cos\varphi) e^{i\omega_n t} \tag{6.65}$$

Here, we introduced the factor $1/r_0$ for convenience to retain the dimension of a pressure for the amplitude $\delta p_{n,s}$ of the perturbation at the surface caused by the nth mode. Analog to Eq. (6.58), the amplitude of the perturbation is thus maximum at the surface and decays upon moving toward the center of the drop. Yet, the decay is algebraic rather than exponential, and the exponent increases with increasing mode number. We find

$$\begin{aligned} \delta p_n(r) &= \delta p_{n,s} \left(\frac{r}{r_0}\right)^n \\ \delta v_n(r) &= -\frac{n}{i\omega_n r} \delta p_n(r) = \delta v_{n,s} \left(\frac{r}{r_0}\right)^{n-1} \end{aligned} \tag{6.66}$$

Next, we relate the amplitude of the pressure perturbation to the amplitude of the shape mode using the boundary condition at the interface, i.e. we equate again $P_s = P_0 + p(r_0, \varphi, t) = P_0 + \gamma\kappa$, which will allow us to express $\delta p_{n,s}$ in terms of δr_n. While formally the same step, the explicit evaluation of this condition is a lot more tedious for oscillating drops than for capillary waves. For the calculation, we make use of the general formula for the capillary equation in spherical coordinates with cylindrical geometry, which reads

$$\kappa = \frac{2r^2 + 3(\partial_\varphi r)^2 - r\partial_{\varphi\varphi} r}{S^3} - \frac{\cot\varphi\, \partial_\varphi r}{r\, S} \tag{6.67}$$

Here, $S = \sqrt{r^2 + (\partial_\varphi r)^2}$. Inserting the ansatz for $r(\varphi)$ from Eq. (6.63) and keeping only terms up to first order in $\delta r_n / r_0$, we find

$$\kappa = \frac{1}{r_0}\left(2 + \sum_{n\geq 2} \frac{\delta r_n}{r_0}(-2P_n - 2\cos\varphi\, P_n' + \sin^2\varphi\, P_n'')\right) \tag{6.68}$$

The right-hand side of the equation can be simplified by making use of the specific identity of Legendre polynomials:

$$n(n+1)P_n = 2\cos\varphi P_n' - \sin^2\varphi P_n'' \tag{6.69}$$

Taking into account $P_0 = 2\gamma/r_0$, this leads to the relation $\delta p_{n,s} = \frac{\gamma}{r_0^2}\,(n(n+1) - 2)\,\delta r_n$. Using the kinematic boundary condition $v_r = \partial_t r$ and $\rho\,\partial_t v_r = -\partial_r p(r = r_0)$, Eq. (6.56), we find

$$\begin{aligned} \rho\,\partial_{tt} r &= -\rho \sum_{n\geq 2} \omega_n^2\; \delta r_n\, P_n(\cos\varphi) e^{i\omega_n t} \\ &= -\sum_{n\geq 2} \gamma \frac{n(n-1)(n+2)}{r_0^3}\; \delta r_n\, P_n(\cos\,\varphi) e^{i\omega_n t} \end{aligned} \tag{6.70}$$

From this equation, we can read the eigenfrequency of the nth mode as

$$\omega_n = \sqrt{\frac{\gamma}{\rho r_0^3} n (n-1)(n+2)} \tag{6.71}$$

This famous result was first obtained by Rayleigh in 1879. For the lowest mode, we thus find an eigenfrequency of $\omega_2 = \sqrt{8\gamma/\rho r_0^3}$, which is again close to but somewhat smaller than our initial estimate. As expected, the eigenfrequency thus increases with increasing mode number because less fluid is involved in the actual motion. (This can be seen from the radial dependence of the velocity perturbation in Eq. (6.66).) The result also implies that the liquid can respond locally very quickly, while the global response is dominated by the lowest P_2 mode. In EW experiments, this is readily seen in high-speed video images: Close to the contact angle, the contact angle changes quickly, while the rest of the drop lags behind (see Figure 6.4).

It is also interesting to note that Eq. (6.71) is consistent with our earlier result for the capillary waves on an infinite surface (Eq. (6.61)). The perturbation that is described by the Legendre polynomials can be seen as a wave propagating along the liquid surface. The requirement of uniqueness of the perturbation δr and a function of the polar angle φ requires that the circumference of the drop must be an integer multiple of the wavelength, i.e. $n\,\lambda = 2\pi\, r_0$. Inserting this relation into Eq. (6.71), we indeed recover Eq. (6.61), as expected.

Furthermore, it is interesting to consider a few extensions of the present scheme. If the drop is surrounded by an immiscible ambient fluid of density ρ_a, that ambient medium also contributes to the flow, and its mass contributes to the inertia of the system. The derivation then involves solutions for the pressure and flow fields both inside and outside the drop that have to be matched by the same boundary conditions on the drop surface as above. As a result, the eigenfrequency of each mode is reduced to

$$\omega_n^* = \sqrt{\frac{\gamma}{\rho r_0^3} \frac{n(n-1)(n+1)(n+2)}{(n+1)+n\hat{\rho}}} = \omega_n \sqrt{\frac{n+1}{(n+1)+n\,\hat{\rho}}} \tag{6.72}$$

Here, $\hat{\rho} = \rho_a/\rho$ is the density ratio between the ambient phase and the drop phase. Secondly, it is also possible to consider the damping due to the finite viscosity of the liquid. Without any proof, we simply provide the formula

$$\widetilde{\omega}_n^* = \omega_n^* - \frac{\mu}{\rho r_0^2} F \sqrt{Re_{d,n}} \tag{6.73}$$

where $Re_{d,n} = \rho r_0^2 \omega_n^*/\mu$ is the droplet Reynolds number and μ is the viscosity of the drop.

$$F = \frac{(2n+1)^2 \sqrt{\hat{\rho}\hat{\mu}}}{4\sqrt{2}\pi\,(n\hat{\rho}+n+1)(1+\sqrt{\hat{\rho}\hat{\mu}})} \tag{6.74}$$

is a viscous correction factor that depends on $\hat{\rho}$ and on the viscosity ratio $\hat{\mu} = \mu_a/\mu$. For a detailed derivation, see [8]. It is interesting to note that the quantitative shifts due to the damping are rather moderate (see Problem 6.6).

6.4.3 Example: Electrowetting-Driven Excitation of Eigenmodes of a Sessile Drop

A very spectacular and at the same time very instructive and practically relevant phenomenon in dynamic EW is the excitation of eigenmodes of sessile drops upon driving them by an AC voltage at variable drive frequency $f = \omega/2\pi$. The key observation is that the amplitude of the response varies strongly with the applied frequency and goes through a series of maxima whenever f coincides with one of the resonances of the drop. The resonances have a finite width Δf, typically of the order of 10–20 Hz for millimeter-sized drops in air and somewhat larger in ambient oil due to the additional damping of the surrounding medium. Figure 6.18a shows snapshots of a drop for the second, third, and fourth eigenmodes as recorded by a camera at slow frame rate such that several phases of the oscillation are superimposed in each of the images. First of all, it is clear that the shape modes of the sessile drop indeed resemble the eigenmodes that we just analyzed for the free drops. Counting the number of nodes on the surface for each of the modes, we can identify that the shapes in Figure 6.18a correspond to the eigenmodes $P_4, P_6,$ andP_8. (The lowest P_2 mode (not shown) has a resonance frequency of $\approx 30\,\text{Hz}$.) For typical conditions with Young's angle $\theta_Y \approx 90^\circ$, we thus excite only even eigenmodes. For these modes, the sessile drop simply corresponds to the top half of an oscillating free drop, as illustrated in the inset of Figure 6.18b – with the only difference of course that a no-slip boundary condition should be imposed at the drop–substrate interface instead of a zero-normal flux condition for the free drop. Apparently, it does not significantly affect the overall character of the even eigenmodes. For uneven modes, however, the top half and the bottom half of the drop are not symmetric with respect to the drop surface. To oscillate in an uneven mode, the center of mass of the drop has to oscillate substantially in order to compensate for the exchange of mass between the top and the bottom half of the drop in different phases of the oscillation. Because different uneven eigenmodes have different frequencies, such combinations of modes are suppressed – at least for not too large amplitudes.

We will discuss general aspects of drop oscillations in Section 8.3. Here, we only mention briefly the very elegant analysis of the EW-driven excitation of drop eigenmodes by Oh et al. [9]. The key idea of their analysis is that the electric driving force (per unit length) $f_{el} = \frac{\epsilon_0 \epsilon_d}{2d} U^2$ experienced by the contact lines can be decomposed into the same eigenmodes (Legendre polynomials) similar to the drop surface. As a consequence, each eigenmode in Eq. (6.70) is driven by the corresponding component of the electric driving force $f_n = (2n+1)P_n(0)\frac{\epsilon_0 \epsilon_d}{2dR} U^2$, where $U = U_0 \cos(\omega_{el} t)$ is the applied AC voltage. The resulting response of the drop as a function of the drive frequency including viscous dissipation is shown in Figure 6.18b. It turns out that the calculated eigenfrequencies overestimate the experimental ones by 10–20%. Part of this deviation is caused by the large oscillation amplitudes in typical EW experiments, which can reach up to 50%

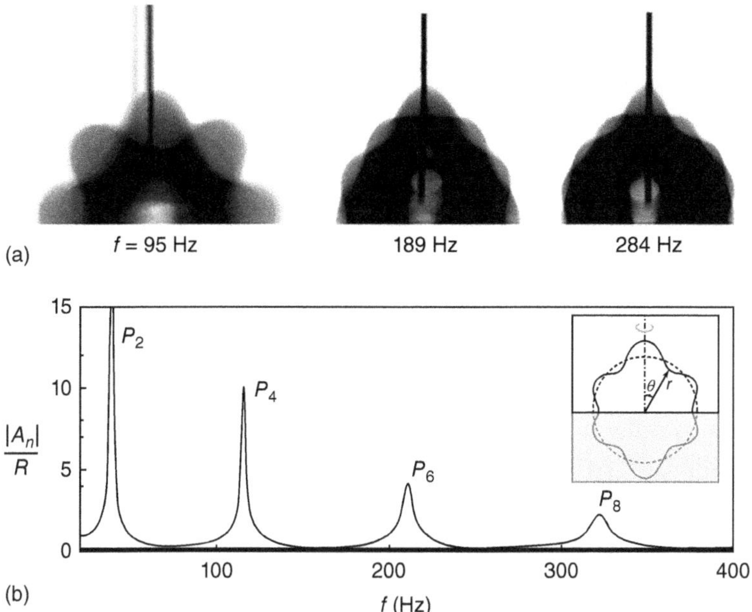

Figure 6.18 Eigenmodes of a sessile drop excited by AC voltage in electrowetting. (a) Snapshots corresponding to the excitation of the P_4 (95 Hz), P_6 (189 Hz), and P_8(284 Hz) mode. (b) Calculated drop response of the modes $P_2, \ldots, P_8$ versus excitation frequency $f = \omega/2\pi$. Inset: illustration of the symmetry condition imposed by the substrate in combination with the 90° average contact angle. Source: Oh et al. 2008 [9]. Adapted with permission of ACS.

of the drop radius, while the validity of the linear analysis of drop oscillations is limited to amplitudes $< 10\%$. The boundary condition at the drop–substrate interface that we already mentioned above is also expected to affect the absolute values of the resonance frequency. For more details, the reader is referred to [9].

6.4.4 General Consequences

The present subsection showed that the response time of drops (or surfaces) is ultimately limited by the inertia of the liquid. On small scales, corresponding to large wavenumbers or high eigenmodes, the liquid can respond more quickly than the entire system on the global scale. This has important consequences for the minimum response time or – vice versa – the maximum operation speed of EW-driven devices such as optofluidic lenses and EW-driven displays. For any given size L of the system, operation faster than the lowest corresponding Rayleigh's eigenfrequency is generally prohibited by inertia. In fact, the useful maximum operation frequency is generally lower than the lowest Rayleigh mode, because operating at the maximum frequency implies that the liquid oscillates at that frequency. In applications such as tunable lenses, the relevant time scale is the time required for the system to settle in the desired new configuration rather than an oscillation around the final state. This means that the inertial oscillations should be damped out by viscous dissipation. For harmonic oscillations such as the eigenmodes of the drops described here, the settling to the new equilibrium is fastest for so-called critical damping conditions, which – for aqueous drops – generally requires the addition of small amounts of viscosity modifiers such as polyethylene oxide (PEO). We will come back to such practical considerations in the chapters on EW applications.

Another important aspect is the separation of time scales between the fluid response and the typical time required to activate an electrode. The former is typically of the order of a few (tens) of milliseconds, as we just derived. The latter is much faster for the capacitances and output powers of amplifiers typically used in EW. As a consequence, the variation of the equilibrium contact angle upon applying a step voltage to an electrode is typically instantaneous on the time scale of the response of the fluid. Hence, the electrical response of the system and the fluid dynamic response are separate. The physics that limits the operation speed of EW systems is therefore generally governed by the fluid dynamic response. As a consequence, in order to describe the dynamic response of EW devices, it is usually sufficient to prescribe the local equilibrium contact angles on each electrode according to the voltage applied at any moment in time and to calculate the subsequent hydrodynamic response of the system.

Problems

6.1 Consider liquid flowing in steady state through a channel of length L bounded by two parallel walls at a distance h. The flow is driven by a pressure gradient ΔP between the inlet and the outlet of the channel. (a) Calculate the total drag force transmitted by flowing liquid to the solid walls. Why does the result not depend on the viscosity of the fluid? (b) Calculate the elements of the viscous stress tensor for any arbitrary position within the channel. Why does the stress tensor vanish in the center of the channel?

6.2 Consider two parallel plates moving with respect to each other at a fixed velocity V, as shown in Figure 6.2. The lower part of the channel between $z = 0$ and $z = H$ is filled with liquid 1 with viscosity μ_1, and the upper part for $H < z < h$ is filled with a second immiscible fluid 2 with viscosity μ_2. (a) Calculate the velocity profile in both fluids. (b) Consider the limit $H \ll h$ and $\mu_2 \ll \mu_1$ that is encountered in electrowetting if an aqueous drop moves in ambient viscous oil that is entrained underneath the drop by the completely wetted substrate. (c) Calculate the force required to keep the top plate moving at the speed V.

6.3 Calculate the resonance frequency of the first free eigenmodes of a water drop with radius $r_0 = 1$ mm in ambient air and in ambient silicone oil with a density of 970 kg m^{-3} and for oil with two viscosities of $\mu_1 = 1$ mPa s and $\mu_2 = 100$ mPa s.

6.4 Consider our derivation of the capillary waves on a liquid surface. (a) Show that the result for the pressure field in Eq. (6.58) indeed satisfies the Laplace equation $\nabla^2 p = 0$. (b) Calculate the components of the resulting flow field in two dimensions, and show that the tracer particles of fluid follow circular trajectories with a diameter that decreases exponentially with increasing distance from the surface.

6.5 Use the dispersion relation of capillary waves, Eq. (6.61), to calculate the phase velocity of the waves and to verify that a perturbation of a reasonable length scale indeed requires approximately 1 ms to travel toward the apex of a mm-sized drop, as shown in Figure 6.4.

6.6 Calculate the eigenfrequencies for the four lowest eigenmodes of a drop of water in air with a diameter of 1 mm. Use Eqs. (6.72)–(6.74) to calculate the frequency shift of these four modes upon immersing the drop in a (i) density-matched ambient oil of the same viscosity and (ii) in a density-matched oil with a viscosity 10 times higher than water.

6.7 Consider two networks of microfluidic channels that are represented by their electric equivalent circuits in Figure 6.19. Make use of the analogy between electric current and hydrodynamic flux and potential and pressure to calculate for the left diagram: (i) the pressure at the left intersection where the incoming flux splits in two and (ii) the fluxes through the top and the bottom channel with the hydraulic resistances R_{H2} and R_{H3}. For the right system, (i) derive a general expression of the flux Q_c through the connecting channel as a function of the difference of the inlet pressures P_{1in} and P_{2in} assuming that the inlet and outlet channels in the top and bottom branches are the same (i.e. $R = R_{1in} = R_{2in} = R_{1out} = R_{2out}$) and $R_c \ll R_{1in}$. (ii) Calculate Q_c for $P_{1in} = P_{2in}$ and $R = R_{2in} = R_{1out} = R_{2out}$ and $R_c \ll R$ but now for $R_{1in} = R - \Delta R$ to first order in ΔR. (Such a device is known as hydrodynamic comparator. See, e.g. Ref. [10].)

6.8 Liquid films with a thickness h_0 in the range of the molecular interactions can be destabilized by the action of molecular forces such as van der Waals forces. As discussed in Section 1.6, the energy (per unit area) of a thin wetting film due to van der Waals forces is given by $\Phi(h) = A_H/12\pi h^2$, where A_H is the Hamaker constant. This energy contribution gives rise to a disjoining pressure $\Pi(h) = -\partial_h \Phi(h)$, which enters in Eq. (6.37) as a contribution to the pressure term. Perform a linear stability analysis for a thin liquid film under the influence of van der Waals interaction and capillarity.

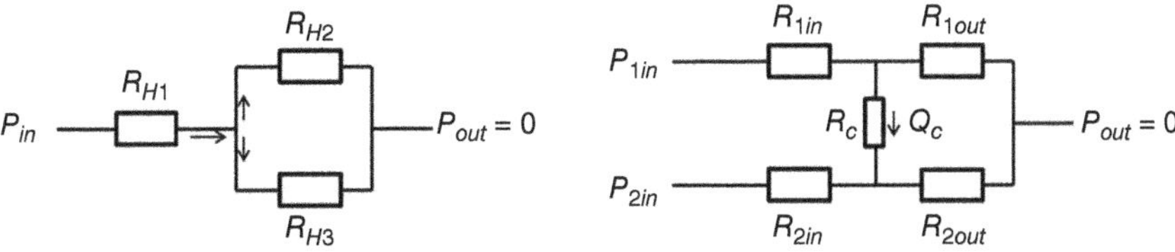

Figure 6.19 Sketch of two resistive microfluidic networks.

(i) Show that the film is unstable against sinusoidal perturbations for wavenumbers shorter than the critical value $q_c = \sqrt{|\partial_{hh}\Phi|/\gamma}$, provided that $A_H < 0$. (ii) Show that the growth rate of the fastest-growing mode is given by $\tau = 12\mu/(h_0^3 q_c^4 \gamma) \propto h_0^5$.

6.9 Derive Tanner's law $r(t) \propto t^{1/10}$ for the spreading dynamics of a liquid drop on a completely wetting substrate (i.e. $\theta_Y = 0$) by balancing the imbalanced Young forces, Eq. (6.47), and the viscous dissipation at the contact line, Eq. (6.50). Make use of the limit of small contact angles for spherical caps, which allows you to express the conserver drop volume as $\Omega = \pi h^2(3R - h)/3 \propto \pi r^3\theta/4$, where R is the radius of the spherical cap in three dimensions, h is its maximum height, and $r = R \sin\theta$ is the radius of the drop–substrate contact area.

References

1 Oron, A., Davis, S. H., and Bankoff, S. G. (1997). Long-scale evolution of thin liquid films. *Rev. Mod. Phys.* 69 (3): 931–980.

2 Staicu, A. and Mugele, F. (2006). Electrowetting-induced oil film entrapment and instability. *Phys. Rev. Lett.* 97 (16): 167801.

3 Sun, B. and Heikenfeld, J. (2008). Observation and optical implications of oil dewetting patterns in electrowetting displays. *J. Micromech. Microeng.* 18 (2): 025027.

4 Howison, S. (2005). *Practical Applied Mathematics*. Cambridge University Press.

5 Bonn, D., Eggers, J., Indekeu, J. et al. (2009). Wetting and spreading. *Rev. Mod. Phys.* 81 (2): 739–805.

6 Blake, T. D. (2006). The physics of moving wetting lines. *J. Colloid Interface Sci.* 299: 1–13.

7 Blake, T. D., Clarke, A., and Stattersfield, E. H. (2000). An investigation of electrostatic assist in dynamic wetting. *Langmuir* 16: 2928.

8 Miller, C. A. and Scriven, L. E. (1968). The oscillations of a fluid droplet immersed in another fluid. *J. Fluid Mech.* 32: 417.

9 Oh, J. M., Ko, S. H., and Kang, K. H. (2008). Shape oscillation of a drop in ac electrowetting. *Langmuir* 24 (15): 8379–8386.

10 Vanapalli, S., Banpurkar, A. G., van den Ende, D. et al. (2009). Hydrodynamic resistance of single confined moving drops in rectangular microchannels. *Lab Chip* 9: 982.

7

Electrowetting Materials and Fabrication

7.1 Practical Requirements

This chapter provides both applied and theoretical perspectives on the impact of materials choices on electrowetting (EW) behavior and on device performance. This chapter also reviews common materials issues that result in experimental results which deviate from ideal EW. Furthermore, best practices are described for creating devices that are both reliable and operable at low potentials. These materials and device fabrication methods are driven by practical requirements that can vary widely based on application. *Lab-on-a-chip* (now a product, Advanced Liquid Logic/Illumina) devices are in some ways the easiest and in other ways the most difficult to satisfy from a materials standpoint [9]. Typically the EW lab-on-a-chip *cartridges* themselves are low cost and disposable and are connected with an electronic control unit that can be of higher cost and greater sophistication (not disposable). Therefore, lab-on-a-chip devices in many ways are the simplest to fabricate because the electronics can provide high potential control (>10 V) and allows for use of very thick and reliable EW dielectrics in the disposable cartridge. Furthermore reliability is also easier to satisfy since the occurrence of EW on a given surface could be several tens of times at most, before the chip is disposed of. The EW materials challenges for lab-on-a-chip are therefore primarily limited to coatings (surface fouling; Chapter 3), fluids (surface fouling, biocompatibility, cross-diffusions between phases, etc.), and possible surfactant incompatibilities. *Optics* (now a product, Varioptic/Parrot) or optical EW devices can also use higher potentials and thicker dielectrics because in most cases only two or very few electrodes can be directly addressed. However, fluid development can be challenging because typically the fluids must provide uncommon refractive indices for desired optical focusing power, low viscosity for fast switching speeds, wide temperature ranges, and proper fluid density matching to reduce the effects of gravity or vibration. There has long been an argument that EW lenses have utility in smartphones. However, the materials challenges here are more difficult in this application, driven by the need for very small package size, increased operating temperature range, and batch fabrication of multiple lenses on a single substrate. *Displays* (in development, Liquavista/Amazon) or EW displays have generated significant excitement by promising to improve the readability of a display in sunlight

Electrowetting: Fundamental Principles and Practical Applications,
First Edition. Frieder Mugele and Jason Heikenfeld.

and/or reduce display power consumption to very low levels [11]. However, EW displays pose a significant challenge because the pixels are limited to ~15–20 V operation that can be provided with conventional active matrix transistors [13]. Furthermore, a display product operating at 60 Hz and for 10 000 hours requires a challenging 36 million voltage cycles without pixel degradation. Furthermore, this requirement holds for having no pixel failures over very large areas of tens to hundreds of square centimeters. In terms of reliability, the only advantage of the displays is that compared with lab-on-a-chip, they can use nonaqueous conducting fluids that improve reliability. However, one of the fluid phases in displays has to dissolve or suspend a high weight percent of a colored dye or pigment, which has required intensive research and development (R&D) to provide suitable combinations of colorants and fluids.

A final introductory comment should be made on published results that can be misleading to those new to EW. There have been numerous reports of ultralow potential (several volts) EW with very thin dielectrics and/or very low interfacial surface tensions. These examples will not be discussed herein, as currently they most often have very little practical relevance due to poor reversibility (large hysteresis) and very slow response times. For current commercial applications, generally the minimum required interfacial surface tension between conducting and insulating fluids is ~5–10 mN m^{-1} or greater, which in combination with existing reliable dielectrics still requires ~15–20 V for operation.

7.2 Electrowetting Deviation: Caused by Non-obvious Materials Behavior

Ideal EW behavior on a dielectric was discussed in Chapter 5. In terms of real behavior for the EW response versus potential (voltage), there are two major types of nonideal EW behavior. To describe these, we examine evidence of a nonideal EW response as shown in Figure 7.1 and will adopt terminology provided by Fan et al. [10]. In the first region, we will refer to simply as the *deviation region* or *EW deviation* where contact angle reduces with potential but at weaker response than the theoretical response. With a well-planned approach, EW deviation can be minimized and closely follows the theoretical response

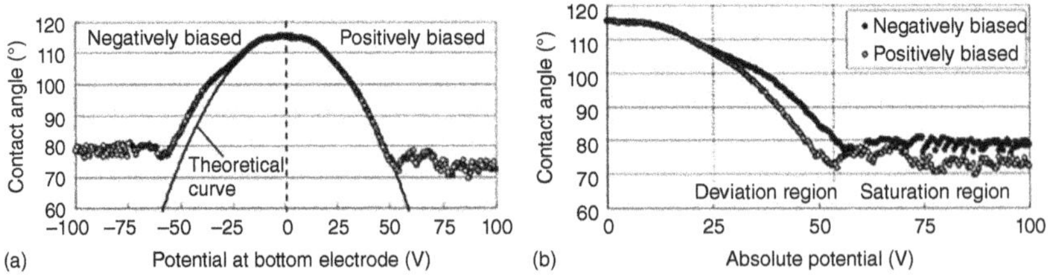

Figure 7.1 Plots of a sessile drop electrowetting experiment in air (no oil) showing asymmetric electrowetting. (a) Measured contact angle from a potential of −100 to 100 V, including theoretical curve. (b) Data replotted versus absolute potential showing the contact angle differences caused by polarity and with labeling of the deviation and saturation regions. Source: Fan et al. 2007 [10]. Reproduced with permission of Royal Society of Chemistry.

up to a second region, which we will refer to as the *saturation region*. In the saturation region, the contact angle nearly or completely ceases to decrease, or even reverses, as the potential is increased. The saturation region will be discussed in greater detail in Section 7.3. Simply, saturation should be avoided. EW into saturation has no applied value. Therefore we will focus the rest of our discussion in this section on the deviation region.

Turning our discussion back to the deviation regime, we can further separate some of the known effects for nonideal behavior into those that are *temporal* (changing with time) and *nontemporal* (invariant with time). In this book, we have chosen to separate effects into these categories of temporal versus nontemporal effects, because one can observe if deviation in an experiment is temporal or nontemporal and then much more quickly diagnose the physical origins of the EW deviation. Consider a simple example for temporal deviation. Let us say you applied a constant (static) potential to achieve a certain contact angle, and even though the potential is kept static, the observed contact angle begins to increase (reverse): this example would be temporal deviation (dynamic with time). Several effects known to cause temporal deviation are shown in Figure 7.2, and a more exhaustive list is provided below with references to articles for further details (e.g. see [4] that includes numerous experimental examples).

We will now present the most common forms of temporal and nontemporal deviation. For each subtype of deviation, we also provide a common method to diagnose if each subtype of deviation exists or not in an experiment (bold text at the end of each subsection). Typically, EW deviation is a combined effect of many subtypes of deviation, and optimization to remove the overall deviation involves careful diagnosis of each subtype. The reader should also use this section as a practical guide only and not as a theoretical review. Each subtype of deviation has rich and complex physics behind it, which can be better understood by careful study of Chapters 1–6 of this book.

7.2.1 Commonly Observed Temporal Deviations

7.2.1.1 Dielectric Failure (Leakage Current)

The dielectric fails to provide insulation between the conducting fluid and the EW electrode (Figure 7.2b). For severe dielectric failure with water, gas bubbles can even be observed due to water electrolysis at the electrode.

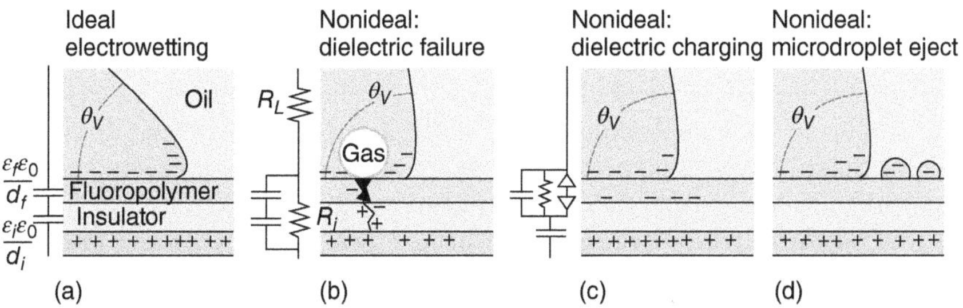

Figure 7.2 Zoom-in diagrams of ideal and nonideal electrowetting near the triple point (contact line) including labeling of the contact angle as a function of applied potential (V).

It is first worthwhile to reiterate that the absolute amount of charge density and instantaneous current required to change the contact angle in EW is very low. Consider a typical EW capacitor of 10^{-5} F m^{-2}; the charge density required to change the contact angle by 90° is only about 10^{-3} C m^{-2} or 10^{-9} C mm^{-2}. 10^{-9} C s^{-1} is only 1 nA! This gives the reader some greater perspective of just how sensitive EW systems can be to any deviation from a perfectly insulating dielectric.

Dielectric leakage or complete failure will cause a DC current flow, and cause a portion of the applied potential to drop partially or fully, across electrically resistive materials in the pathway of this DC current flow. Often, the most electrically resistive material is the conducting fluid itself (R_f), and this is often true even if the conducting fluid is an aqueous salt solution. A potential drop across the conducting fluid of course does not contribute to EW. To account for this potential drop, the EW equation (Eq. (5.1)) can be substituted with a reduced potential $U_i = U \times R_i/(R_f + R_i)$, where R_i is the resistance of the dielectric layer:

$$\cos\theta(U) = \cos\theta_Y + \frac{c(U \times R_i/(R_f + R_i))^2}{2\gamma} \tag{7.1}$$

It should be noted that most dielectrics exhibit a nonlinear response in electrical resistance versus applied potential and a simple value for R_i is typically only accurate for a short period of time at a constant potential that is well below the breakdown potential of the dielectric. Furthermore, if the leakage current through a dielectric system is significant enough to cause noticeable EW deviation, then the leakage current will very likely get worse with time (lower R_i). This is especially true for aqueous solutions where electrochemical degradation can increase the rate of failure of a dielectric that otherwise may seem stable with dry (solid) electrodes. It is for these reasons that we generally classify dielectric failure as a temporal effect.

The temporal nature of dielectric failure can also be nearly instantaneous if complete dielectric failure occurs. This is typically observed in an experiment as abrupt partial or full dewetting, and for water significant electrolysis and gas bubble formation is also often observed (Figure 7.7). In EW systems, the term *complete dielectric failure* is more appropriate than *breakdown field* because electrical failure in EW systems is dominantly due to dielectric pores or defects through which the conducting fluid can propagate (a liquid electrode that penetrates!). Therefore in EW systems, complete dielectric failure is commonly observed well below the bulk breakdown field that one would typically see with dry metal electrodes. That is to say, complete dielectric failure with conductive fluids does not require a classical avalanche breakdown process where electrically conductive charges are impact ionized or thermally (joule) created within the dielectric itself.

Quantifying dielectric failure can be performed by measure of faradaic (direct) current and/or imaging the density of gas bubbles generated for aqueous fluids (see Figure 7.7) [24]. Even as little as picoampere of DC electrical current for an millimeter-sized droplet is evidence of dielectric failure that will very likely get worse over time.

7.2.1.2 Dielectric Charging

This is one of the most commonly observed mechanisms for EW deviation (Figure 7.2c). The applied electric field exceeds the electrical breakdown field of the materials, and/or the materials are partly conductive to defects or pores though which fluid and ions can propagate [4]. In a multilayer dielectric like that shown in Figure 7.2, the layer that is least resistant to charging is often the fluoropolymer topcoat, and the E-field for any individual layer can be derived from a simple series capacitance of layers as

$$E_{layer} = (U \times c_{total})/(c_{layer} \times d_{layer}) \tag{7.2}$$

where d is the layer thickness and c is the total capacitance per unit area of the series of capacitances in a multilayer dielectric or for just a single layer itself. The equations for calculations of these capacitances, for an example, two-layer system, are shown at left of Figure 7.2 (f = fluoropolymer, i = insulator).

If charge is injected from the conductive fluid or the electrode into any portion of the dielectric, then that injected charge will electrically screen (counteract) the remaining charges in the fluid or electrode and decrease the electromechanical force that causes EW. The ideal EW equation can then be modified with a simple trapped charge potential (U_T) as first described by Prins and coworkers [28]:

$$\cos\theta(U) = \cos\theta_Y + \frac{c(U - U_T)^2}{2\gamma} \tag{7.3}$$

If the charging is at or near the surface (which is typically the case), then the amount of charge density Q_T (coulombs per unit area) can be predicted by a simple calculation of the charge-to-potential relation $Q_T = cU_T$, where c is the capacitance per unit area and U_T is the trapped potential. Charging near the surface is typically dominant, especially so with use of thin fluoropolymer topcoats that charge quite easily. A modified EW equation can therefore also be presented in terms of the trapped charge:

$$\cos\theta(U) = \cos\theta_Y + \frac{c(U - Q_T/c)^2}{2\gamma} \tag{7.4}$$

Interestingly, Eq. (7.4) is valid regardless of the depth of charge. This is both counterintuitive and intuitive. It is counterintuitive because we think of charge on a capacitor as due to $Q = cU$, and if charge is halfway through the dielectric, then it is only half a dielectric thickness away from the lower electrode (half capacitance). It is intuitive though, because regardless of position only one point charge in the dielectric is needed to screen an opposite polarity point charge in the electrode beneath the dielectric. The appendix at the end of this chapter contains a complete derivation that supports Eqs. (7.3) and (7.4) and the fact that charge depth does not matter.

The temporal nature of dielectric charging can be quite complex. Any EW system, if left with a static (DC) potential for long enough time, will experience dielectric charging because in practice there is no such thing as a perfect electrical insulator. As shown in Figure 7.2c, this slow leakage of charge into the fluoropolymer layer can be approximated as a shunt resistor (a resistor in parallel with the capacitor). A shunt resistor also implies that the charging can be slowly removed with no potential (slowly discharges with time) or more quickly with a reversing of the potential. Prevention of a net charge with reverse of potential is why alternating potentials are strongly preferred for achieving ideal EW. In fact, the shunt resistance R can be experimentally determined by increasing the alternating potential frequency to a frequency at which the amount of charging is reduced to $1/e$ of the DC charging condition, corresponding to the RC time constant of the shunt resistance and capacitance network of Figure 7.2c.

For higher potentials, the current feeding the charging can become nonlinear with potential because of complete dielectric failure of the fluoropolymer (see previous subsection), approximated by the two back-to-back zener diodes of Figure 7.2c. In practice, this *fast charging* of the fluoropolymer is all too often caused by applying potentials well into the saturation region of operation. The back-to-back zener diode approximation also reveals what is commonly observed – that injected charge can stay even permanently even as the external potential is removed from the experiment. *A homework Problem 7.9 is provided at the end of this chapter on this very topic and is solved simply by treating the trapped charge density Q_T as a built-in potential as taught above. This very problem shows how one can quantify the effect of dielectric charging using simple contact angle measurements.* Lastly, we note how the charging shown in Figure 7.2c appears most strongly near the contact line because of higher electric field produced by the sharper profile of the conducting fluid at the contact line (see Figure 5.5). This localization of the charging has been modeled in detail by the Papathanasiou group [8] and is easily pictured in earlier reports as well [27].

7.2.1.3 Charges into the Oil

Similar to dielectric charging, charges can be injected into the oil, which also partially screens the charges remaining in the conducting fluid. In addition, when a very strong (and impractically high) EW potential is applied, the conducting and insulating (oil) fluid interface can become unstable due to the accumulation of a high density of identically charged species, and microdroplets of charged conducting fluid can be ejected and cause screening of remaining charges (Figure 7.2d) [19]. The reader may refer back to Section 5.3 for a deeper theoretical consideration of this effect. Like dielectric charging, oil charging is dominantly found near the contact line. One therefore has to be careful of oil selection, as even low surface tension silicone oils such as Dow Corning OS oils can be very hygroscopic, which will typically increase the amount of oil charging as more water is dissolved into the oil. *Other than the extreme cases of visible microdroplet ejection, it is difficult to experimentally and conclusively observe the effect of oil charging because these charges in an oil fluid are not "trapped" locally like they are with dielectric charging (they can easily drift with electric field).*

7.2.1.4 Oil Relaxation

An increase or *relaxation* of the contact angle with time is often observed in EW displays and optics when using insulating fluids that consist of molecules with a nonzero dipole moment such as chloronaphthalene or that contain a large amount of polar additive in the fluid such as a dye or surfactant. This effect is not well understood nor published with any significant data or experiments beyond that for displays [30]. Regardless, the effect is easily observable and typically occurs in the range of tens of microseconds such that AC potentials of 10–100 Hz are often required to prevent this nonideal behavior. This effect can also be observed if charged impurities already exist in the oil. *To help in diagnosis if oil relaxation is an issue, a control experiment should be tested with an oil such as tetradecane, which exhibits no such relaxation. However, confirming oil relaxation is challenging because other effects such as oil charging must be ruled or separated out.*

7.2.1.5 Surfactant Diffusion (Interface Absorption)

If a surfactant is used and the surfactant concentration is near the critical micelle concentration (CMC point), then when the droplet is electrowetted, the increased surface area of the droplet is not immediately repopulated with surfactant. Therefore the instantaneous response of the droplet is that of one with a higher apparent surface tension or interfacial tension [9, 21, 25, 33–36]. Again, the reader is directed to Chapter 3 for deeper theory on diffusion of a molecule to a surface or interface where it becomes stable. Such behavior is possible for an abruptly electrowetted droplet in air (surfactant inside the droplet) or in oil (surfactant can be in the droplet or in the oil phase). This form of a temporal response is opposite of all those described above, as the contact angle deviation is immediate, but over a period of microseconds to seconds, the contact angle *decreases* and approaches its theoretically predicted value. A dynamic surface or interfacial tension $\gamma(t)$ can simply be substituted into the EW equation (Eq. (5.1)) and will asymptotically decrease over time as $\gamma(t) = (\gamma_0 - \gamma_e)e^{-t/\tau} + \gamma_e$, where γ_0 is the surface tension at $t = 0$ and τ is the characteristic relaxation time for the surfactant. τ generally increases with increasing size of the surfactant molecule and decreases as the concentration of the surfactant molecule is increased. The resulting modified EW equation is

$$\cos\theta(U,t) = \cos\theta_Y + \frac{cU^2}{2((\gamma_0 - \gamma_e)e^{-t/\tau} + \gamma_e)} \tag{7.5}$$

Raccurt measured $\tau \sim 13$ s for a 0.2% concentration of Tween 20 surfactant in NH_4HCO_3 surrounded by silicone oil [21]. For small surfactants at high concentrations, the effect is nearly unnoticeable in a typical EW experiment. *In most situations, this temporal deviation is easily minimized or avoided if surfactant concentrations are kept well above the CMC point and easily diagnosed because the temporal trend is the opposite of nearly all other forms of temporal deviation.*

7.2.1.6 Oil Film Trapping

Young's angle of 180° implies a continuous film of oil between the conductive fluid and the dielectric surface. Imagine a rapidly advancing contact line or fast droplet transport in lab-on-a-chip. The faster the advancement or transport, the thicker the layer of oil that will become trapped (similar to the physics of dip coating; see Section 6.2). The oil acts a dielectric as well and therefore initially reduces the EW effect substituting a dynamic capacitance c_{dyn} into the EW equation: $c_{dyn} = (c_{oil} \times c)/(c_{oil} + c)$.

$$\cos\theta(U, t = 0^+) = \cos\theta_Y + \frac{(c_{oil} + c)\, U^2}{2\,(c_{oil} \times c)} \tag{7.6}$$

where c_{oil} is the oil film very quickly after the potential is applied. Use of the term *very quickly* is overly simplified, because as soon as the oil film is entrapped, it begins to break up into smaller islands of oil (Figures 6.10 and 8.16) [26]. As the oil film then breaks up, the capacitance approaches the expected value of nearly the dielectric layer by itself minus the small effect of the remaining islands of oil, and accordingly EW contact angle nears its expected value. The time evolution of this breakup is also quite complex and is covered in detail in Section 6.2. The experimentally seen profile of the entrapped and broken up oil film is reserved for the next chapter.

Oil trapping is easily diagnosed experimentally by measuring the total electrical capacitance versus time (see Figure 8.16a). Problem 7.9 at the end of this chapter gives an experimental example of diagnosing the thickness of entrapped oil. Oil trapping is like surfactant diffusion in that the contact angle will *decrease* over time.

7.2.2 Commonly Observed Nontemporal Deviation

7.2.2.1 Unexpected Young's Angles: Gravity Effects

Numerous factors can cause deviation of Young's angle, and in many cases simple gravitational force and limitations of contact angle imaging/measurement software can lead to Young's angles of 180° (in oil) to appear decreased by as much as tens of degrees. To be clear, gravity *cannot* change the actual contact angle, but it can only change the apparent contact angle that is simply an artifact of the limitations of commonly used measurement techniques such as goniometers. *The effects of gravity can be diagnosed simply through calculation of Bond number (Section 1.4) and mitigated through density matching of the oil and conducting fluid. The effects of imaging and software limitations on contact angle calculation are far more difficult to diagnose and typically require secondary measurements such as electrical capacitance and careful confirmation of the interfacial tensions used to calculate Young's angle. For example, typically one measures with a goniometer Young's angle of 160–170° even for the case of a continuous oil film (180°).*

7.2.2.2 Unexpected Young's Angles: Surface and Interface Fouling

In addition, surface fouling can quickly alter the expected Young's angle, and is included here as a nontemporal effect, because once it happens it is usually permanent in nature. The reader is encouraged to revisit Chapter 3 of this book for greater detail on adsorption at interfaces and how it can change the observed Young's angle. *The effects of surface fouling can be diagnosed with advanced surface analysis tools, but a quicker method is to repeat an experiment with fluids that are pure and free of solutes that could foul a surface.* However, it should be noted that oils even if bought as *reaction grade* typically contain surface-active contaminants. Lead author Mugele's laboratory has seen this quite extensively even for commonly used alkanes. Without additional cleaning, the low interfacial tension as controlled by pendant drop measurements has a value that is about $\sim 10\,\mathrm{mN\,m^{-1}}$ too low, along with an observable time dependence. A general recommendation for precision measurements is to clean the solvents by running them through adsorption columns. Even after the solvents are expected to be *clean*, they should be verified for interfacial tension one second after pendant droplet formation to one hour (or even longer time scales for commercially intended applications). Lastly, it should also be noted that for biological applications, especially those using whole or partial biofluids that are full of amphiphilic solutes, such cleaning is typically impractical and some degrees of interfacial surface tension change or surface fouling are inevitable (see last section of this chapter for techniques to mitigate fouling in lab-on-a-chip).

7.2.2.3 Unexpected Young's Angles: Dielectric Charging

Dielectric charging was listed previously in this section as a temporal effect but is included here as well because it can also cause a nearly permanent decrease in the apparent Young's angle when potential is set to 0 V. At 0 V the charges remaining in the dielectric impart an electromechanical force on the conducting droplet (as explained previously for Figure 7.2c). The key word here is nearly, because most of the charges will be released back into the conducting fluid over long periods of time. *In experiment, therefore, Young's angle should always be measured before any potentials have been applied in order to eliminate the possibility of an altered apparent Young's angle due to dielectric charging. Dielectric charging will also create an offset potential in the EW experiment (Eq. (7.3)).* Nontemporal dielectric charging is, however, easily avoided with proper materials and by avoiding the onset EW saturation, but even in well-prepared EW experiments, if DC potentials are used, then temporal dielectric charging is more difficult to avoid.

7.2.2.4 Wetting Hysteresis

Wetting hysteresis will cause the EW angle to appear larger as potential is being increased and to appear smaller as potential is being decreased. It appears in equilibrium for a system and is therefore listed a nontemporal type of deviation. However, it certainly will effect fluid geometry or motion in a dynamic system as well (e.g. reducing droplet transport speed in a lab-on-a-chip system). Hysteresis is an obvious topic for many but is mentioned here for the purpose of completeness for readers who are new to wetting science. Fundamental details were provided toward the end of Chapter 1, and here we introduce two simple hysteresis-adapted EW equations:

$$\cos\theta(U) = \cos(\theta_Y \pm \alpha) + \frac{cU^2}{2\gamma} \tag{7.7a}$$

$$\cos\theta(U) = \cos(\theta_Y) + \frac{c(U - U_H)^2}{2\gamma} \tag{7.7b}$$

where Eq. (7.7a) $+\alpha$ is the hysteresis angle for a contact line that is advancing over a surface and $-\alpha$ is the hysteresis angle for a contact line that is receding over a surface and Eq. (7.7b) U_H is a threshold voltage induced by contact line pinning (again, see end of Chapter 1). Either is appropriate in simple form, but neither equation is fully complete. For example, Eq. (7.7a) assumes a contact line is already moving (there is no threshold voltage), whereas Eq. (7.7b) assumes the contact line is pinned by hysteresis and has not yet begun to move. A fully complete equation would be quite complex for other reasons as well, such as device configuration; hysteresis is dependent on voltage types (DC versus AC, which can locally depin and lubricate the contact line) [16]; with structured surfaces, hysteresis can be dependent on the local contact line profile that changes with applied voltage. For an application specific example, see Eq. (8.5) and related discussion. Dielectric charging or damage, surface fouling, and other factors can cause appearance of hysteresis-like effects. Therefore, to diagnose hysteresis in an EW system, it is recommended to use fluids of high purity and utilize *a syringe and syringe pump to inflate or deflate the volume of a droplet in an EW system before and without any potential applied. An approximate amount of contact angle hysteresis can be calculated as the taking half of the advancing contact angle minus the receding contact angle.*

7.2.3 Deviation That Is Often Both Highly Temporal and Nontemporal

7.2.3.1 Chemical/Surface Potentials

In very low potential EW, chemical potentials can impart a noticeable deviation in EW response, typically tens of millivolts or more (see Chapter 3). Teflon, for example, has surface potential at neutral pH of −40 mV for 0.1 mM KCl solution and −100 mV for 0.1 mM K_2SO_4 solution [31]. These potentials do have a temporal response as fluid comes into contact with a solid surface, and the response can be fast (hard to observe with time) or surprisingly slow depending on the surface and/or chemical species involved.

Recently (submitted for publication at the time of writing this book), coauthor Mugele and colleagues have identified significant and near-permanent charging of Teflon surfaces by OH^- absorption for Teflon that was simply immersed in highly purified water. For the results shown in Figure 7.3, it should first be noted that the results are for a 4 μm thick (low capacitance) Teflon dielectric layer. Therefore the large potential shifts are simply the result of a fixed amount of charge leading to a large potential according to $U = Q/c$. If, for example, the same dielectric material were only 200 nm thick, then the 20 V shift (Figure 7.3a) would instead only be 1 V (but that is still a significant shift for a low voltage 200 nm

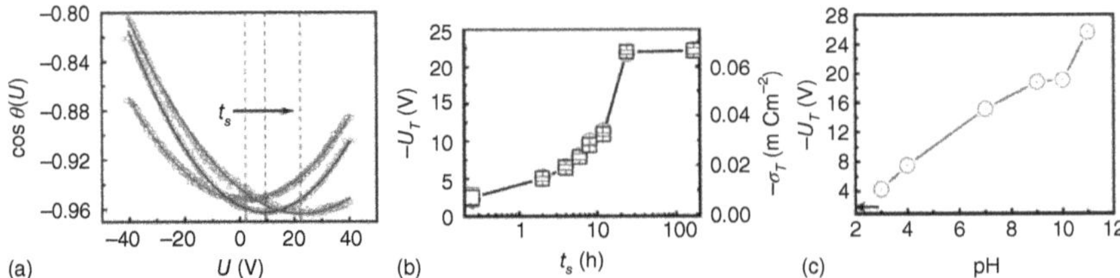

Figure 7.3 (a) Chemical surface potential shift for electrowetting of water on 4 μm thick Teflon after immersion in the highly purified water for 0 hour (bottom curve at −40V), 8 hours (middle curve at −40V), and 24 hours (top curve at −40V). (b) The potential shift is gradual and (c) due to OH^- absorption on the Teflon as evidenced by strong pH dependence for 6 hours of immersion. Source: [32]. Reproduced with permission of Royal Society of Chemistry.

dielectric system). As shown in Figure 7.3b, the process of charging is clearly temporal and slow, and as shown in Figure 7.3c, it is pH dependent, which confirms OH^- absorption as the dominant effect (data in Figure 7.3c is after six hours of water immersion). Even after substrates were dried and retested, the accumulated chemical surface potentials persisted. This should leave the reader with no doubt that surface absorption must be considered in EW experiments (Chapter 3), even if so-called high-purity fluids are utilized. *Chemical surface potential is most easily diagnosed by a potential shift in the EW curve at low potentials (i.e. eliminate the possibility of dielectric charging by not applying high potentials). In some cases, chemical surface potentials can be diagnosed by adjusting parameters such as pH of the conducting fluid. Chemical surface potentials are a complex topic, and the reader is encouraged to revisit Chapters 3 and 4 for additional details.*

7.3 Electrowetting Saturation

Referring back to Figure 7.1, the second region, which we have not discussed yet, is the *saturation region*. In the saturation region, the contact angle nearly or completely ceases to decrease as the potential is increased. At very high potentials, several of the electric field-dependent mechanisms discussed for the deviation region can become so severe that they actually cause the contact angle to increase with applied potential. Therefore, in some cases with use of poor materials, poor device design, or impractically high potentials, the saturation region can be understood in terms of EW deviation mechanisms described in the previous section. However, as we will teach in the next section, even with good materials and proper experimental conditions, saturation still exists, and there is no universal theory that has been shown to explain why.

Typically, with aqueous conducting fluids, saturation occurs around 60–80° for both EW in air or oil (see Figures 7.1 and 7.4). Other studies have seen that lower *saturation* angles down to 15–30° are possible using ionic liquids [20] or glycols

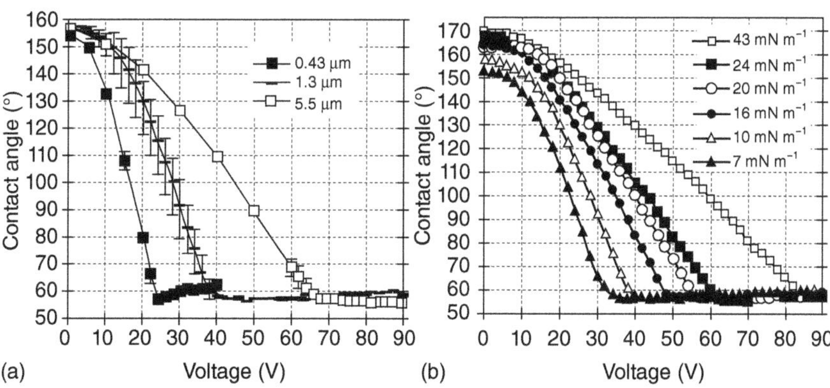

Figure 7.4 (a) Influence of dielectric thickness on saturation for glass/ITO/Parylene C/50 nm Fluoropel; test fluid 0.02 wt% NaCl aqueous solution with 0.028 wt% Triton X-102 in silicone oil; DC potential steps of 2 V/2 s. (b) Effect of interfacial tension on saturation for glass/ITO/1.3 μm Parylene C/50 nm Fluoropel; test fluid 0.02 wt% NaCl aqueous solution with Triton X-102 ranging from 0 to 0.055 wt% in silicone oil; DC potential in steps of 2 V/2 s. Source: [5]. Reproduced with permission of Taylor & Francis.

(unpublished data, Heikenfeld group). However, these same works clearly show the onset of *deviation* in the range of 60–80°. This apparent invariance of the onset deviation or saturation is discussed in greater detail in the next section.

7.4 The Invariant Onset of Deviation or Saturation and Lack of a Universal Theory for This Invariance

From a theoretical perspective, it is clear that EW leads to diverging electric fields in the immediate vicinity of the contact line. While the actual divergence will be circumvented in reality by some microscopic cutoff scale such as the Debye screening length, these conditions nevertheless expose the materials used in EW experiments to rather extreme conditions. In that sense, it should not be surprising that the materials cease to respond ideally and linearly as implicitly assumed in basic EW theory (see Section 5.3). In fact, macroscopic calculations in two-dimensional (and cylindrically symmetric) systems that enforce linear material response display ideal EW response down to contact angles as low as 5° [33], suggesting that contact angle saturation is indeed based on some sort of material response that is unaccounted for in current theory. More detailed molecular dynamics simulations show that in the extreme case of very high voltages but otherwise ideal material response, contact angle saturation eventually occurs because individual ions get pulled out of the drop by the enormous electric fields within nanometer distance of the contact line (Section 5.3). These ions subsequently settle on the substrate and screen the electric field. *However, these findings discussed above and the deviation factors discussed in Section 7.2 are presently only clues at best, and a universal theory predicting saturation is currently lacking. In addition to the lack of a universal theory for saturation, this section will also show that in some cases, none of the factors discussed in Section 7.2 for deviation can be attributed to causing saturation.* At best, scientists have produced multiple disconnected hypotheses (droplet ejection, charge injection, thermodynamic limit, etc.) that do not satisfactorily hold for the large body of EW experimental results.

7.4.1 The Invariance of Saturation for Aqueous Conducting Fluids

In practice, the effect of contact angle saturation *for aqueous solutions with* $\gamma_{ci} \geq 5\,\text{mN}\,\text{m}^{-1}$ has always been observed to typically limit the contact angle modulation down to an electrowetted contact angle of 60–70° for both systems in air (~110–120° Young's angle; Figure 7.1) and oil (~160–180° Young's angle; Figure 7.4).

It has further been demonstrated experimentally that when using DC potential and with aqueous solutions with $\gamma_{ci} > 5\,\text{mN}\,\text{m}^{-1}$, EW contact angle saturation is invariant with electric field, contact line profile, interfacial tension, choice of nonpolar insulating fluid, and type of polar conductive fluid or ionic content in the polar fluid [4]. One experimental example of many reported by Chevalliot et al. [4] is shown in Figure 7.4. In Figure 7.4a, the dielectric thickness is decreased, and therefore according to the EW equation, the electric field strength under the conducting fluid and near the contact line increases for a given contact angle (see Eq. (7.12)). However, the saturation angle of approximately 60° is constant. Therefore saturation hypotheses that should be dependent on electric field such as dielectric charging, oil charging, and microdroplet ejection are all invalidated for this particular experiment. Similarly, in Figure 7.4b, the dielectric thickness is kept constant, and the fluid interfacial surface tension decreased with surfactants to alter the required EW potential and therefore the electric field. *It is clear that for aqueous solutions with practically relevant values for γ_{ci}, existing theories fail to provide a universal theory for EW saturation.*

7.4.2 The Invariance of the Onset of Deviation or Saturation for All Types of Conducting Fluids with $\gamma_{ci} > 5\ \mathrm{mN\,m^{-1}}$

Other studies have seen that lower *saturation* angles down to 15–30° are possible using ionic liquids [20] or glycols (unpublished data, Heikenfeld group). However, these same works clearly show the onset of *deviation* in the range of 60–80°. *Therefore, for all the all EW experiments shown to date, the onset deviation or saturation is invariant (e.g. one of them always begins to appear at no lower than a range of 60–80°).*

7.4.3 Summary

Therefore, at present time, only one universal conclusion can be drawn with regard to the onset of deviation or saturation: if high quality dielectrics, normal operating potentials, and normal fluid interfacial surface tensions are utilized (~5 mN m^{-1} or higher), then *either deviation or saturation always begins to appear at or before a range of 60–80°*. This invariance does not change based on choice of materials and therefore not explained by known factors that were discussed in Section 7.2 on EW deviation.

Conversely, if poor materials are chosen (such as dielectrics that are prone to charging or simply too thin), then saturation *can* be mapped to effects also causing EW deviation. However, this type of experimentation is inherently impractical from an applied perspective, because behavior such as significant dielectric charging inevitably leads to degradation of EW performance over time.

From an applied perspective, it does not appear that you can overcome saturation, and EW down to very small angles is currently not possible with practical materials systems. From a theoretical perspective, a universal theory for saturation is still lacking. The leading speculation points to electrohydrodynamic instability near the contact line, which is beyond the scope of this book at this time (see Section 5.3 for further discussion). Perhaps clues also lie in the choice of conducting fluid, because as previously stated ionic liquids and glycol liquids can achieve saturation angles down to 15° even with interfacial surface tensions of >5 mN m^{-1}.

If wetting to zero degrees (a continuous film) is desired for a particular application, then the reader should visit Chapter 9 where dielectrowetting is presented. Dielectrowetting is not EW, but it is a special case of dielectrophoresis with a potential-squared response like EW. In a sense, dielectrowetting is also an extreme example of avoiding field divergence by a suitable choice of electrode geometry and conductivity. This provides an additional clue that the divergence of electric field near the contact line may ultimately be directly responsible for saturation.

7.5 Choosing Materials: Large Young's Angle and Low Wetting Hysteresis

EW applications typically require a large Young's angle (for maximum change in modulated contact angle) and low wetting hysteresis (for more accurate modulation and modulation without a threshold hysteresis to overcome; Eqs. (7.7a) and (7.7b)). An expanded version of the EW equation (Eq. (5.1)) includes the cosine of Young's angle in terms of interfacial surface tension forces:

$$\cos\,\theta_U = \cos\,\theta_Y + (CU^2)/(2\gamma_{ci}) \rightarrow \cos\,\theta_U = (\gamma_{id} - \gamma_{cd})/\gamma_{ci} + (CU^2)/(2\gamma_{ci}) \tag{7.8}$$

where C is the dielectric capacitance and the γ terms are the interfacial tensions between the conducting fluid (c), insulating fluid (i), and the dielectric (d), as discussed in greater detail in Chapter 1.

Most commercial or near-commercial EW applications use an oil for the insulating fluid (not air) because it provides a very large Young's angle and low wetting hysteresis. Oil also reduces the effect of gravity or vibration in sealed devices. In particular, Young's angle of 180° is often ideal for two reasons:

1) $\theta_Y = 180°$ minimizes hysteresis (the conducting fluid never actually touches the dielectric surface, it always rests on a very thin and lubricating film of insulating fluid).
2) It is speculated that $\theta_Y = 180°$ protects the dielectric surface (the conducting fluid is never brought into electrochemical contact with the dielectric, again separated by a thin film of the insulating fluid).

Achieving $\theta_Y = 180°$, which is the same as $\cos(\theta_Y) \leq -1$, requires $(\gamma_{id} - \gamma_{cd})/\gamma_{ci} \leq -1$ according to $\cos\,\theta_Y = (\gamma_{id} - \gamma_{cd})/\gamma_{ci}$. In practice, this is typically achieved by having the (γ_{id}) near or equal to zero and ($-\gamma_{cd}$) that is nearly equal to or greater than the interfacial tension between the conducting and insulating fluids (γ_{ci}). In a first approximation, this can be achieved by selecting insulating fluids and dielectrics that have similar surface tensions/energies. For this same reason, it can be seen why in air $\theta_Y = 180°$ is impossible on a planar surface because air has effectively zero surface tension and there is no material with a zero surface energy. The only way to increase contact angles in air is then to partially replace significant portions of the dielectric surface air pockets, which is exactly how superhydrophobic surfaces work, a topic reserved for the last portion of Section 7.5.

For a more detailed discussion of the effect of interfacial tensions on Young's angle, the reader may also refer back to Chapter 1. Conceptually, the interfacial tensions can assume any value. If the ratio of interfacial tensions on the right-hand side of Young's equation yields something greater than 1 or less than −1, this simply means that one of the two phases is completely wetting the substrate. In fact, in some EW systems, the data shows a threshold for wetting explained only by having the case $(\gamma_{id} - \gamma_{cd})/\gamma_{ci} < -1$.

We now present three ways how hydrophobicity is typically achieved, which include (i) low surface energy coatings (fluoropolymers), (ii) interfacial tension matching the oil with the surface, and (iii) superhydrophobic surfaces.

7.5.1 Conventional Ultralow Surface Energy Coatings (Fluoropolymers)

The most widely utilized and conventional approach to achieve a large Young's angle is to use fluoropolymer dielectrics or a thin topcoat of fluoropolymer. The high electronegativity of fluorine reduces the polarizability of the atom and therefore also reduces the subset of van der Waals forces that give rise to the intermolecular attractive forces that contribute to a material's surface energy (again, see Chapter 1 for more details). Generally fluoropolymers can be deposited in vacuum using plasma reaction chemistry or solution deposited, with the latter being the dominantly used technique. The most widely used solution-deposited fluoropolymers are dissolved at ~0.1–1 wt% in a fluorosolvent, spin- or dip-coated, and then baked at manufacturer-specified temperatures into a smooth film that is typically tens to hundreds of nanometers thick. Unless made very thick (several micrometers), the solution-deposited fluoropolymer films are inherently poor EW insulators because of a high defect/pore density arising from two causes: (i) the fluoropolymer chains themselves are weakly interactive, and they sublimate before they melt, such that significant film densification and melting reflow cannot be achieved; (ii) solution deposition requires that the solvent be removed during a bake or drying step, and that inherently requires a porous pathway through the polymer film for the solvent to escape as a gas. The most commonly studied fluoropolymers are listed in Table 7.1 – given by their trade names: DuPont Teflon AF, Asahi Cytop, and Cytonix Fluoropel 1600V series.

Table 7.1 Example methods to achieve a large Young's angle in air or in insulating oil.

Material	Insulating fluid (mN m^{-1})	Conducting fluid	Surface (mJ m^{-2})	Young's angle (°)	Dielectric constant
Teflon AF	None, air	Water	16	~115	1.9
Cytop 809 M	None, air	Water	19	~110	2.1
Fluoropel 1600V	None, air	Water	16	~115	2.0
Any fluoropolymer	Silicone oil (17)	Water	16–20	~180	—
Any fluoropolymer	Hexane (18)	Solution [17]	16–20	~180	—
Parylene C	Hexane (18)	Solution [17]	40	~120	3.1
Parylene C	Dibromooctane (38)	Solution [17]	40	~180	3.1
Teflon AF	Dibromooctane (38)	Solution [17]	16	~120	1.9
Parylene HT	Tetradecane (25)	Water	26	~180	2.2

Source: Values as taken from manufacturer data sheets or from [6, 15, 17].

Fluoropolymer coatings typically provide Young's angles ranging from 100° to 120° in air with ~10° of wetting hysteresis. The range of surface energies, 16–19 mJ m^{-2}, for common EW fluoropolymers is found in Table 7.1. Depending on the choice of the ambient oil, fluoropolymers can also provide Young's angles for water up to 180° and effectively no hysteresis if the right combination of fluids is used. Recalling our previous discussion for Eq. (7.8) and further examining Table 7.1, it is not surprising that silicone oils and alkanes with surface tensions in the range of 16–24 mN m^{-1} provide large values of $\theta_Y = 160–180°$ on fluoropolymers that have a similar surface energy to the oils (a rough but not fully correct predictor of likelihood of low interfacial surface tension). A hybrid approach of soaking a fluoropolymer in oil and then drying it is known to reduce hysteresis in air down to several degrees as the oil is trapped in pores or at the rough features that are typically the dominant cause for hysteresis in air.

Lastly, it should be noted that although most fluoropolymers are inherently chemically resistant, not all will allow reversible EW. For example, Cytonix's Fluoropel 1600A series is noted as not being *UV light resistant* like the 1600V series (hence the V in the name), and upon EW with water, the surface becomes irreversibly hydrophilic due to electrochemical reaction. Even for the best fluoropolymers, electrochemical stability must be monitored. Consider recent data in Figure 7.5 [15] that shows loss of hydrophobicity (decreased θ_Y) and loss of electrical insulation (increased θ with potential applied) after repeated EW actuations in air. The loss of EW performance in Figure 7.5 could also be due to often-observed electrochemical degradation of the SiO_2 layer underneath the

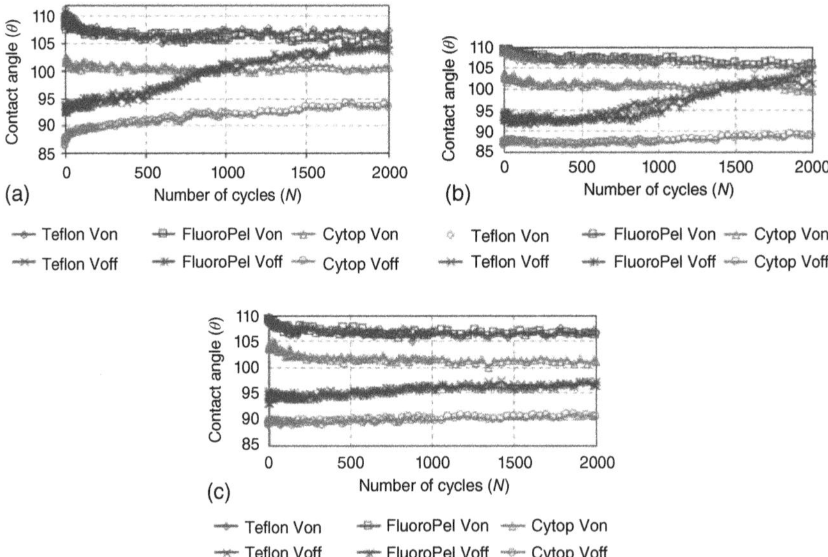

Figure 7.5 0 and 30 V electrowetting responses for (a) contact angle data with + DC V on the conducting fluid, (b) contact angle data with −DC V on the conducting fluid, and (c) contact angle data with 1 kHz AC V for a 0.1 M KCl aqueous solution on a dielectric stack of 50 nm fluoropolymer and 565 nm SiO_2. Source: Koo and Kim 2013 [15]. Reproduced with permission of IOP.

fluoropolymer that occurs more rapidly if the fluoropolymer does not act as a strong electrochemical barrier. As shown in Figure 7.5, AC operation is clearly the most reliable, because a net movement of ions or charge is more difficult. With AC operation the potential reverses before a charge or ion reaches its destination for electrochemical reaction or charging. Also noticeable in Figure 7.5 and commonly observed in most other EW experiments, positive potential on the electrowetted droplet exhibits the fastest degradation. This polarity-dependent degradation can be speculated to be due to a negative surface potential in aqueous solutions [31], which allows mobile positive ions to propagate through the diffuse layer inside pores in the fluoropolymer (see Chapter 4 and also more details in the conducting fluid section of 7.7). Conversely, the negative surface potential in the pores will screen out mobile negative ions, impeding or blocking negative ion entry from the aqueous solution when negative potential is applied to the aqueous solution. Lastly, it should be mentioned that the aging data of Figure 7.5 was performed in air and would have likely been improved substantially if air was replaced with an insulating fluid promoting $\theta_Y = 180°$ for the conducting fluid, such that the conducting fluid never comes into electrochemical contact with the fluoropolymer surface.

7.5.2 Hydrophilic Coatings Made Hydrophobic Through Proper Choice of Insulating Fluid

In 2009, Maillard published a detailed study on EW using two immiscible liquids on dielectric coatings that were not fluoropolymers [17]. The study achieved $\theta_Y = 180°$ by using insulating fluids and dielectrics that provide an interfacial tension (γ_{id}) near or equal to zero (see Eq. (7.8) and related discussion). This study confirmed that the surface energy of the dielectric was not the most critical parameter, but rather a low γ_{id} was needed to obtain a large Young's angle for the conducting fluid. This and similar results are illustrated beautifully in Table 7.1. Notice how even hydrophobic materials such as Teflon fail to provide $\theta_Y = 180°$ if the insulating fluid and dielectric are not carefully matched for a low γ_{id}. Practically, this finding is very powerful as high quality insulators can be used instead of fluoropolymers, which are often porous and defective and prone to electrical charging (see previous section on fluoropolymers). The only significant challenge for this approach is that the insulating fluids used typically are more polar (higher surface tension), which in some cases can make them incompatible with the conducting fluid due to miscibility of the two fluids or mutual solubility for solutes.

7.5.3 Superhydrophobic Coatings: Larger Young's Angle in Air but Small Modulation Range

EW devices with an initial superhydrophobic water contact angle ($\theta_Y > 150°$) have now been demonstrated on a variety of structured substrates [12]. These substrates are more complex than a conventional superhydrophobic surface since EW further requires an electrical conductor that is coated with a high performance hydrophobic dielectric. Example substrate structures that have

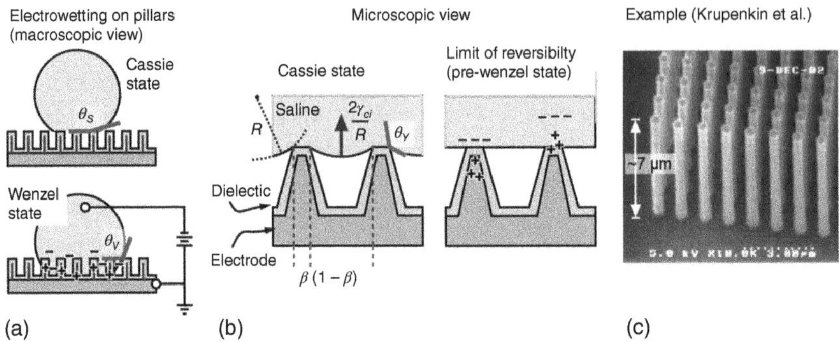

Figure 7.6 Electrowetting on superhydrophobic (structured or textured) surfaces. Source: [12]. Reproduced with permission of Taylor & Francis.

been studied include silicon nanoposts and nanowires, carbon nanofibers and nanotubes, and polymer microposts. These techniques can be further distinguished by whether or not the structure surface is EW itself (e.g. Figure 7.6, [35]) or the case where the conductive EW electrode is buried beneath a structured layer that is entirely insulating [36]. Regardless of all the various methods utilized, there are significant commonalities for all types of EW on structured surfaces, as will be discussed next.

A commonly perceived goal of such work has been to achieve greater hydrophobicity in air and less wetting hysteresis; however it has been determined theoretically and experimentally that little improvements can be realized from an applied perspective. Even though the demonstrated structured surfaces are geometrically very diverse, there are several consistencies in EW behavior for all these platforms. As an EW bias of tens of volts is applied between the droplet and the substrate, the macroscopically observed contact angle is typically decreased from >150° to ~100°. This magnitude of contact modulation is similar to that achieved for planar surfaces in air, and significant applied value has yet to be demonstrated. As the potential is increased, the electromechanical force eventually causes wetting between the substrate structures and the droplet transitions from the Cassie state to the Wenzel state (Figure 7.6a, Section 1.3). The Wenzel state presents a new energy minimum for the system, and the wetting is irreversible unless an additional external stimulus like heat (boiling) or oil (dewetting) is added. With reference to Figure 7.6, EW behavior on superhydrophobic surfaces involves modifying Young's equation with the reduced fraction of area for fluid–solid contact β ($0 < \beta < 1$) to achieve the structured superhydrophobic angle θ_S (see Figure 7.6):

$$\cos\theta_S = \beta(1 + \cos\theta_Y) - 1 \tag{7.9}$$

Also due to β, the electrical capacitance for EW is also greatly diminished. Therefore, the EW equation is also modified by β, which further causes a greatly diminished electromechanical force (EW number) for small values of β:

$$\cos\theta_U = \cos\theta_S + \frac{1}{2} \times \frac{\beta \times c \times U^2}{\gamma} \tag{7.10}$$

This equation is an approximation that allows ease of understanding of how the electromechanical force is weakened. In reality, the exact calculation requires a deeper treatment as shown elsewhere [34].

It is interesting to think of superhydrophobic surfaces similar to the discussion on creating $\theta_Y = 180°$ by interfacial surface tension matching of the oil and the dielectric to reduce γ_{id}. With superhydrophobic surfaces, you are essentially doing the same thing by having the insulating fluid as air and the dielectric surface as voided and composed dominantly of air as well (obviously and air-to-air interface has zero interfacial tension). However, as a final reminder, this approach is far less effective than using an oil insulating fluid for two reasons: (i) the EW force is diminished by β and (ii) in air it is impossible to achieve $\gamma_{cd} < \gamma_{ci}$ because γ_{cd} always involves partial contact with a physical surface, and it is therefore impossible to achieve $-\gamma_{cd}/\gamma_{ci} \leq -1$, which is the requirement for $\theta_Y = 180°$. Therefore, if large ranges of contact angle modulation and low hysteresis are required, use of an insulating fluid that is oil is clearly a superior approach for EW (if oil is allowable).

Lastly, it should be noted that in the use of trenches [18] or corrugations [2], a reversible Cassie-to-Wenzel transition can be enabled for wetting in one dimension. However, this case is limited in functionality and is still practically inferior for most any application where oil could instead be utilized.

7.6 Choosing Materials: the Electrowetting Dielectric (Capacitor)

There are two ways to approach creating high performance EW dielectrics. The simplest approach is to make the dielectric very thick (low capacitance), typically in the range of several micrometers thick for common polymers coated with a fluoropolymer or by use of a thick stand-alone fluoropolymer film itself. As the dielectric thickness increases, according to the EW equation, the required electric field (E) across the dielectric of thickness d actually *decreases*. Equation (7.11) is simply the EW equation solved for voltage:

$$U = \sqrt{\frac{2\gamma \times d(\cos\theta_U - \cos\theta_Y)}{\varepsilon}} \tag{7.11}$$

Now, in the equations shown in Eq. (7.12), we clearly see from Eq. (7.11) that voltage U_{EW} required to achieve a certain contact angle variation is proportional with the square root of dielectric thickness. We know that the required electric field for EW (E_{EW}) is always the potential U_{EW} divided by the thickness of the capacitor (dielectric) material (d). Therefore E_{EW} actually *decreases* as the dielectric thickness is increased:

$$U_{EW} \propto \sqrt{d}, \quad E_{EW} = U_{EW}/d \quad \therefore E_{EW} \propto \sqrt{1/d} \tag{7.12}$$

This is a somewhat counterintuitive result for beginner practitioners of EW, who initially turn to a reduced dielectric thickness to allow reduced potential to improve reliability. In reality, when you thin down the dielectric, things get dramatically worse because (i) the electric field required to achieve the same contact angle reduction is actually increased and (ii) the thinner dielectric itself more likely to have a catastrophic defect. This segues this discussion into the next paragraph, which explains why thinner dielectrics are often more defective.

Thicker EW dielectrics typically have less catastrophic defects than thinner ones (see the catastrophic defect example illustrated in Figure 7.7). The submicron defects that might plague a very thin dielectric might not span the entire thickness of a thicker dielectric, reducing the likelihood of complete dielectric failure. Therefore, if high potentials of 60–200 V are not of concern, and because the energy consumption is equal (except for transformer loss, e.g. Durel D388A), one can simply use 2–5 μm of Parylene C dielectric with a ~50 nm fluoropolymer topcoat to achieve defect-free and reliable EW.

The more challenging and interesting discussion is related to *low* potential operation (<20 V). As discussed above, simply reducing the thickness of a conventional dielectric to reduce the required potential is insufficient and typically results in dielectric failure as shown in Figure 7.7. As stated in the introduction for this chapter, there have been numerous reports of ultralow potential (several volts) EW with very thin dielectrics and/or very low interfacial surface tensions. However, these examples will not be discussed herein, because other than for basic exploratory research, currently they have very little practical relevance due to poor reliability or very slow response times. For most applications, the current commercially viable state of the art requires fluid interfacial tensions no less than ~5–10 mN m^{-1}, which in combination with reliable dielectrics still requires ~15–20 V for operation. Companies such as Amazon/Liquavista have clearly made progress in reliable operation with 15–20 V and millions of cycles

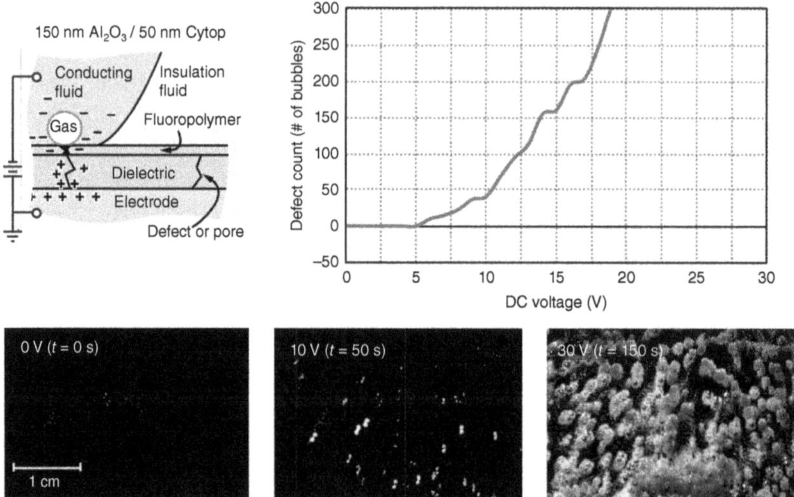

Figure 7.7 Defectivity measurement technique for single-layer dielectrics using a voltage increase of 1 V at every five seconds. The defects are counted by taking top-view photographs and using software such as image J to calculate the number of bubbles due to electrolysis. Source: [24]. Reproduced with permission of Elsevier.

of reliable contact angle modulation from 180° down to 90–120°. However, none of their findings have been made publicly available. Recently, there have been several publications on reliable dielectrics by coauthor Heikenfeld's group, for which the three of the most impactful approaches will now be presented:

1) The use of more slowly deposited Parylene HT as a thin single-layer dielectric, resulting <15 V operation with a 300 nm thick film. Parylene HT has less defectivity largely because of reduced gas-phase reaction compared with Parylene C (gas-phase reaction causes granules of Parylene to deposit and therefore a *granular* and defective film) [7]. For this same reason, it is speculated that Parylene N may give similar results because it deposits very slowly as well.
2) The use of *self-healing* dielectrics through an acidic conducting fluid and an electrode such as Al that is anodizable (forms a passivating oxide dielectric at any regions of dielectric failure) [6]. This approach is limited to DC operation or AC with a DC offset potential, which currently limits the maximum number of actuation cycles due to eventual dielectric charging.
3) The multilayer dielectric approach [24] (Figure 7.8), where instead of making a perfect single insulator, multiple thinner defective layers are allowed so long as their individual defects do not overlap/connect. This approach was inspired by the multilayer inorganic/organic film work performed to make robust hermetic *barrier* sealing for moisture-sensitive organic electronics and displays. Amazon/Liquavista has referred to their undisclosed dielectric stack as a *barrier* layer, suggesting similar structure. There are several major findings for this multilayer work, which will now be discussed in greater detail.

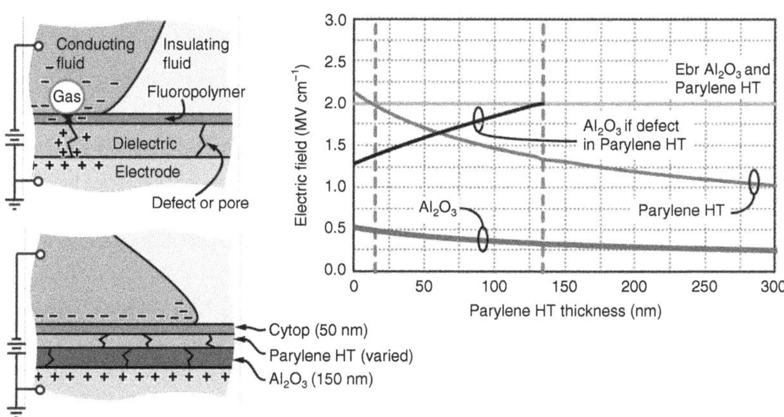

Figure 7.8 Reliability improvement over the results shown in Figure 7.7 by adding a second dielectric (Parylene HT). Over >10 cm^2 area, no dielectric failure was observed even at >2 times the required electrowetting potential needed to reach $\theta(U) = 70°$. The dotted lines in the theoretical plot show the reliable ranges of Parylene HT. Source: [24]. Reproduced with permission of Elsevier.

7.6.1 Current State of the Art for Low Potential Electrowetting: Multilayer Dielectrics

Before the value of a multilayer dielectric can be understood, the effect of defectivity of a single-layer dielectric should again be emphasized. As shown in Figure 7.7, a single ~150 nm layer of anodization-grown Al_2O_3 exhibits a high level of defectivity under even low potentials of 5–10 V (this device would require 10–20 V for the full range of EW modulation). This is considered a *single-layer* dielectric, because as mentioned previously the solution-deposited Cytop fluoropolymer topcoat is a very poor electrochemical barrier itself.

Now, as shown in Figure 7.8, a second dielectric coating of Parylene HT is added (Parylene C was also demonstrated to be effective). With only 50 nm of Parylene HT added, it was observed that all defects were completely passivated. The same test shown in Figure 7.7 was performed, and zero electrolysis bubbles were observed even at 30 V. This observation was for >10 cm^2 area, and even at greater than two times the required 15 V EW potential needed to reach $\theta = 70°$. Individual tests of Parylene HT or Al_2O_3 alone with 50 nm of Cytop coating showed high densities of defects and failure, clearly demonstrating that the improvement is not merely the conventional benefit of a *thicker combined dielectric.*

At this time, it speculated that in multilayer systems, the multilayer dielectric stack is effective if defects in the individual dielectric layers do not overlap or connect in the combined dielectric stack. This however assumes that if one of two individual layers is defective, then the other of the two layers will experience the full applied potential at the site of that defect and must reliably hold the full potential without failure (remember help from the fluoropolymer layer can be ignored due to extremely high defectivity). The plot provided in Figure 7.8 considers what happens at such single-layer defects and reveals some non-obvious insights into multilayer design. The plot in Figure 7.8 assumes a challenging and practically relevant case: $\gamma_{ci} = 30\,\mathrm{mN\,m^{-1}}$ and EW from $(\gamma_{id} - \gamma_{cd})/\gamma_{ci} = -1$ $(\theta_Y = 180°)$ to $\theta(U) = 70°$. The theoretically required electric field is plotted across a 150 nm Al_2O_3 layer and across a Parylene HT layer as the Parylene HT thickness is varied. The electric field across the Al_2O_3 layer is also plotted for the case of a defect in the Parylene HT layer. 2 $\mathrm{MV\,cm^{-1}}$ is assumed as the breakdown field for both dielectric layers with a liquid electrode, indicated by the horizontal line. Vertical dotted lines represent a limiting window to avoid dielectric failure. Two key conclusions are as follows. Firstly (left dotted vertical line), the Parylene HT layer cannot be made too thin or else it will experience >2 $\mathrm{MV\,cm^{-1}}$ and will no longer contribute as an insulator in the stack. Secondly (right dotted vertical line), if the Parylene HT layer is made too thick, then this lowers the stack capacitance, raising the required EW potential, and if the Parylene HT has a defect in it, the Al_2O_3 layer will experience the entire applied potential at that defect (bottom solid line). This would cause the Al_2O_3 layer to exceed its breakdown field, thereby causing complete dielectric failure. Therefore, in multilayer dielectric systems, the individual electric fields must be accounted for and optimized by control of layer thicknesses. Interestingly, if the data in Figure 7.8 were replotted for $\gamma_{ci} = 50\,\mathrm{mN\,m^{-1}}$ (not shown), as is the case for some lab-on-a-chip applications with low surface tension silicone insulating fluids, there would be no reliable operating window and all layers would need to be increased in thickness to reduce the electric fields.

7.6.2 A Note of Critical Importance for the Topcoat in a Multilayer System

A non-obvious result is that the required electric field in the layer with the lowest dielectric constant (usually the fluoropolymer topcoat) will be the same for constant values of U, γ_{ci}, θ_Y, $\theta(U)$. This result is regardless of the dielectric constant or thicknesses of the dielectric materials placed around or below the layer with the lowest dielectric constant. Derivation of this result is saved as a challenging homework problem at the end of this chapter.

Therefore, for a given potential and contact angle range, the E-field across the fluoropolymer topcoat is predetermined and should carefully be considered. See the solution for homework Problem 7.6 at the end of the chapter for a full derivation. This non-obvious result also highlights why attempts to use *high permittivity* dielectrics alone with a thin fluoropolymer topcoat are largely a flawed approach, which leads the conversation to the next section.

7.6.3 Carefully Choosing the Best Materials for Each Individual Layer of the Dielectric Stack

Generally, EW stacks *are deposited in the order listed below* for two reasons: (i) because of fabrication process compatibilities with layers already deposited onto the substrate and (ii) because of increased materials similarity, having the fluoropolymer/organic layers adjacent generally minimizes interfacial charge trapping compared with the fluoropolymer/inorganic case.

7.6.3.1 First Layer: Inorganic Dielectrics

Inorganic dielectric choice is the most confusing to make, as there are many possibilities.

Firstly, it should be clearly stated that high-ε dielectrics (ε of tens to thousands, like $BaTiO_3$) provide no benefit to EW systems unless the dielectrics are found to be more reliable (only if they can be made thicker than the length of problematic dielectric defects). A quick way to compare performance for potential inorganic dielectrics is by calculation of the maximum charge density Q_{max} they can sustain:

$$Q_{max} = \varepsilon\varepsilon_0 U_{max}/d \tag{7.13}$$

where U_{max} is the voltage at which the dielectric will electrically fail. Calculating U_{max} for various types of dielectrics quickly reveals that even dielectrics with permittivity in the hundreds or even thousands still have U_{max} values that are not dramatically different than those of more common low permittivity dielectrics.

Furthermore, in EW systems the inorganic dielectrics are already ~100 nm or more in thickness, and high-ε dielectrics with thickness >1 μm are difficult to deposit economically and without significant buildup of thin film strain, which can cause delamination and cracking. Therefore the *thicker* inorganic dielectrics are typically impractical.

Secondly, when the conducting fluid is, or partially comprises, a reactive liquid like water, the electrochemical inertness of the inorganic dielectric must be considered. For this reason, oxide dielectrics are often readily attacked and degrade, whereas more inert dielectrics such as Si_3N_4 exhibit better resistance to attack. Furthermore, oxides such as SiO_2 hydrate quite easily, forming pathways for ionic current. As a general rule (but not without exception), inorganic dielectrics that are difficult or impossible to wet-chemical-etch, like Si_3N_4, are highly preferred. However, for convenience in basic research or for applications where the conducting fluid is not water, thermally grown silicon oxides can be attractive.

7.6.3.2 Second Layer: Organic Dielectrics

Generally, vapor-deposited organic dielectrics are preferred. The leading class at present time is clearly the family of Parylene dielectrics. These dielectrics deposit conveniently at room temperature using equipment that is relatively inexpensive to purchase (R&D scale coaters, which can coat a dozen or more 12″ diameter parts in a single run). At this time, there have been no published reports of organic dielectrics showing superiority to the Parylenes. With the exception of Parylene HT (fluorinated Parylene), Parylene dielectrics degrade (oxidize) during the bake cycle of the fluoropolymer, generally as soon as temperatures increase above 80–100 °C. Baking in vacuum or N_2 can partially alleviate potential damage. This baking issue is mentioned not because the Parylene needs to be baked, but because many fluoropolymer topcoats require a final bake step.

7.6.3.3 Third Layer: Fluoropolymer

Fluoropolymer topcoats are solution deposited and baked. Cytop has demonstrated the most reliable operation in air, which is a highly challenging environment due to the absence of a protecting oil layer. It can also be purchased with built-in adhesion promoters to reduce blistering (peeling) of the fluoropolymer from surfaces it is coated on. However, a word of caution is that over time adhesion promoters can migrate to the surface and reduce hydrophobicity. Cytop is also quite expensive and has limited shelf life, though, and therefore many researchers have utilized lower-cost Fluoropel products with significant success in noncommercial devices. Fluoropel can also be baked as low as 90–120 C, which is much lower than Cytop (180 C) and Teflon AF (160 or 240 C). The low baking temperature of Fluoropel is therefore further attractive because an underlying Parylene layer will experience less oxidation during the fluoropolymer baking step.

7.6.3.4 The Simplest Approaches Available to Electrowetting Practitioners

For research or applications where low potential is not required, the simplest approach at this time is a single Parylene layer that is thick (>1 μm) and coated with ~50–100 nm of spin- or dip-coated fluoropolymer. Alternately, if Parylene is not available, a >1 μm thick layer of fluoropolymer can be utilized. The simplest approach if low potentials are needed is to room-temperature anodize a metal such as Al or Ta and then add a thin Parylene coating, followed by a fluoropolymer to achieve a multilayer dielectric system [24].

7.7 Choosing Materials: Insulating and Conducting Fluids

Let us start this section with a very clear statement: in order to achieve reliable EW results, the choice of fluids is arguably the most important factor to consider and traditionally the most ignored by EW practitioners. It is also important to note up front that for any high quality experiment, fluid purity is critical and fluids should be purchased with certified high purities or purified prior to use. Fluid purity is especially important considering the surface adsorption effects discussed in Chapter 3.

7.7.1 The Insulating Fluid

The insulating fluid is any neutral dielectric fluid at room temperature that is immiscible with the polar fluid. In EW, the conducting fluid is not always water, and in some cases insulating *oils* will have undesirable miscibility with conducting fluids such as glycols. So not all *oils* will satisfy the materials requirements. Most often in advanced or applied work, a mixture of insulating liquids is utilized, sometimes with solutes such as dyes; hence the most appropriate term is insulating *fluid*.

The choice of insulating fluid was previously discussed in terms of achieving a value of $\theta_Y = 180°$, which minimizes hysteresis and provides a protective insulating fluid film between the conductive fluid and the top surface of the dielectric. Interfacial tension requirements for the insulating fluids were discussed in Section 7.5. Commonly used insulating fluids include alkanes and silicone oils. Silicone oils typically have wider operating temperature ranges for commercial products. However, a warning is that some silicone oils, although having very low surface tensions, can be highly hydroscopic. An exemplary non-silicone and non-alkane insulating fluid can be found by examining popular insulating oils used for electrophoretic devices, namely, the Isopar series of oils. The Isopar series of oils are synthetic isoparaffins that provide low viscosities, wide temperature ranges, and very importantly low miscibilities with many nonaqueous conducting fluids (the topic that will be discussed next). Hence, Isopar oils are strongly recommended. The most challenging cases for insulating fluid selection are applications like EW displays, where the insulating fluid has to also dissolve a blend of dyes at 5–20 wt%. Insulating fluids are also challenging for EW optics, where typically the insulating fluid must also support the highest possible refractive index. Surfactants can also be added to the insulating fluid instead of the conducting fluid, but surfactants in the insulating phase can cause EW deviation by oil relaxation (see Section 7.2).

7.7.2 The Conducting Fluid

Water is the most convenient conducting fluid for EW due to immediate availability, high surface tension, low viscosity, immiscibility with a wide range of insulating fluids, and compatibility with bio-applications. *However, from a reliability perspective, water is one of the worst possible conducting fluids to utilize.*

Unless one is using very thick dielectrics, water has never been shown to allow reliable EW operation. Water is a superior solvent that hydrates a wide range of polymers and inorganic dielectrics and is electrochemically highly reactive. For these reasons and others, water is one of the worst fluids to utilize if you are trying to eliminate many types of the EW deviation discussed in Section 7.2.

Let us start our discussion with water anyway. We will begin with water because for some applications it is often essential (lab-on-a-chip) and because for researchers its properties, solutions, and EW behavior are also the most well characterized in the literature. So, starting with water, there are two major steps that can be taken to improve reliability.

7.7.2.1 Ionic Content

EW devices often utilize conducting fluids with ionic surfactants and inorganic salts to modify the EW response. It has been observed in low potential EW devices (thin dielectric of <12 V) that a frequent onset of dielectric failure (electrolysis) occurs with the use of ionic solutes such as KCl or sodium dodecyl sulfate. More detailed current potential investigations reveal less dielectric failure for larger-size ions. Specifically, improved resistance to failure is seen for surfactant ions carrying a long alkane chain. Therefore, a catanionic surfactant (in which both ions are amphiphilic) was custom synthesized, and elimination of dielectric failure was observed in both negative and positive potentials [22]. Experimental results showing the ion-size dependency of dielectric failure are shown in Figure 7.9. Two general conclusions can be made: (i) if ionic content needs to be added to provide conductivity, use the largest possible ions that are compatible with the application and (ii) use the minimum ionic content necessary (if any ions at all) to satisfy the needed electrical charging time (RC time constant).

It is worth to expand on the value of minimizing ionic content. Preferential anion adsorption exists on fluoropolymers (Figure 7.3, Chapter 3, and [31]). These adsorbed ions do not easily move unless under very high electric field, and even when they do move, they are not as mobile as ions positioned further away from the surface. The surface-adsorbed anions locally deplete additional anions via

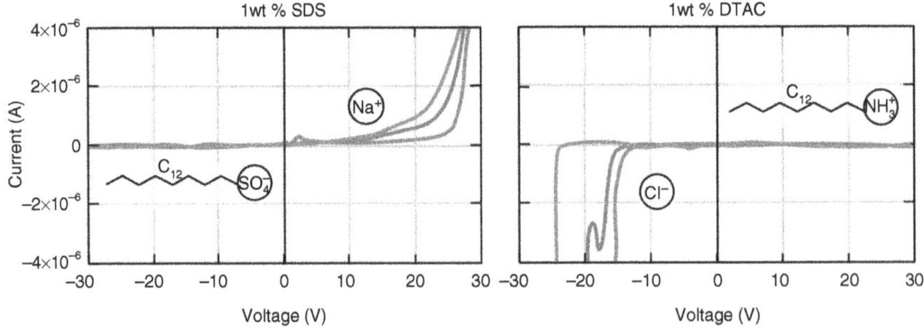

Figure 7.9 Dielectric failure current versus potential for 1 μL drops of two different aqueous solutions on a fluoropolymer/Al_2O_3 dielectric stack. Every test and individual positive and negative polarity potential sweeps were measured at independent dielectric sites. The data is presented without any averaging, allowing one to appreciate that the data is fairly repeatable even for dielectric failure tests. Source: [22]. Reproduced with permission of ACS.

Debye screening. This could explain why a stand-alone thin fluoropolymer film often exhibits water electrolysis at lower potential for positive bias (because only the positive ions comprise the mobile Debye sheath inside a small pore). Now, the Debye screening length increases as ionic concentration is reduced, thereby reducing leakage current through a dielectric. Furthermore, a more resistive fluid can act as a *current limiter* and therefore slow the electrochemical degradation rate of the dielectric stack. However, remember the fluid must be conducting enough to not present a series resistance such that the RC time constant limits the operation to only DC potentials that would likely cause more severe device degradation and EW deviation. A large RC time constant could also impair the EW speed needed for some applications.

7.7.2.2 Don't Use Water!

Ultimately, the simplest way to achieve reliable operation is to not use water. A detailed analysis of nonaqueous conducting fluids for lab-on-a-chip applications has been performed for the use in droplet transport in lab-on-a-chip applications [3]. More recently, an analysis of greater detail was performed on nonaqueous conducting fluids for EW devices and displays [4]. Emphasis was provided on real-world testing parameters, including contact angle response and immiscibility with oil, environmental range, interfacial tension, ionic content and influence on dielectric reliability, compatibility with additives such as soluble dyes or particle dispersions, and cross-diffusion of fluids or solutes. Out of 16 preselected fluids, 6 exhibited EW performance comparable with the best aqueous–surfactant solutions (see Table 7.2). Of these, an exemplary nonaqueous conducting fluid was found to be propylene glycol and ethylene glycol, both readily available and propylene glycol being nontoxic as well. The effect of the use of glycols on device reliability is significant.

Table 7.2 Downselected set of nonaqueous conducting fluids for robust electrowetting actuation.

Fluid	m.p. (°C)	b.p. (°C)	μ (cP) at 25 °C	σ (nS cm^{-1})[a]	TBAB conc. (wt%)[b]	γ_{ci} (mN m^{-1})[a]	SDS conc. (<wt%)[b]
DMSO	18.5	189	1.99	0.62	0.018	8.5	0.026
Ethylene glycol	−13	197.3	16.1	0.03	0.072	15.6	Not studied
Formamide	2.6	220	3.34	1500	Not needed	23.8	Not studied
γ-Butyrolactone	−43.3	204	1.7	0.84	0.016	7.0	0.033
N-Methyl formamide	−3.8	199.5	1.68	160	Not needed	6.8	Not needed
Propylene carbonate	−48.8	242	2.53	5.3	0.027	8.6	0.039
Propylene glycol	−60	187.6	40.4	~0	0.305	8.5	0.372
2-Pyrrolidone	25	251	13.3	4.6	0.086	7.8	0.135

The insulating fluid phase was the same in all tests and was a silicone oil blend (80 wt% Dow Corning OS-20–17.5 mN m^{-1}, 10 wt% OS-10–17.2 mN m^{-1}, 10 wt% OS-30–17.3 mN m^{-1}).

a) Data were measured or imported from Handbook of Organic Solvents. Boca Raton, FL: CRC, 1995.

b) Tetrabutylammonium bromide and sodium dodecyl sulfate were added to reach a conductivity of 20 > S cm^{-1}.

Source: [4]. Reproduced with permission of Taylor & Francis.

Low potential (thin dielectric) devices that degrade in as little as several minutes with aqueous EW solutions last weeks or longer when glycols are utilized. For commercial product development, if reliable low potential operation is required (e.g. optics and displays), at this point, nonaqueous fluids seem to be required. However, the insulating fluids utilized typically must be of low surface tension (very nonpolar), which prevents miscibility with the conducting fluids that are not as polar as water. Many of the higher surface tension (more polar) insulating fluids, even alkanes, are miscible with many nonaqueous conducting fluids. Isopar is an exemplary insulating fluid that also is an excellent complement to propylene glycol. Ionic liquids also represent an alternative nonaqueous solution but typically have higher viscosity, higher costs, and very high conductivity (which can decrease reliability) compared with nonaqueous fluids such as glycols.

7.8 Summary of General Best Practices

A summary of best practices is as follows. *Firstly*, understand your application and all the unique challenges it requires (temperature, speed, potential, etc.). *Secondly*, firm up your understanding of EW deviation and saturation regimes and causes, such that experimental results can be properly understood in the context of theoretically ideal operation. *Thirdly*, choose a combination of fluids and surfaces that provide the contact angle modulation range that is needed, typically with insulating fluids that have very small or zero interfacial tension with the dielectric surface. *Fourthly*, if the amount of applied potential is not a concern, then simply use several micrometers thick Parylene dielectrics, and if low potential is needed, then use multilayer dielectric stacks where the electric field distributions have been carefully calculated. *Fifthly*, if maximum reliability is desired to avoid water as the conducting fluid, minimize the ionic content and conductivity of the conducting fluid, and if ionic content is required, use ions with the largest possible physical sizes. *Lastly*, use AC potentials that prevent or reduce a net ion migration through the dielectric and that also reduces dielectric charging and wetting hysteresis (see Section 7.2).

7.9 Mitigating Surface Fouling in Biological Applications

Surface fouling and unexpected Young's angles were discussed in a subsection of Section 7.2 and in Chapter 3. In biological applications such as EW lab-on-a-chip (also commonly referred to as digital microfluidics), surface fouling is particularly challenging. Biological solutes are so challenging because structures such as proteins, peptides, and large biomolecules have a structure that is very heterogeneous in terms of hydrophilic/hydrophilic features/groups. This same heterogeneity is essential for selectivity of binding events in living systems. However, in a microfluidic system, these affinities can become much less selective with the solutes *sticking* or *fouling* solid surfaces. Similarly, such interactions can cause surfactant-like behavior at liquid–liquid interfaces (see Section 7.2 and Chapter 3). Liquid–liquid interfaces are not the primary concern however (and are not discussed further in this subsection). The main challenge is that many solutes have one or multiple locations with hydrophobicity that could have low interfacial energy with a hydrophobic surface such as a fluoropolymer. As they foul the fluoropolymer surface, hydrophobicity is at least partially lost, wetting hysteresis increased, and ability for EW transport of droplets is diminished.

There are several approaches to mitigating fouling in biofluids. It is important to understand the application in mind when considering which approaches to adopt, in particular asking for how much time does significant fouling need to be prevented. Some applications only require few droplet manipulations before reaching the end result of fluid transport. Some applications can tolerate imprecise droplet transport or splitting caused by pinning and hysteresis at fouled surfaces. However, other applications may be more challenging, as a droplet may need to be incubated in a stationary position for hours or more and/or the need for highly predictable transport or splitting.

One of the simplest improvements is to include immersion of biofluid droplets surrounded by oil or thin oil films: $(\gamma_{id} - \gamma_{cd})/\gamma_{ci} = -1$, $\theta_Y = 180°$. Use of oil immersion allows EW of whole tears, sweat, urine, saliva, plasma/serum, or even something as complex as whole blood [25]. The approach is simple and somewhat effective, because it is more difficult for a solute to foul a solid surface if there is a thin layer of oil between the solution and the solid surface. However, gradually over time, even in oil, surfaces will foul as fouling solutes can partition into the oil phase and onto the solid surface. Other approaches include careful control of pH and applied potential to keep fouling solutes away from a surface (Figure 7.10, adapted from [29]). Many solutes take on a net charge, and for many lab-on-a-chip applications, dielectric degradation is less of a concern, and a DC potential can be utilized on the EW electrode, which is the same polarity as the charged solute for which fouling is to be prevented (e.g. simple electrostatic repulsion). Another approach is solution additives. Some additives such as surfactants, along with a moving droplet, can be used to partially *scrub* or clean a surface-free fouling solute. However, of greater interest is to use additives to prevent the fouling in the first place.

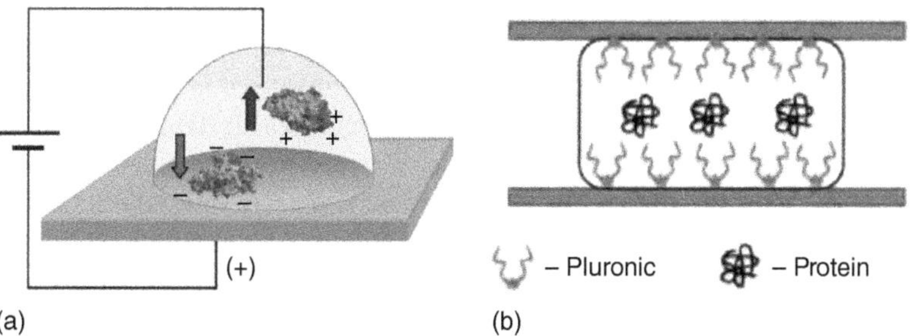

Figure 7.10 Biofouling mitigation strategies including (a) polarity dependence for BSA and lysozyme at pH 7.2. Source: [29]. Reproduced with permission of ACS. (b) use of pluronic additives to prevent surface fouling. Source: [1]. Reproduced with permission of ACS.

An additive approach known to work in microfluidics and also adapted to EW is to use pluronic additives in the aqueous droplet [1]. Pluronic additives are long, amphiphilic, and water-soluble molecules such as polyethylene oxide. However, they are not so heterogeneous that they would cause surface fouling. The physics is simple: if these long molecules are added at above their CMC (~0.1 wt%), they densely populate the inner surface of the aqueous droplet and therefore block other solutes from reaching that surface. These additives do have limits, however, because at high concentrations they can be toxic to living cells that are suspended in the aqueous droplet in some applications. In biological applications pluronic additives are useful both above and below the CMC, but again the *anti-fouling* effects are only improved above the CMC. Therefore an even more desirable approach would therefore be additive-free.

Beyond simply immersing the droplet in oil, the only other additive-free option left in an EW system would be the solid surfaces themselves. The only major improvement reported for the solid surface is the use of copolymer coatings activated by electric potential [23]. These coatings have been shown up to more than a fivefold improvement of mitigating fouling compared with the state of the art. The physics is quite complex and beyond the scope of this book. Generally modifying the EW surface is a hard problem. Surface chemists will frequently instruct to *modify with PEG* (or some other hydrophilic uncharged materials capable of resisting fouling), but that simplistic strategy fails, as EW requires a hydrophobic surface to enable manipulation of water droplets. Thus, the challenge is one of *threading the needle*, in which you try to use a polymer that is overall hydrophobic, but with a few functionalities added to reduce adhesion. Wheeler's 2015 work [23] is not a perfect manifestation of such a polymer, but rather a starting point that can likely be improved upon in the future.

7.10 Additional Issues for Complex or Integrated Devices

There are numerous other fabrications issues that are beyond the scope of this chapter. Two comprehensive example studies are included here by reference. Reading these studies reveals that a large amount of work and numerous challenges are faced for complete integration of displays or optical EW devices [14, 30]. Integration issues for lab-on-a-chip include sample introduction, biofouling (see previous section), and sensor integration, to name a few. Arrayed devices such as displays often present challenging large-area fabrication and fluid dosing methods. Most devices must be at least partially sealed and sometimes need to be sealed with the fluids inside them, making epoxy selection challenging. This chapter has provided the basis for one to begin to immediately create state-of-the-art EW performance in any university or company lab. However, the path is much more difficult to achieve functional and self-contained devices for real-world applications.

Acknowledgement

Special acknowledgement is given to Dr. Stein Kuiper who has collaborated with the second author of this book (Heikenfeld) on numerous EW materials and reliability studies. This chapter would not have been possible in its present form without his involvement and knowledge stemming from years of expert work in EW systems, including his time while working at Philips Research labs. Therefore an acknowledgement for this chapter should also extend to the many research staff at Philips Research labs who pioneered much of the early EW materials work. The same could be said of Varioptic as well (Berge and colleagues).

7.A Trapped Charge Derivation

In the presence of trapped charge, the EW curve is asymmetric, and the contact angle becomes maximum not at zero voltage but at a finite voltage U_T. As first derived by Prins and Verheijen, $\theta(U)$ follows the relation shown previously in Eqs. (7.3) and (7.4).

The derivation of this expression is in principle straightforward. Yet, it requires a careful consideration of all contributions to the electrostatic energy. The laterally homogeneous trapped charge density σ_T is assumed to be located at a fixed depth δ below the solid surface as shown in Figure 7.A.1.

For a dielectric with a homogeneous dielectric constant ϵ, the basic laws of electrostatics (see Chapter 2) provide relations between the charge densities on the drop, σ_d, the trapped charge, σ_T, and the charge density on the electrode both under the solid–liquid interface, σ_{el}^L, and next to the drop under the solid–vapor interface, σ_{el}^V, and the corresponding electrostatic potentials ϕ_i^j:

$$\sigma_d = \epsilon\epsilon_0 E_1 = -\epsilon\epsilon_0 \frac{\phi_T^L - \phi_d}{\delta} \tag{A.1}$$

$$\sigma_T = \epsilon\epsilon_0 (E_2 - E_1) \rightarrow \sigma_T + \sigma_d = \epsilon\epsilon_0 E_2 = -\epsilon\epsilon_0 \frac{\phi_{el} - \phi_T^L}{d - \delta} \tag{A.2}$$

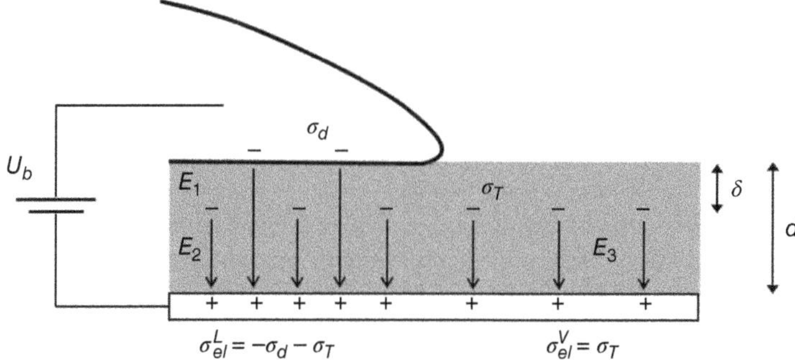

Figure 7.A.1 Electrowetting in the presence of a trapped charge density σ_T at a depth δ.

$$\sigma_{el} = -\epsilon\epsilon_0 E_2 = +\epsilon\epsilon_0 \frac{\phi_{el} - \phi_T^L}{d - \delta} \tag{A.3}$$

$$\sigma_T = +\epsilon\epsilon_0 E_3 = -\epsilon\epsilon_0 \frac{\phi_{el} - \phi_T^V}{d - \delta} \tag{A.4}$$

The equilibrium contact angle is obtained as usual by considering the virtual work upon displacing the contact line by δx toward the right and requiring the free energy to be stationary, i.e.

$$\delta W = \delta W_{surf} + \delta W_{el} = 0 \tag{A.5}$$

where $\delta W_{surf} = (\gamma_{sl} - \gamma_{sv} + \gamma \cos\theta)\delta x$, as usual in the derivation of Young's law. Moreover,

$$\delta W_{el} = \delta W_{batt} + \delta W_{drop} \tag{A.6}$$

consists of the gain of electrostatic energy upon transferring charge from the battery to the drop. Specifically, we find

$$\delta W_{batt} = -(\phi_d - \phi_{el})\sigma_d\, \delta x = -\frac{1}{\epsilon\epsilon_0}(\sigma_d\, d + \sigma_T(d - \delta))\sigma_d\, \delta x \tag{A.7}$$

and

$$\delta W_{drop} = \frac{\epsilon\epsilon_0}{2}\, \{E_1^2\, \delta + (E_2^2 - E_3^2)(d - \delta)\}\, \delta x = \frac{1}{2\epsilon\epsilon_0}\, \{d\, \sigma_d^2 + 2\,(d - \delta)\sigma_d \sigma_T\}\, \delta x \tag{A.8}$$

For the right-hand part of the equation, we made use of a linear combination of Eqs. (A.1), (A.2), and (A.4). Combining Eqs. (A.5) and (A.6), we find

$$\delta W_{el} = -\frac{d\, \sigma_d^2}{2\epsilon\epsilon_0}\, \delta x = -\frac{\epsilon\epsilon_0}{2d}(\phi_d - \phi_T^V)^2 \tag{A.9}$$

Rewriting $\phi_d - \phi_T = (\phi_d - \phi_{el}) - (\phi_T^V - \phi_{el}) = U - U_T$ and inserting it into Eq. (A.5), we recover Eq. (7.3).

Note also that a positive trapped charge density σ_T implies a positive voltage U_T and hence requires a *positive* voltage $U = U_T$ on the drop (with respect to the electrode on the substrate) to achieve the maximum contact angle. Similarly, a negative trapped charge requires a negative applied voltage. This result may seem counterintuitive at first glance. Yet, Eq. (A.9) implies that the electric contribution vanishes if the potential of the drop equals the potential of the trapped charge. In this case, the E_1 vanishes and the wetting state of the drop is not affected by the electric fields at all. Consequently, we find $\theta(U = U_T) = \theta_Y$. A second remarkable aspect of Eq. (A.9) (and hence the resulting contact angle response in Eq. (7.3)) is that it is independent of the location δ of the trapped charge. However, the amount of trapped charge will likely depend on the dielectric thickness, because for thick dielectrics the required EW voltage is very high and the required charge density very low. If the energy stored in the EW capacitor is $cU^2/2$, and using $Q = cU$ that is equivalent to $Q^2/2C$, then it is clear that for a given EW number (contact angle), if capacitance increases, then the required charge decreases.

Problems

7.1 Consider an electrowetting system with fluids of 10 mN m^{-1} interfacial surface tension, Young's angle of 180° (is in oil), and a two-layer dielectric stack consisting of 200 nm Si_3N_4 (permittivity of $\varepsilon_r \sim 8$) and 100 nm of Teflon (permittivity of $\varepsilon_r \sim 2.1$). Calculate (a) the electrical capacitance, (b) the potential required to electrowet down to 70°, and (c) the E-field across the Teflon if there is an electrical defect in the Si_3N_4 layer and comment on the reliability of the device if the breakdown electric field for the Teflon is 200 V μm^{-1}.

7.2 If you keep all other parameters constant and you increase dielectric thickness by four times, how much will the required operating potential increase? You may assume a simple single-layer dielectric system.

7.3 For the electrowetting experiment of Problem 7.1, now assume that the Si_3N_4 is defect-free and that the Teflon is defective such that when you electrowet down to 90° with positive potential applied to the conductive fluid, 15% of the total charge that would normally be in the conductive fluid is permanently injected into the Teflon layer. Calculate the deviation in potential for wetting to 90° as caused by this charge injection.

7.4 Alkanes are often used with water on hydrophobic fluoropolymer surfaces to achieve Young's contact angles near 180°. Conversely, chlorinated or bromated alkanes can also achieve Young's contact angles near 180° on non-hydrophobic surfaces such as Parylene C. For this three-phase systems, which of the interfacial surface tension components is being carefully designed to achieve the large Young's angle, and to what value is it approaching in the best possible design?

7.5 You perform an electrowetting experiment with deionized water and you add a significant amount of benzalkonium chloride surfactant (do an online search for it). This surfactant dominates the conductivity of the aqueous solution. Generally speaking, do you expect your device to operate more reliably in positive or negative potential?

7.6 Perform a simple theoretical derivation or numerical calculation that proves that using high dielectric constant is useless for reducing electrowetting potential, because the required electric field in the layer with the lowest dielectric constant (usually the fluoropolymer topcoat) will be equal regardless of the dielectric constant of the dielectric material placed below it. To simplify your derivation, you may assume constant values for the required U and constant values for γ_{ci}, θ_Y, θ_V, while only the thickness of the fluoropolymer may change as may the capacitance of other dielectrics in the dielectric stack.

7.7 Numerically calculate the electrical energy density needed to sustain electrowetting to 90°, which is the same for a dielectric that is 1 μm thick or 100 μm thick, even though the potential is higher for the latter.

7.8 See the images below. The time and potential (on or off) for each droplet is labeled. Describe what is likely happening for each experiment (a) and (b), and note any differences that you see between (a) and (b). In your analysis consider the deviation region also seen in Figure 7.1.

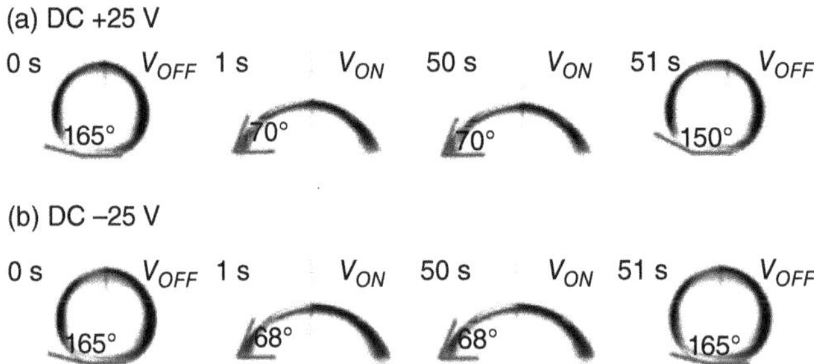

7.9 See the temporal data from a lab-on-a-chip experiment [9]. The data is for a droplet being quickly advanced onto an adjacent electrode (the voltage is applied at $t \sim 5$ seconds). The droplet is then moved away from the electrode at $t \sim 10$ seconds. There is a clear temporal response of electrical capacitance for the advancing droplet with the electrode it advances onto. This temporal response is due to trapping of an oil film beneath the droplet, which slowly dewets (breaks up) with further time from 5 to 10 seconds. Assume for simplicity the dielectric has a thickness of 1 μm and has a dielectric constant of 3. Assume the oil also has a dielectric constant of 3. Using the data below, calculate the initial thickness of the oil film formed at $t \sim 5$ seconds.

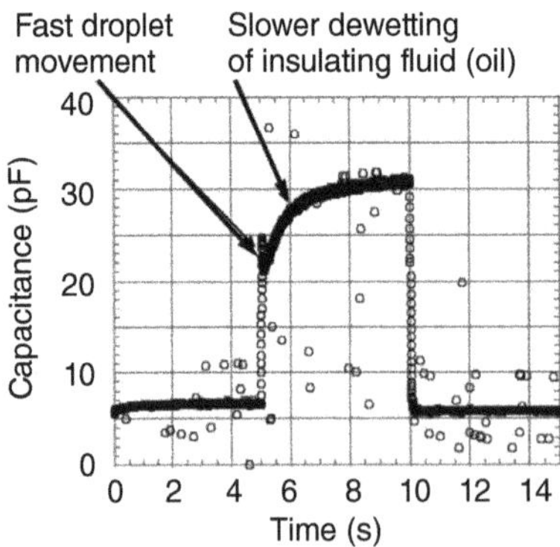

7.10 A liquid metal droplet such as GaInSn or Hg has a surface tension of ~500 mN m^{-1}. You are limited to 200 V for actuation, and you have access to a very high quality dielectric with a breakdown field of 4 MV cm^{-1}

and a dielectric constant of $\varepsilon_r = 3$. Find the optimum thickness for the dielectric that will give you the most contact angle change in a reliable manner (meaning that you do not exceed 50% of the breakdown field of the dielectric, or $2\,\mathrm{MV\,cm^{-1}}$).
Hint: You will discover why liquid metals are currently not robustly electrowettable.

7.11 Consider a superhydrophobic surface with a value of $\beta = 0.05$, $\theta_Y = 115°$ and the superhydrophobic features being pillars (cylinders). Assume that the pillars themselves are coated with an electrode and electrowetting dielectric of $0.5\,\mu\mathrm{m}$ thickness and dielectric constant of 3. You may ignore the possibility of electrical breakdown. Calculate (a) θ_S and (b) the minimum θ_V before the electrowetted contact angle on the posts reaches 90° causing irreversible wetting into the Wenzel state; lastly compare the amount of contact angle modulation you were able to achieve (total range) to the same electrowetting surface if it were a flat/planar surface, and comment on how much practical value a superhydrophobic surface provides.

References

1 Au, S. H., Kumar, P., and Wheeler, A. R. (2011). A new angle on pluronic additives: advancing droplets and understanding in digital microfluidics. *Langmuir* 27 (13): 8586–8594.

2 Berry, S., Fedynyshyn, T., Parameswaran, L., and Cabral, A. (2012). Reversible electrowetting on dual-scale-patterned corrugated microstructured surfaces. *J. Microelectromech. Syst.* 21 (5): 1261–1271.

3 Chatterjee, D., Hetayothin, B., Wheeler, A. R. et al. (2006). Droplet-based microfluidics with nonaqueous solvents and solutions. *Lab Chip* 6 (2): 199–206. http://xlink.rsc.org/?DOI=b515566e.

4 Chevalliot, S., Heikenfeld, J., Clapp, L. et al. (2011). Analysis of nonaqueous electrowetting fluids for displays. *IEEE/OSA J. Disp. Technol.* 7 (12): 649–656.

5 Chevalliot, S., Kuiper, S., and Heikenfeld, J. (2012). Experimental validation of the invariance of electrowetting contact angle saturation. *J. Adhes. Sci. Technol.* 26 (12–17): 1909–1930.

6 Dhindsa, M., Heikenfeld, J., Weekamp, W., and Kuiper, S. (2011). Electrowetting without electrolysis on self-healing dielectrics. *Langmuir* 27 (9): 5665–5670.

7 Dhindsa, M., Kuiper, S., and Heikenfeld, J. (2011). Reliable and low-voltage electrowetting on thin Parylene films. *Thin Solid Films* 519 (10): 3346–3351.

8 Drygiannakis, A. I., Papathanasiou, A. G., and Boudouvis, A. G. (2009). On the connection between dielectric breakdown strength, trapping of charge, and contact angle saturation in electrowetting. *Langmuir* 25 (1): 147–152.

9 Fair, R. B. (2007). Digital microfluidics: is a true lab-on-a-chip possible? *Microfluid. Nanofluid.* 3: 245–281.

10 Fan, S.-K., Yang, H., Wang, T.-T., and Hsu, W. (2007). Asymmetric electrowetting – moving droplets by a square wave. *Lab Chip* 7 (10): 1330–1335. http://xlink.rsc.org/?DOI=b704084a.

11 Hayes, R. A. and Feenstra, B. J. (2003). Video-speed electronic paper based on electrowetting. *Nature* 425 (6956): 383–385.

12 Heikenfeld, J. and Dhindsa, M. (2008). Electrowetting on superhydrophobic surfaces: present status and prospects. *J. Adhes. Sci. Technol.* 22 (3–4): 319–334.

13 Heikenfeld, J., Drzaic, P., Yeo, J.-S., and Koch, T. (2011). Review paper: a critical review of the present and future prospects for electronic paper. *J. Soc. Inf. Disp.* 19 (2): 129. doi: 10.1889/jsid19.2.129.

14 Hou, L., Zhang, J., Smith, N. et al. (2010). A full description of a scalable microfabrication process for arrayed electrowetting microprisms. *J. Micromech. Microeng.* 20 (1): 015044.

15 Koo, B. and Kim, C. J. (2013). Evaluation of repeated electrowetting on three different fluoropolymer top coatings. *J. Micromech. Microeng.* 23 (6): 067002.

16 Li, F. and Mugele, F. (2008). How to make sticky surfaces slippery: contact angle hysteresis in electrowetting with alternating voltage. *Appl. Phys. Lett.* 92 (24): 244108.

17 Maillard, M., Legrand, J., and Berge, B. (2009). Two liquids wetting and low hysteresis electrowetting on dielectric applications. *Langmuir* 25 (11): 6162–6167.

18 Manukyan, G., Oh, J. M., Van Den Ende, D. et al. (2011). Electrical switching of wetting states on superhydrophobic surfaces: a route towards reversible Cassie-to-Wenzel transitions. *Phys. Rev. Lett.* 106 (1): 014501.

19 Mugele, F. and Baret, J. C. (2005). Electrowetting: from basics to applications. *J. Phys. Condens. Matter* 17: R705–R774.

20 Paneru, M., Priest, C., Sedev, R., and Ralston, J. (2010). Static and dynamic electrowetting of an ionic liquid in a solid/liquid/liquid system. *J. Am. Chem. Soc.* 132 (24): 8301–8308.

21 Raccurt, O., Berthier, J., Clementz, P. et al. (2007). On the influence of surfactants in electrowetting systems. *J. Micromech. Microeng.* 17 (11): 2217–2223.

22 Raj, B., Dhindsa, M., Smith, N. R. et al. (2009). Ion and liquid dependent dielectric failure in electrowetting systems. *Langmuir* 25 (20): 12387–12392.

23 Sarvothaman, M. K., Kim, K. S., Seale, B. et al. (2015). Dynamic fluoroalkyl polyethylene glycol co-polymers: a new strategy for reducing protein adhesion in lab-on-a-chip devices. *Adv. Funct. Mater.* 25 (4): 506–515.

24 Schultz, A., Chevalliot, S., Kuiper, S., and Heikenfeld, J. (2013). Detailed analysis of defect reduction in electrowetting dielectrics through a two-layer "barrier" approach. *Thin Solid Films* 534: 348–355.

25 Srinivasan, V., Pamula, V. K., and Fair, R. B. (2004). An integrated digital microfluidic lab-on-a-chip for clinical diagnostics on human physiological fluids. *Lab Chip* 4 (4): 310–315.

26 Staicu, A. and Mugele, F. (2006). Electrowetting-induced oil film entrapment and instability. *Phys. Rev. Lett.* 97 (16): 167801.

27 Vallet, M., Vallade, M., and Berge, B. (1999). Limiting phenomena for the spreading of water on polymer films by electrowetting. *Eur. Phys. J. B* 11 (4): 583–591.

28 Verheijen, H. J. J. and Prins, M. W. J. (1999). Reversible electrowetting and trapping of charge: model and experiments. *Langmuir* 15 (20): 6616–6620.

29 Yoon, J. Y. and Garrell, R. L. (2003). Preventing biomolecular adsorption in electrowetting-based biofluidic chips. *Anal. Chem.* 75 (19): 5097–5102.

30 Zhou, K., Heikenfeld, J., Dean, K. A. et al. (2009). A full description of a simple and scalable fabrication process for electrowetting displays. *J. Micromech. Microeng.* 19 (6): 065029.

31 Zimmermann, R., Rein, N., and Werner, C. (2009). Water ion adsorption dominates charging at nonpolar polymer surfaces in multivalent electrolytes. *Phys. Chem. Chem. Phys.* 11 (21): 4360–4364. http://xlink.rsc.org/?DOI=b900755e.

32 Banpurkar, A., Sawane, Y. B., Wadhai, S. M. et al. (2017). Spontaneous electrification of fluoropolymer-water interfaces probed by electrowetting. *Faraday Discuss.* 199: 29–47.

33 Mugele and Buehrle (2007). Equilibrium drop surface profiles in electric fields. *J. Phys.: Condens. Matter*. 19 (37).

34 Staicu, Manukyan, and F. Mugele (2008). Microscopic structure of electrowetting-driven transitions on superhydrophobic surfaces. arXiv:0801.2683.

35 Tom N. Krupenkin, Ashley Taylor J., Tobias M. Schneider, and Shu Yang (2004). From rolling ball to complete wetting: the dynamic tuning of liquids on nanostructured surfaces. *Langmuir* 20 (10): 3824–3827.

36 Dale L. Herbertson, Carl R. Evans, Neil J. Shirtcliffe, Glen McHale, and Michael I. Newton (2006). Electrowetting on superhydrophobic SU-8 patterned surfaces. *Sens. Actuat A: Phys.* 130–131: 189–193.

8

Fundamentals of Applied Electrowetting

8.1 Introduction and Scope

This chapter is far more than a simple review of the various applied demonstrations for electrowetting. Instead, this chapter presents the applied nature of electrowetting in terms of *fundamental methods and capabilities*. This approach is taken because by understanding the fundamental functions that electrowetting can impart on fluids, the reader is better equipped for both work in existing applications and for discovery of new applications. Each fundamental capability is presented in terms of example applications, the underlying physics, and the achieved effect. In addition, important considerations such as materials, control electronics, and other factors are provided where relevant. This chapter will not cover simple contact angle reduction on a sessile droplet or wetting into a capillary, which is covered extensively in previous chapters of this book. The chapter only touches on major topics, and is not comprehensive (e.g. [18]). The chapter also excludes currently impractical topics such as superhydrophobic surfaces. Advanced readers are also reminded that the basic effects introduced in this chapter are non-obvious to newcomers to electrowetting. The reader should consider that many effects/devices were discovered and reported more than 10 years after electrowetting became broadly known!

8.2 Droplet Transport

Droplet movement has example applications including, but not limited to, lab-on-chip, simple indicator displays, and active cooling [29, 36, 41, 42].

8.2.1 Basic Force Balance Interpretation of Droplet Transport

Causing movement of a droplet obviously requires a force imbalance on the droplet. For a sessile droplet (no horizontal movement) on a single large electrowetting plate, the electromechanical force imbalance imparted by electrowetting is uniform along the entire contact line perimeter. Therefore the system is able to achieve an equilibrium state by only reducing the contact angle as the interfacial surface tension forces are repositioned to achieve that equilibrium (Section 5.1). For a sessile droplet on a single large electrowetting

Electrowetting: Fundamental Principles and Practical Applications,
First Edition. Frieder Mugele and Jason Heikenfeld.

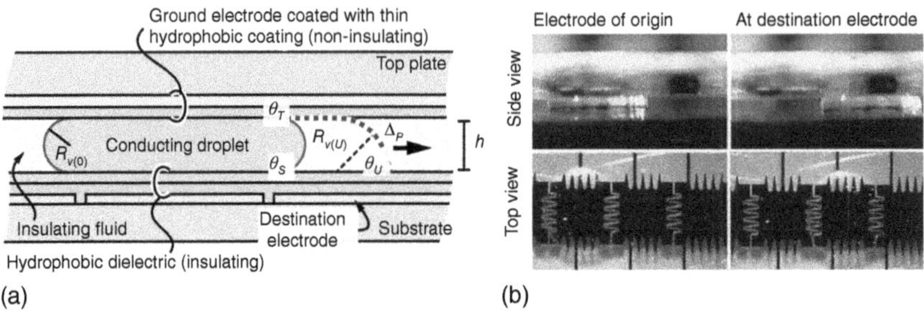

Figure 8.1 Simple droplet transport. (a) Diagram of conducting fluid droplet (solid line, shading) and movement onto the destination electrode by electrowetting (dotted line). (b) Experimental photos, including a top view showing the electrode interdigitation. Source: [28]. Reproduced with permission of Royal Society of Chemistry.

plate, the contact angle change results in a geometry change for the droplet, but does not result in any net horizontal movement of the droplet.

For droplet movement (transport), the foremost requirement is that the electromechanical force *be nonuniform* around the droplet perimeter. Therefore, a droplet only partially overlapping an electrowetting surface will experience increasing force to move the droplet onto that surface as potential is increased on the electrowetting electrode. From a fluid mechanics perspective, this droplet motion is most easily understood and designed for not in terms of force, but in terms of force per unit area (pressure), which we will help describe with the help of Figure 8.1.

Figure 8.1 illustrates a conventional electrowetting microchannel for droplet movement. The substrate contains a series of electrodes coated by a hydrophobic dielectric, thereby creating independent controllable destinations for a conducting droplet. The top plate has a ground electrode, with only a hydrophobic coating that is so thin that the ground electrode can be approximated as being in direct electrical contact with the conducting droplet. The Laplace pressure ($\mathrm{N\,m^{-2}}$) within the conducting fluid is calculated using a version of Eq. (1.5):

$$\Delta P = \gamma_{ci}\left(\frac{1}{R_1} + \frac{1}{R_2}\right) \tag{8.1}$$

This general equation takes into account the interfacial surface tension between the conducting and insulating fluids (γ_{ci}, $\mathrm{N\,m^{-1}}$) and the principal radii of curvature for the conducting fluid meniscus (R_1, R_2, m), which are determined by the contact angles and any solid bounding surfaces around the fluid. In the microchannel of Figure 8.1, one of these radii is in the horizontal plane (not labeled) and traces the perimeter (circumference) of the droplet as observed from the top view. We will refer to this horizontal radius as R_H. This horizontal radius is not substantially changed by electrowetting and relative to other radii is very large, and therefore this horizontal radius weakly contributes to the Laplace pressure. As a result horizontal radius R_H will be ignored in our discussion.

Now, instead take a vertical cross section through the conducting droplet (side view, Figure 8.1a), and consider the contact angles with the top plate and substrate, θ_T and θ_S, along with a channel height h. These factors predict the vertical radius of curvature of the droplet $R_{V(0)}$ at no applied potential. This vertical radius $R_{V(0)}$ is typically much smaller than the horizontal radius R_H discussed in the previous paragraph and therefore will dominate the Laplace pressure predicted by Eq. (8.1). Using simple geometry, the resulting Laplace pressure inside the confined droplet of Figure 8.1 can therefore be expressed as

$$\Delta P = \gamma_{ci}\left(\frac{1}{R_V(0)}\right) \tag{8.2}$$

$$R_V = \frac{-h}{(\cos\theta_T + \cos\theta_S)} \tag{8.3}$$

In Figure 8.1, an insulating fluid such as oil is used to promote Young's angles of 180°, and therefore $R_{V(0)} = h/2$ (both mathematically and visually obvious in Figure 8.1). Now applying electrowetting (Eq. (5.1)) voltage U to the destination electrode, the new droplet meniscus is shown as the dotted line in Figure 8.1a for the example case of electrowetting to $\theta_U = 90°$. This results in $R_{V(U)} = h$, which can also be seen visually in Figure 8.1. The trailing edge of the droplet still has a pressure of $\Delta P = \gamma_{ci}2/h$, and the advancing edge has an electromechanically lowered pressure of $\Delta P = \gamma_{ci}/h$, for a net pressure acting on the droplet of $\Delta P = \gamma_{ci}/h$. As a result of this pressure (force) imbalance, the droplet moves forward onto the destination electrode. The droplet will cease forward movement when either (i) the potential is removed or (ii) the droplet reaches the far edge of the destination electrode where the availability of electromechanical force also ends. Droplet motion could also cease if the destination electrode were larger than the droplet and both the advancing and receding ends of the droplet became electrowetted on such a larger electrode ($\Delta P = 0$) (Eq. (8.5)). The above description is a simple one, allowing one to easily visualize the exact change in pressure balance for the system. Substituting the electrowetting equation into Eqs. (8.2) and (8.3) provides a prediction of the pressure imbalance as a function of electric potential:

$$\Delta P = -\gamma_{ci}(\cos\theta_T + \cos\theta_S + cU^2/2\gamma_{ci})/h \tag{8.4}$$

If both the top and bottom electrowetting plates are electrowetted with identical dielectrics and voltages, then in Eq. (8.4) the electrowetting number can simply be doubled to cU^2/γ_{ci} (this assumes that the droplet is electrically grounded by a third electrode, or else the required potential doubles due to two capacitors in series). The reader should note, however, that Eqs. (8.2)–(8.4) predict that a droplet will begin to move immediately as even very low electric potentials are applied. In practice, this is not observed due to factors such as contact angle hysteresis (Eq. (7.7)) and viscous drag. These same forces that create a threshold voltage for movement also slow the velocity of the droplet. Therefore the discussion now turns to threshold and velocity for transport.

8.2.2 Advanced Droplet Transport Physics: Threshold and Velocity

The speed (velocity) at which a droplet can be moved brings into play forces that were neglected in the previous section. Furthermore, in practice the droplet does not move at very low voltages, clearly indicating the presence of a threshold that is not captured in Eqs. (8.2)–(8.4). To accomplish a more complete model, we need

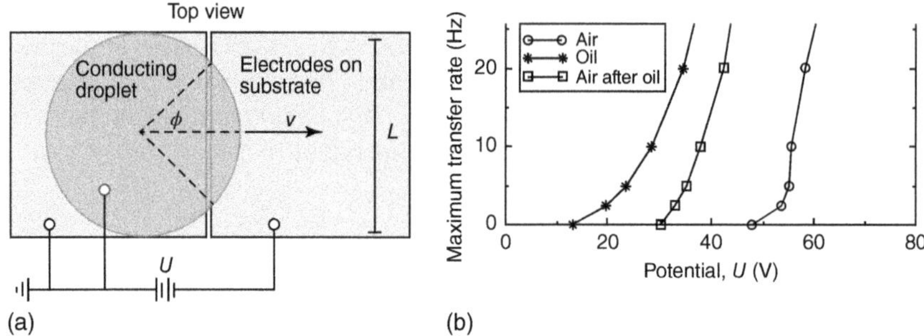

Figure 8.2 (a) Electrowetting actuator top view. Source: [31]. Redrawn with permission of Taylor & Francis. (b) Transfer rate (velocity) profiles for several different conditions. Source: [40]. Reproduced with permission of Duke University.

to revert back to equations based on force, not pressure, since some of the factors opposing droplet transport are forces imparted at surfaces (are not pressures).

Using the same terms adopted in the previous subsection, a more detailed model of the force pulling a droplet forward onto a square electrowetting electrode with side length L inside a microchannel is provided below [31]:

$$F_U = L \sin \varphi[\cos \alpha U^2/2 - \gamma_{ci} \sin \alpha[\sin \theta_U + \sin \theta_S]] \tag{8.5}$$

where α is the amount of contact angle hysteresis. As shown in Figure 8.2, the term in front L sin ϕ simply allows the integration of the electrowetting force per unit length over the electrowetted electrode, where L is the side length of the square electrowetting electrode and the diameter of the droplet (assumed equal for simplicity, they typically are similar). sin ϕ nicely captures the actuation force during the entire transport process, where, for example, sin ϕ is small when the process starts and reaches a maximum when $\phi = 90°$. The total actuation force diminishes as ϕ approaches 180° because the trailing edge of the droplet also begins to electrowet as it moves onto the destination electrode and therefore creates a force in the opposite direction of the force on the advancing edge of the droplet. Also notice how this force equation lacks the channel height h of the pressure equation in Eq. (8.3), which serves as a further reminder that the driving electromechanical force is localized to near the contact line.

Obviously, there are forces opposing the motion in the channel, or else a droplet could reach infinitive velocity. The dominant motion-opposing forces in a planar and open channel are drag forces at the droplet–solid interface (F_{ds}) and drag at the droplet/insulating fluid interface (F_{di}) if oil is utilized in place of air (which is the most common and practical case). Understanding these drag forces is not only important to predicting droplet velocity but as will be seen in the next section drag forces cause internal droplet flows that can assist in rapid droplet mixing (recall Figure 6.6 and related discussion). The generic drag force equation and the specific drag forces of interest to electrowetting droplet transport are

$$F_{ds} = 2C_v \mu v L^2/h \tag{8.6}$$

$$F_{di} = C_D A_D \rho_i v^2/2 \tag{8.7}$$

where v is the droplet velocity, C_D is the drag coefficient for a microchannel, and C_v is an empirical constant, typically with a value near 10, discussed in greater detail elsewhere [31]. Other new terms include viscosity of the droplet μ, the density of the insulating fluid (oil) ρ_i, and the projected area of the droplet in the direction of transport A_D.

A force balance between Eqs. (8.4)–(8.6) then yields a transport velocity equation. Assuming the droplet diameter is similar to L and that $A_D = Lh$, the droplet transport velocity is

$$v = \frac{F_U}{L(12\mu h/L + 2C_V \mu L/h)} \tag{8.8}$$

The terms in the denominator represent the solid surface drag and insulating fluid drag and have oppositely arranged h/L and L/h ratios. This clearly shows that channel height h and electrode dimension L must be optimized, which is especially true when the droplet and oil viscosities are similar. Equation (8.8) as expected shows that transport velocity has a U^2 dependence, which is also visible in the plots of Figure 8.2b. However, notice also in Figure 8.2b the threshold dependence for droplet motion that changes dramatically with the presence of oil. This threshold dependence is largely due to the pinning forces that cause contact angle hysteresis α. Simple dipping treatment of a fluoropolymer surface with oil causes the oil to fill in rough spots on the surface and reduce hysteresis. Performing the electrowetting while in oil reduces the roughness and hysteresis even further. However, even for the case of $\theta_Y = 180°$, hysteresis still exists in practice and opposes the onset of droplet transport.

One last note on droplet velocity is γ_{ci}. In the previous chapter, it was shown that in most cases electrowetting deviation and saturation are independent of γ_{ci}. Therefore, for a higher γ_{ci}, it requires a larger U, and more electrowetting force is generated at a given contact angle (Eq. (8.4)). Therefore to maximize transport speed, generally one also maximizes γ_{ci} [23].

8.2.2.1 Advanced Droplet Transport Physics: Flow Field

In the previous subsections, the bulk velocity of the droplet was developed. However, bulk droplet velocity is only an *external* picture of velocity, and there is also an *internal* flow that must be considered. In the previous section, it was shown that drag forces oppose F_d, the bulk droplet motion. The drag force is experienced near the droplet interface with the substrate and the surrounding environment of oil. Now recall Section 6.1, where it was shown that at the interface of the droplet with a solid surface, the velocity of the fluid is zero (stationary with the solid surface). Therefore, as the droplet moves forward, the substrate interface causes a drag force in the opposite direction of transport. This drag force then locally results in an *internal* flow field (velocity) near the solid and oil interfaces, which is opposite of that of *external* droplet velocity. Such a flow field for a moving droplet is illustrated beautifully in the microparticle image velocimetry data shown in Figure 8.3 [20]. The plot in Figure 8.3a includes horizontal (x–y-axis) streamline components of flow inside the droplet at a height of $z = 148$ μm inside a 388 μm channel ($z = 0$ μm is the surface of the bottom electrowetting substrate). Because the center of the droplet (along $y = 400$ μm) has the longest path length for drag across the bottom substrate, it experiences the most net drag force, and therefore fluid in the center of the droplet has a negative *internal velocity* with respect to the x-axis. Note, however, that the *external net velocity* along the x-axis during

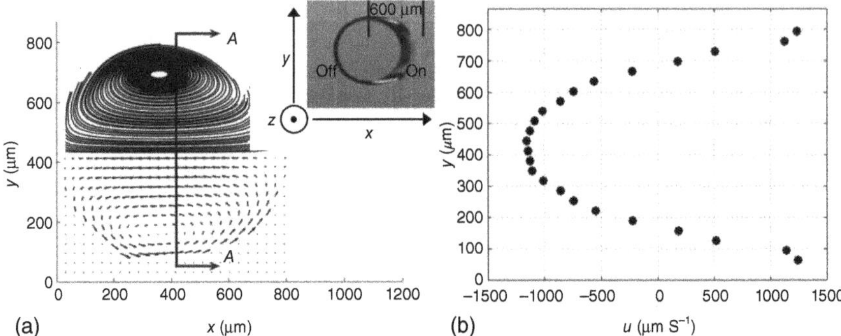

Figure 8.3 (a) Two-dimensional (horizontal x–y) velocity field at $z = 148$μm (bottom half) and the streamlines (top half). (b) The axial velocity component, u, along the cross-sectional *AA*. Source: [20]. Adapted with permission of Royal Society of Chemistry.

transport is still positive. The edges of the droplet (near $y = 100$ and 700 μm) also experience drag in the same direction, but a reduced path length such that fluid flow is forced into the positive x-direction by the more powerful negative x-direction flow in the center of the droplet. The oil also imparts a drag force, but its effects on internal fluid velocity are weaker since in this experiment the area of contact with oil is much smaller than the area of contact with the substrates. For calculation of drag force, the reader may return to the previous subsection on droplet velocity. The flow field in a moving droplet has practical application in microfluidic mixing, which will be covered in Section 8.3.

8.2.3 Additional Practical Notes on Implementation of Basic Droplet Transport

Material preferences for droplet transport and lab-on-chip were partly discussed in Chapter 7. Droplet movement typically involves several to dozens of electrodes arranged in series (1D transport) or 2D arrays (2D transport). Example electrode sizes are typically about 1 mm^2 in area with channel heights being 100 μm or more (if droplets are wider, their horizontal Laplace pressure component can become too weak to stabilize the droplet shape during transport). The top plate typically uses a transparent electrode with a very thin fluoropolymer coating (typically 50 nm or less). As noted in this section, generally a high γ_{ci} is desired for maximum transport velocity and also for increasing splitting accuracy (next section). Because high voltage sources are more easily integrated with lab-on-chip devices, the use of thick dielectrics and high γ_{ci} is not of significant practical concern for the required drive electronics.

8.3 Droplet Transport for Splitting, Dosing, Merging, and Mixing

Droplet splitting and merging is primarily of interest to lab-on-chip [29, 36, 41, 42].

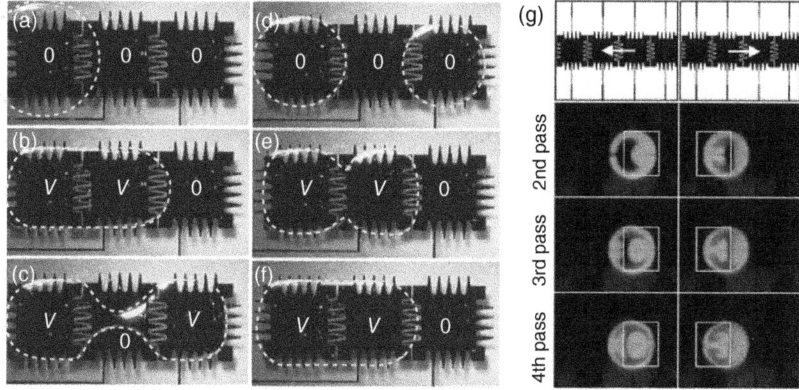

Figure 8.4 Example photographs of (a–c) droplet splitting, (d–f) remerging, and (g) time-lapse images of mixing of fluorescent dye by repeated movement of a droplet between two electrodes. In (a–f) the transparent droplet is outlined, and the electrodes are labeled as either no potential *0* or as with potential *V*. Source: Adapted from [27], [Paik 2003].

8.3.1 Simple Experimental Examples

The basic processes of splitting, merging, and mixing (but not dosing/dispensing) are demonstrated in Figure 8.4. Droplet splitting is more complex than conveyed by the simple time-lapse photographs of Figure 8.4c and will be presented in greater detail in the next subsection. For example, splitting accuracy (unequal volumes between split droplets) can be several volume percent or more and highly dependent on how the splitting is performed [2, 36]. Droplet dosing, also referred to as dispensing, refers to the process of introducing fluid droplets by electrowetting into a microchannel. As will be seen in subsequent sections, droplet dosing relies on physics similar to that for droplet splitting. Mixing by transport or splitting/merging (Figure 8.4g) involves internal flow fields similar to those discussed in the previous section, but is more complex in that they must be implemented in a manner such that internal flow is not fully reversible.

8.3.2 Fundamentals of Droplet Splitting

Splitting of a droplet in a microchannel obviously requires oppositely directed electrowetting forces to overcome surface tension that is always working to instead minimize droplet surface area. As shown in Figure 8.4c, the simplest way to do this is with three electrodes, two of which are electrowetting and destinations for the split droplet and one electrode in between at the original droplet location with no electrowetting. Although this splitting technique is simple, when potentials are abruptly applied, the splitting is not perfectly precise with errors in split droplet volumes of at least several percent [36]. The culprit is hydrodynamic instabilities. A simple way to eliminate these hydro*dynamic* instabilities is simply to slow things down. The instabilities can be dampened by more gradual application of potentials (seconds) for splitting to the point where $<1\%$ variation can be achieved [2].

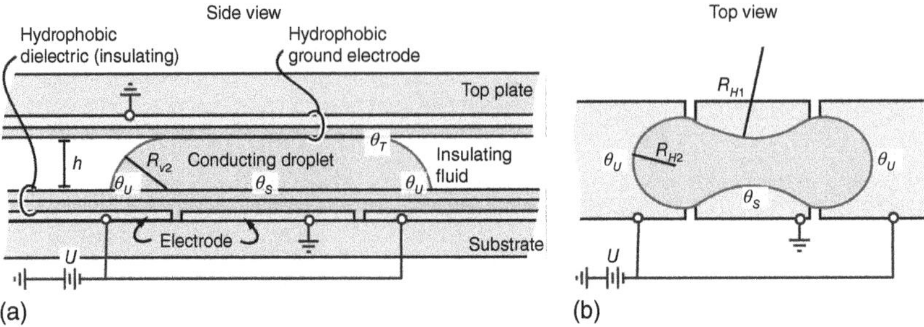

Figure 8.5 Side-view and top-view diagrams of basic parameters affecting droplet splitting. Note that θ_S is labeled in its proper location on the center electrode but is not visible along the cross-sectional axis used for visualization in this diagram. Adapted from: [7]. Reproduced with permission of Elsevier.

The basic splitting process increases the total surface area and therefore requires input energy to be completed. The total energy required is even greater than that, as during splitting the surface area is even larger than that of the two droplets after splitting. Fortunately in a microchannel the pressure holding the droplet is weakened, as the droplet is smashed between the plates, which increases the radius or curvature for the only unbounded surface (the horizontal surface). Figure 8.5 provides side-view and top-view diagrams of droplet splitting where competing forces can be easily seen.

8.3.2.1 Influence of Vertical Radii of Curvature

As potential is applied to the destination electrodes, the contact angle is decreased on those electrodes to θ_U as labeled in Figure 8.5a. Using Eqs. (8.2) and (8.3), this can be interpreted as a decrease in net pressure acting on the droplet surface due to an increased vertical radii of curvature R_{V2}. For simplicity in Figure 8.5a, θ_U is shown to be around 90°, which would result in a value for R_{V2} that is twice the vertical radius of curvature over the center electrode (R_{V1}, not visible in the diagram) because there is no potential applied and $\theta_T = \theta_S \sim 180°$. As a result based on vertical radii of curvature alone, there is a pressure imbalance that will split the droplet apart. However, recalling Eq. (8.1), all radii of curvature must be considered, and in our discussion we have neglected the horizontal radii or curvature that will oppose droplet splitting.

8.3.2.2 Influence of Horizontal Radii of Curvature

As visible in Figure 8.5b, in the horizontal plane the radii of curvature are not in favor of splitting. Curvature R_{H1} is concave (negative), resulting in pressure pulling the droplet surface outward, and curvature R_{H2} is convex (positive), resulting in an inward pressure. Both of these pressures are opposite of what is needed to split the droplet. However, recall that this system is in a microchannel and that the horizontal radii are much larger (weaker) than the vertical radii, and accordingly in Eq. (8.1) the vertical radii of curvature will dominate such that splitting can occur.

With the horizontal and vertical influences on pressure clearly in competition, a minimum criterion for splitting can be expressed as follows, as slightly modified from [7]:

$$\frac{1}{R_{H1}} - \frac{1}{R_{H2}} > -\frac{\cos\theta_U - \cos\theta_S}{h} \tag{8.9}$$

In this equation, the influence of horizontal radii of curvature (left side of the equation) is negative in value, and the influence of electrowetting (right side of the equation) is also negative in value and must be *more negative* such that electrowetting can cause the splitting to occur. cos θ_U can be taken directly from the electrowetting equation to bring in terms such as potential and capacitance. Notice how a smaller channel height h increases how negative the right-hand side of Eq. (8.9) is, which therefore makes splitting easier. This was noted in the beginning of our discussion where it was said that increased smashing of the droplet decreases h and makes splitting easier because it weakens the influence of horizontal radii of curvature (makes them larger).

8.3.3 Fundamentals of Droplet Dosing (Dispensing)

The dosing (dispensing) of a droplet into a microchannel is an extension of droplet splitting. Firstly, it is important to note that in conventional lab-on-chip device, for practical reasons of simplicity, only the bottom substrate is electrowetting (Figures 8.1 and 8.5). Therefore it is impossible to pull in by electrowetting the conducting fluid in from the external environment, because $\theta_T \sim 180°$and the vertical radii of curvature can therefore never be concave (negative pressure). In practice, mechanical pressure is used to inject a larger droplet into a reservoir (see the large droplet in Figure 8.6a). This droplet remains and in a ready position for dosing a portion of its volume whenever needed.

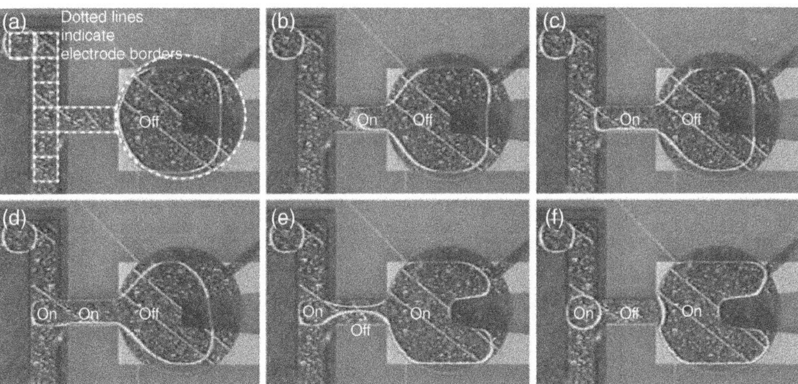

Figure 8.6 Top-view photographs of droplet dosing (dispensing) in a microchannel. The dotted lines in (a) show the locations of independently controllable electrodes, simply marked as *on* or *off* in terms of applied potential. The large droplet and reservoir is also confined by physical vertical side walls (transparent, not visible). Source: [36]. Reproduced with permission of Springer.

As shown in Figure 8.6, the basic process of dosing is similar to droplet splitting except that (i) an elongated liquid slug is created by electrowetting as shown in Figure 8.6c,d and (ii) the splitting electrode is larger. A larger splitting electrode in this example is used because it provides a larger split volume (longer liquid slug before splitting) and possibly because it reduces the voltage required for splitting (larger R_{H1}; see previous discussion).

8.3.4 Fundamentals of Droplet Mixing

Droplet mixing by transport or repeated splitting/merging will be discussed only briefly here, as it will be discussed in greater detail in the next section, and is simply a useful application of the internal flow field that we previously discussed. These complex internal flow patterns formed inside the droplet by forward motion can *unwind* when the motion is reversed. Therefore mixing is facilitated by the formation of complex flow patterns with multiple interfaces across which diffusion-based mixing occurs. Thus, mixing rate depends on the total interfacial area generated between flow patterns. This increase in interfacial area between flow patterns will be visually apparent and described in greater detail in the figures of the next section. Increase in interfacial area that allows increased diffusion is visually observable in Figure 8.4g. Notice the symmetry of the flow paths but how they become more diffuse with each reversal of the motion.

8.4 Stationary Droplet Oscillation, Jumping, and Mixing

This section pertains to the effects caused by a periodic change in the contact angle of a stationary droplet, which imparts nonequilibrium shapes onto the droplet meniscus. Applications include mixing, reducing contact angle hysteresis, and even the possibility to causing the droplet to jump free from a surface [21, 25]. A deeper theoretical treatment of oscillating droplets can be found in Section 6.4.

8.4.1 Droplet Oscillation

Just as the meniscus at the top of the water can create waves when the cup or table beneath the cup is vibrated, alternating (periodic) contact angle modulation can also impart waves of oscillation along the droplet meniscus. As can be seen in Figure 8.7, a simple step function of applying (Figure 8.7b) or removing (Figure 8.7d) the potential temporarily creates a nonspherical meniscus, which eventually due to damping forces settles back to the equilibrium spherical droplet profile. Now, if instead the contact angle modulation is quickly repeated in a periodic fashion (switch rapidly between Figure 8.7b and d), wave propagation across the meniscus can be sustained. Much like creating vibrations on a guitar string, the system vibrates along the meniscus surface and is bound where the meniscus touches the substrate surface, such that a finite number of vibration frequencies (and therefore wavelengths) exhibit resonance. For the case of weak

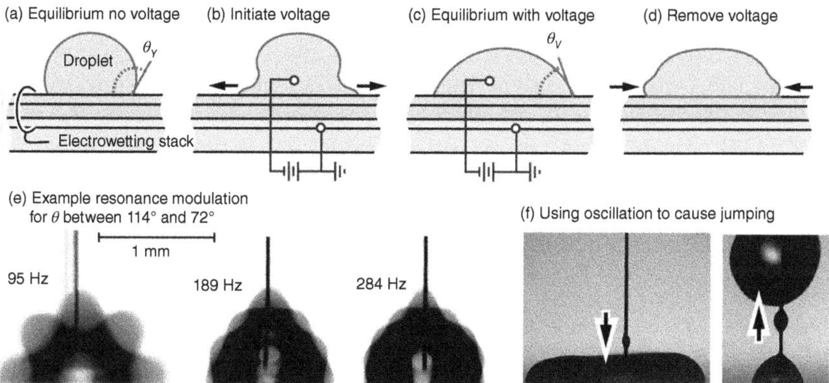

Figure 8.7 Diagrams (a–d) of droplet actuation and temporary nonequilibrium profiles. Photographs of (e) example resonance patterns induced by AC potential. Photographs of (f) resonance strong enough to cause jumping. Source: [25]. Photos adapted with permission of ACS.

viscous damping (e.g. water), the resonance frequencies (ω_n) for the nth mode can be predicted as [25]

$$\omega_n = \sqrt{n(n-1)(n+2)\frac{\gamma}{\rho r_0^3}} \tag{8.10}$$

where the common terms are density ρ, surface tension γ, and r_0, the volume-averaged radius of the droplet (see Section 6.4 for the derivation of this expression). Note that the resonant frequency (droplet) is twice the input frequency (potential) because the electrical force is proportional to the square of potential. At non-resonance frequencies a nonspherical meniscus is also often observable, but a single dominant wavelength of vibration is not observable. As the frequency is increased, damping forces diminish the amplitude of the vibration, and for high frequencies (>1 kHz) and typical droplet sizes (~1 mm base radius), the droplet appears spherical. Such a trend with increasing frequency is easily observable in the photographs of Figure 8.7e.

8.4.2 Droplet Oscillation and Jumping

With an adequately large oscillation, the droplet oscillation can become asymmetric, and the center of mass of the droplet can also begin to oscillate vertically. If the vertical oscillation becomes vigorous enough, the droplet can actually jump free of the substrate ([25], Figure 8.3g). Droplet jumping has potential application in fluid printing.

8.4.3 Droplet Oscillation and Hysteresis

Understanding droplet oscillation is important from several applied perspectives. Firstly, for any electrowetting application, where a device needs to move or rearrange a fluid as quickly as possible into a new equilibrium state, e.g. optics, the damping time for meniscus oscillation typically limits the effective switching speed. This topic will be discussed later in Section 8.8. Also, AC frequency operation can be used to reduce the apparent contact angle hysteresis (Eq. (7.7) and related discussion), an effect that resonance of oscillation will only increase [39]. Simply, a constantly moving contact line is driven by a localized force, and this localized force can easily be made greater than the forces causing pinning. For a sinusoidal voltage source, this time-dependent electrostatic force induced at the contact line can be expressed as

$$f(t) = \frac{cU^2}{2\gamma}(\sin \omega t)^2 \tag{8.11}$$

where ω is the angular frequency and U is the rms potential. As a result, the droplet can freely dewet or slide on a surface as though there were almost zero contact angle hysteresis. For water, frequencies of ~1 kHz can be utilized, which are high enough that a typical-sized droplet still appears spherical. This is another reason why if you must electrowet in air (which has higher hysteresis than in oil) one could again argue that there is no practical need for superhydrophobic surfaces (Section 7.5).

8.4.4 Droplet Oscillation and Mixing

Generally, there are three mechanisms that can effect mixing inside a stationary droplet: Marangoni transport, capillary wave-driven Stokes drift, and chaotic advection patterns [30] similar to those for droplet transport (Figure 8.4g). Marangoni transport relies on local joule heating of the conducting fluid at very highly applied potential frequencies (where the conducting fluid becomes electrically resistive), which further causes surface tension gradients and net flow from regions of low surface tension to high surface tension [16]. This type of mixing is generally impractical in applications such as lab-on-chip due to the very high frequencies required (>100 kHz). Chaotic advection patterns are only a weak influencer of stationary droplet mixing. We will therefore focus our discussion on capillary wave-driven Stokes drift, which can result in robust mixing and which can be implemented at more practical frequencies well below 1 kHz (Figure 8.8).

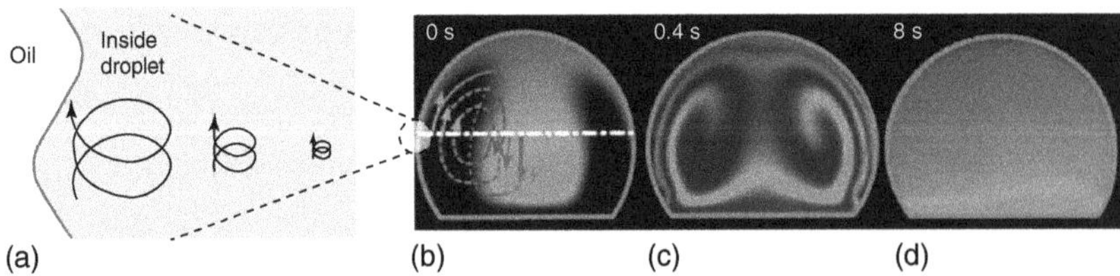

Figure 8.8 (a) Zoom-in diagram of particle trajectories due to a traveling capillary wave on the droplet surface, including a net flow due to Stokes drift. [24]. Reproduced with permission of IOP. (b–d) Photographs of dye mixing at 0, 0.4, and 8 s for 120 Hz sinusoidal potential. Each photograph was captured at the same phase of oscillation corresponding to $U(t) = 0$ V [22]. The experiment was performed in an oil ambient with $\theta_Y \sim 180°$.

Figure 8.8a shows a zoom-in diagram of a capillary wave at the droplet surface and the trajectories of particles (or fluid) inside the droplet [24]. The waves originate at the oscillating contact line, generating a Stokes drift upward along the droplet sides, which then forms a descending flow at the center of the droplet. This trajectory and mixing is visually obvious in the experimental photographs of Figure 8.8b–d (120 Hz potential applied in this experiment). The net flow due to Stokes drift is only possible due to the traveling wave that provides an asymmetry of the droplet shape between the spreading and receding phases of the droplet. Obviously, at very low frequencies (a few Hz or lower), a traveling wave is not sustained, and the Stokes drift diminishes. At very high frequencies (greater than several kHz), the wave becomes significantly damped, and again Stokes drift diminishes. Therefore, the ability to easily enable a strong Stokes drift in the range of tens to approximately hundreds hertz is important because it is practically very achievable (easily compatible conventional drive electronics used in electrowetting lab-on-chip). The time scale for this droplet mixing by oscillation is also similar (several seconds) to that for mixing by droplet transport (Figure 8.4g). However, in a microchannel the free meniscus area is very small (most of the surface is bound by the substrates; Figure 8.1), and mixing by droplet transport is therefore a more practical approach for lab-on-chip.

8.5 Gating, Valving, and Pumping

Electrowetting with droplets is commonly referred to as *digital microfluidics* because the fluids are moved in discrete volumes. It can also have application in gating, valving, and pumping of continuous flows of fluid, which is the dominant commercial paradigm for most microfluidic applications such as lab-on-chip [41].

8.5.1 Fundamentals

First, the concept of gating continuous pressure-driven fluid flow is discussed. There are several possible methods for using electrowetting and Laplace pressure to gate pressure-driven fluid flow. In any scenario, a local increase in Laplace pressure (Eq. (8.1)) can be utilized to halt fluid flow if the Laplace pressure is greater than the pressure driving the flow. Increases in Laplace pressure can be achieved by encountering an increased hydrophobicity of a surface (θ_S in Eq. (8.3)) or by encountering a reduced vertical dimension (h in Eq. (8.3)) or a reduced horizontal channel dimension. As shown in the example of Figure 8.9a, a reduced channel height stops the advancing flow of a conducting fluid under pressure, unless electrowetting is applied such that R_2 is increased to a value and that Laplace pressure is less than the pressure driving the fluid flow. The diagram and photographs of Figure 8.9 show water encountering such reduced channel height gate at $t = 0$. In this specific example of Figure 8.9a, the droplet is like that shown in Figure 8.1, and to move fluid into or past the gate, electrowetting is applied such that R_2 is less than R_1 (remember the trailing edge of the droplet has radius R_1 that can provide a driving pressure to move the droplet into or past the gate).

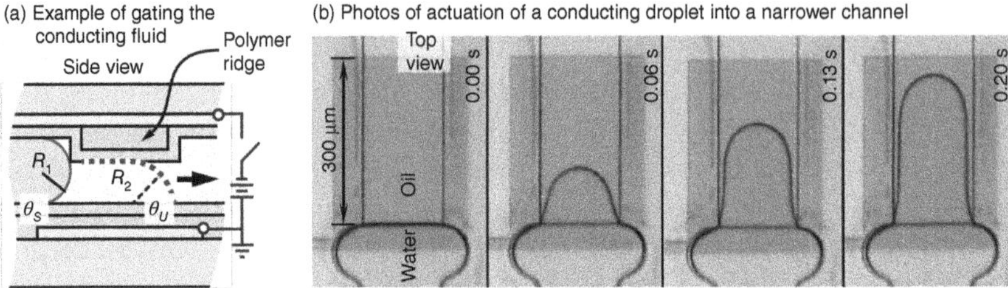

Figure 8.9 Diagrams (a) of an electrowetting gate based on Laplace pressure. Photos (b) of electrowetting actuation of water into a channel of smaller height, similar to that in (a). Potential was applied at $t = 0$ [4]. The film stack in (a) is a modification of that diagramed in Figure 8.1.

As demonstrated in Figure 8.9b, when potential is applied at $t > 0$, the water is then able to electrowet and advance [12]. However, a key requirement to enable an electrowetting gate that will allow continuous flow past the gate is that the electrowetting electrode extend beyond the end of the reduced channel height (otherwise advancement would terminate at the end of the electrowetting electrode where the system would face an even shorter channel height and therefore even higher Laplace pressure).

The specific example used for gating in Figure 8.9 was chosen because it can be extended to electrowetting-driven pumping as shown in Figure 8.10 [12]. This system also utilizes two channel heights: a channel of larger height h_1 and a channel of smaller height h_2. If potential is applied to an electrode, similar to Figure 8.9, conducting fluid can be pulled into the channel of reduced height. If potential is removed, Laplace pressure will then drive the conducting fluid out of the narrow height channel and back into the channel with greater height (and therefore lower Laplace pressure). As shown in Figure 8.10, the fluid that is pumped is the insulating fluid (oil), and the conducting fluid (water) is simply the pump. Potential is first applied to the topmost electrode (see dotted line), which closes off the top oil-inlet port of the pump. Next, the piston (middle electrode) is actuated, which pulls the water into the smaller channel of height h_2 and therefore displaces (pumps) oil toward the outlet of the pump. The lower valve is then actuated to seal the outlet port. Finally, with the lower outlet port sealed, potential is

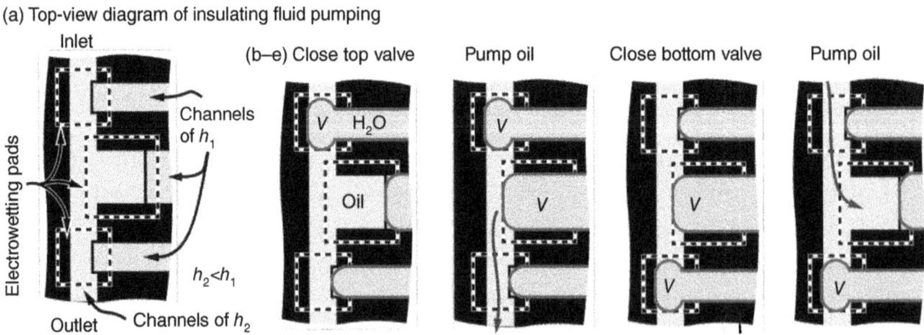

Figure 8.10 Diagrams of electrowetting-driven pumping of the insulating fluid (oil). Source: Adapted from [12].

removed from the piston pump (middle electrode), and the water recedes in the piston pump from the channel (h_2), which pulls oil into the pump at the inlet. The process may then repeat itself to continue to pump additional volumes of oil. Of greater interest is pumping water. The design for this is left as an exercise at the end of this chapter, but a clue to its operation would be to (i) use electrowetting to pull water from a channel of low Laplace pressure into a channel of higher Laplace pressure, (ii) split off that water from the source using splitting similar to that discussed in Section 8.2, and (iii) use electrowetting to gate that droplet into a new low Laplace pressure channel.

8.6 Generating Droplets and Channels

Microfluidic systems have been increasingly applied to study, prepare, and control two-phase fluid emulsions in various fields such as biology, medicine, and food science. Traditionally, droplet size has been controlled through flow rates, but electrowetting can also offer another degree of tunability over droplet size [9]. Furthermore, recently, electrowetting has been shown capable of creating *virtual channels* without solid sidewalls, opening up the possibility for programmable lab-on-chip with continuous flows.

8.6.1 Fundamentals for Droplet Generation

Conventional flow-focusing microfluidics can be used to create a laminar flow of two separate phases, with typically an oil being the continuous phase and a polar fluid such as water being the discontinuous phase. The discontinuous phase is generated as the flow stream of polar fluid experiences capillary breakup. The capillary breakup is a bit different than that for droplet splitting (Section 8.3) because the fluid capillary is typically flowing at such a rate that viscous stresses/shear forces also impact the breakup [34].

Electrowetting can be used to further control this breakup, as illustrated in traditional form (no electrowetting) in the leftmost photograph of Figure 8.11 [9]. Electrowetting has two visible effects shown in the middle and right-hand

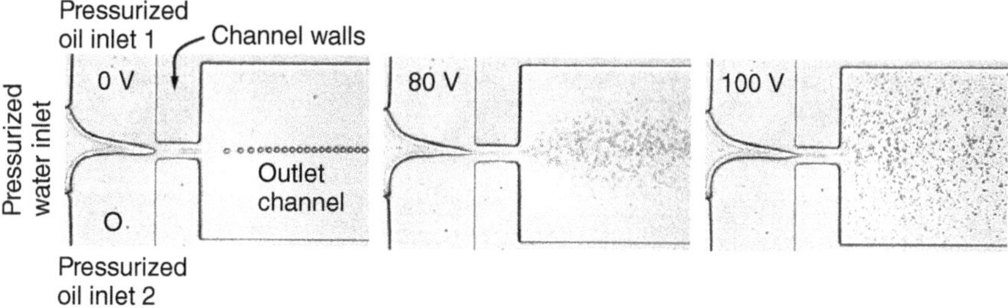

Figure 8.11 Top-view photographs of using potential to control water-in-oil emulsion droplet size. Source: [9]. Reproduced with permission of Royal Society of Chemistry. The entire region where droplets are generated is above an electrowetting surface, and the top plate acts as electrical ground, similar to the configuration of Figure 8.1.

photographs of Figure 8.11. Firstly, the droplets spread out as they enter the outlet channel. Immediately, this suggests self-repulsion, likely due to electrical charge. Secondly, the droplets become smaller in size. Both effects are consistent with the spontaneous generation of microdroplets from an electrowetted droplet as higher potentials cause the electrowetted meniscus to become highly unstable (see Figure 5.11b).

8.6.2 Fundamentals for Channel Generation

Conventional microfluidics utilizes solid sidewalls to confine channels, and conventional electrowetting lab-on-chip uses droplets and a channel that is open (unconfined) in the horizontal plane. Generally, it would be desirable to merge the capabilities of continuous flow/channel microfluidics and electrowetting control. Shown in Figure 8.12 is a top view of a horizontally open channel, filled with oil that has a two inlet electrodes, a central electrowetting channel electrode, and two inlet and two outlet electrodes, each independently controlled [1]. At $t = 0$, the upper inlet, center, and upper outlet electrodes are electrowetted, and a conducting fluid flows along this activated path. The conducting fluid only follows this path because it is back-pressurized with a syringe pump and electrowetting creates a path of reduced Laplace pressure compared with flowing into the surrounding channel. At $t = 4$ s, the upper outlet electrode is turned off, and the lower outlet electrode is electrowetted, switching the flow to a new outlet. In theory, these results could be extended to any number of adjacent electrodes, allowing sophisticated channel formation and routing so long as potential is constantly applied to maintain the channel (lower Laplace pressure, preventing capillary breakup). Creation of channels can also be achieved by having trenches with electrowetting sidewalls, which at low enough electrowetted contact angle exhibit adequate driving pressure for a conducting fluid to drive into the channels [43].

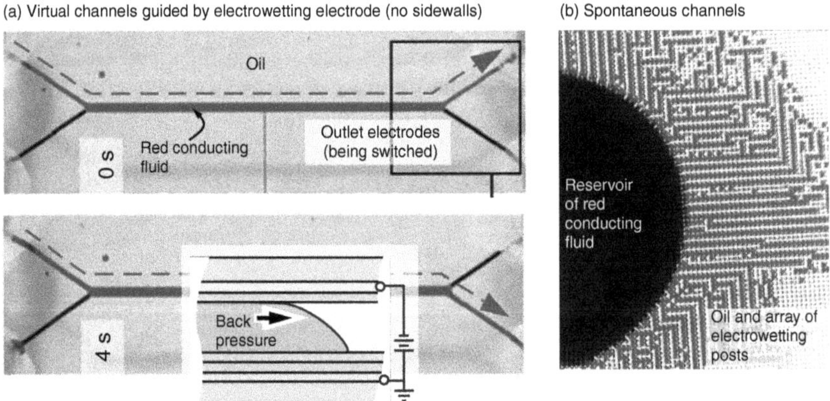

Figure 8.12 Top-view photographs of a red-pigmented conducting fluid (a) using electrowetting electrodes to direct flow of a conducting fluid. Source: [1]. Adapted with permission of Royal Society of Chemistry. (b) Using a porous media and oil to spontaneously create a complex arrangement of channels. Source: [8]. Adapted with permission of Royal Society of Chemistry.

Channels can also be formed without trenches or patterned electrodes, as shown in Figure 8.12b, which provides a top view of electrowetting a red conducting fluid into an electrowetting surface with tightly spaced electrowetting posts [8]. The channel formations form right angles because of the square lattice of posts. The channels are formed not because of a unique aspect of electrowetting, but rather because the posts create porous media and the oil becomes entrapped because it has viscosity similar to or higher than the rapidly advancing conducting fluid [44].

8.7 Shape Change in a Channel

The ability to change the shape of a droplet to a noncircular geometry inside a channel is desirable for efficient liquid storage or for creating channels (Section 8.6) in lab-on-chip but also for potential display applications [47].

8.7.1 Fundamentals

Shape change of fluids in some aspects is related to the discussion of the previous section, but here we cover it specifically for the case of a microchannel and no external stimulus other than electrowetting. As shown in Figure 8.13a, assume a conducting fluid is confined in a channel similar to that shown in Figure 8.1. The conducting fluid can be electrowetted to conform to the shape of the star-shaped electrode, and if potential is removed, the fluid will again minimize its surface energy (and Laplace pressure) by returning to a circular shape in the horizontal plane. The degree of shape change that can be imparted is only limited by Laplace pressure. As shown in Eq. (8.12), the pressure for the fluid with no potential is determined by R_V and R_H. Young's angle of 180°, height h, $R_V = h/2$, and R_H as the horizontal radius of the fluid all together determine the maximum Laplace pressure of the fluid. When electrowetted, the new pressure for the new fluid geometry *must* be less than the starting Laplace pressure with no potential (else there is no driving force for the fluid to change shape):

$$\left(\frac{1}{R_H}+\frac{1}{R_V}\right) \geq \left(\frac{1}{R_{H,min}}+\frac{1}{R_{V,ew}}\right) \rightarrow \left(\frac{1}{R_V}\right) \geq \left(\frac{1}{R_{H,min}}+\frac{1}{R_{V,ew}}\right) \tag{8.12}$$

With applied potential, the Laplace pressure is determined by the new $R_{V,ew}$ (which *increases*) due to electrowetting and the new $R_{H,min}$ (which *decreases* as the horizontal border of the droplet attempts to conform to the new shape). Shown at right in Figure 8.13b, a dotted line reveals the minimum attainable $R_{H,min}$ for that particular experiment and as can be predicted by Eq. (8.12). The photographs of Figure 8.13b also contain noticeable periodic features called Laplace barriers [47]. Unlike the diagram of Figure 8.13a, when potential is removed in Figure 8.8b, the star shape is partially preserved by the Laplace barriers. The Laplace barriers are simply portions of the channel with a reduced height (in Figure 8.13b, an array

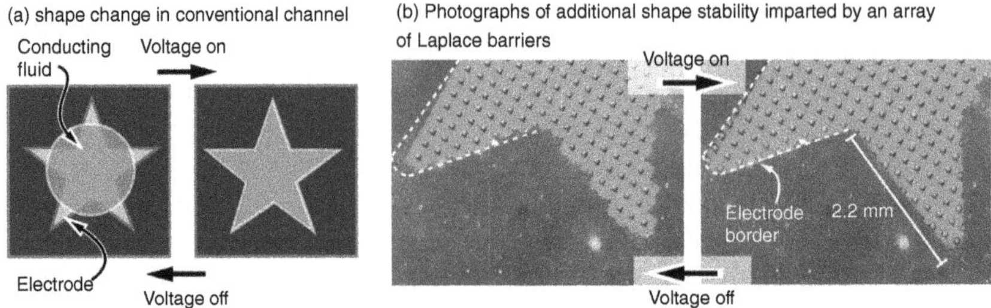

Figure 8.13 Shape change in microchannel, including (a) diagram of conventional operation and (b) photos of operation including the effects of Laplace barriers. Source: Adapted from Kreit 2011 [47].

of square posts). When properly designed, the Laplace barriers provide a gating effect similar to that described for Figure 8.9 and enable noncircular fluid shapes (stars, channels, text shapes) to be changed and maintained even without any applied potential.

8.8 Control of Meniscus Curvature

Controlling the curvature of a meniscus is dominantly utilized for optical applications [3, 15, 45].

8.8.1 Fundamentals

Controlling the curvature of a meniscus of a polar fluid is attractive from an applied standpoint if the meniscus itself imparts some physical effect. A particularly attractive feature of electrowetting is that with single-electrode control, the meniscus always exhibits a curvature that is constant (spherical) and at its limit is infinite in radius of curvature (flat). The pioneering work in modern electrowetting, performed by Berge and Peseux, quickly recognized that if the refractive index inside and outside the meniscus was different, then optical refraction could be achieved and adaptive optics such as switchable lenses could be fabricated [3]. The simplest way to achieve such meniscus shape-curvature control is that shown in Figure 8.14a. A more powerful approach is shown in Figure 8.14b, where the meniscus can be made convex or concave and, if the potential is set for contact angle of 90°, the meniscus can be made flat (infinite in radius). Therefore, for optical applications, various types of plano-concave/convex lenses can be realized for

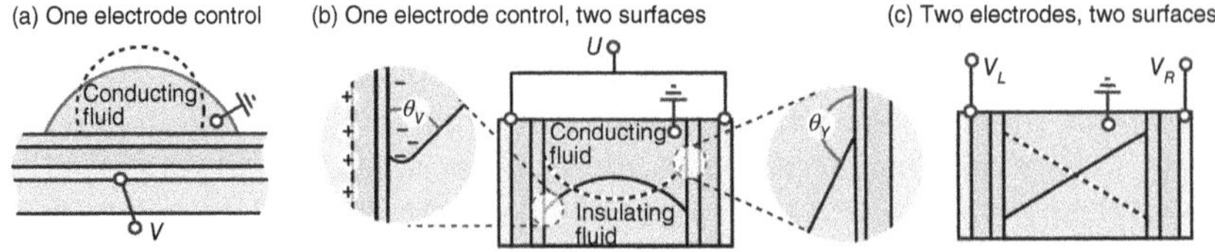

Figure 8.14 Diagrams of several levels of sophistication for meniscus curvature control.

focusing light. To do so, typically the conducting fluid is aqueous in part (due to low viscosity and low refractive index $n \sim 1.33$) and the insulating fluid being a low-viscosity oil with a high refractive index ($n \sim 1.5$ or more). As shown in Figure 8.14c, independent control of two or more electrodes allows the meniscus to be tilted, realizing a tunable optical prism for steering light. For a prism, the primary requirement for a flat meniscus is that the contact angles on the left and right sidewalls sum to 180°. Note that the device in Figure 8.14c shows two electrodes, but four electrodes are also possible, allowing 2D curvature control [48].

8.8.2 Additional Notes on Implementation

Electrowetting optics has several additional challenges. Firstly, the meniscus must maintain an optical quality curvature and not be strongly effected by vibration, and therefore the diameters of the devices are typically limited to several millimeters or less. The refractive index difference between the fluids must be maximized, and typically low viscosities are needed for fast switching times, both of which severely limit the choices of available fluids. Therefore typically the conducting fluid contains water (low viscosity, low refractive index) even though water is a difficult choice from a reliability perspective (see previous chapter). Lastly, switching speeds are typically important (e.g. fast focus), and meniscus vibrations and damping times must be carefully considered in design. Design of a critically damped system (fastest possible damping time) is practically not useful, because the effective spring constant for an electrowetting system changes with contact angle and numerous contact angles are typically needed for variable focusing.

8.9 Control of Meniscus Surface Area/Coverage

The previous section discussed controlling meniscus curvature, which obviously also changes the surface area. In this section, techniques for very large changes in meniscus surface area and coverage are presented, with applications being primarily in display devices [10, 38, 46].

8.9.1 Fundamentals

There are two general ways to control the surface area of the meniscus. The first is to view it from the perspective of the surface area of the conducting fluid. The conventional approach would be to perform this in air (Figure 8.14a), but in air, modulation of surface area is limited to a very narrow range as contact angle changes from 115° to 60–70°(with one exception being superspreading; see Chapter 9). A more widely used approach is to consider the change in surface area and coverage of the oil, particularly for the case where $\theta_Y = 180°$ and the oil forms a complete and continuous film. Recall from Chapter 7 that this requires $(\gamma_{id} - \gamma_{cd})/\gamma_{ci} \leq -1$ and can be achieved by selecting insulating fluids and dielectric surfaces that generally have similar surface tensions/energies (Figure 8.15a,

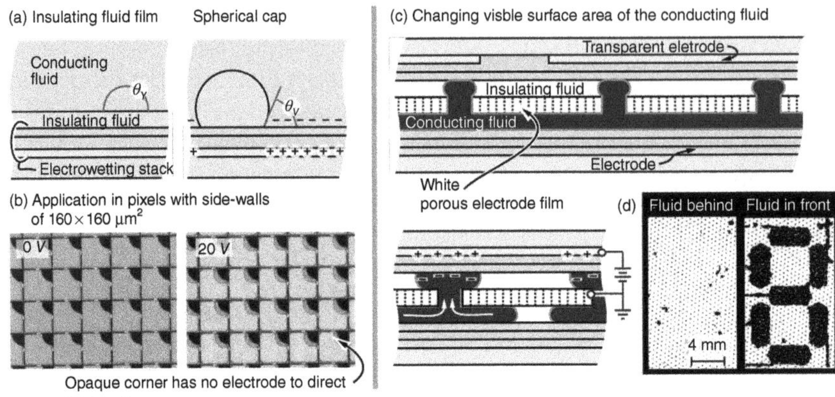

Figure 8.15 Changing the surface area/coverage through one of two methods: (a) changing the coverage of an oil film and (b) display application with dyed oil and (c) changing the coverage of a conducting fluid and (d) display application with pigment in the colored fluid.

Table 7.1). If electrowetting is applied, >100° of contact angle change can be typically achieved, reducing the oil coverage substantially (Figure 8.15a). Now, adding a bit of dye to the oil, and trapping it horizontally with a hydrophilic grid, creates the basis to make a switchable light valve (Figure 8.15b). The pixels in Figure 8.15b are from Philips (Liquavista) [38] and utilize and lower electrode that is reflective in order to make a reflective display device. This approach is attractive because of its simplicity of fabrication and operation but is limited in that surface area change requires constant application of potential and the colored oil cannot be fully hidden. A significantly different and somewhat more complex approach was therefore developed and is shown in Figure 8.15c,d [10]. In Figure 8.15c, the conducting fluid is now colored, and the insulating fluid is a clear oil. The device also includes a white porous film that has a ground electrode coating (not shown). The device of Figure 8.15c simply pulls by electrowetting a colored conducting fluid in front of (visible) or behind (invisible) the white film in order to alter its visible surface area/coverage in the upper channel (upper respective to the film). The pores in the film can be made small, increasing the maximum reflective brightness (area) compared with the approach of Figure 8.15b. Another attractive design feature is that the upper channel and lower channel can have equal channel heights, and as a result, the Laplace pressures in either channel are equal. Therefore various levels of conducting fluid coverage can be maintained in either channel without need for electrical power/potential.

8.9.2 Additional Notes on Implementation

This section on display devices required fluids that are colored. Coloring a conducting fluid is rather straightforward (many soluble dyes or pigment dispersions are satisfactory). Coloring an insulating fluid is far more challenging, as the dyes themselves must be highly nonpolar and free from conductive impurities to ensure maximum electrowetting modulation [46]. For most display applications, switching speed is critical; hence scaling the pixels to small dimensions is important but is also challenging as pixels approach <100 μm sizes.

8.10 Control of Film Breakup and Oil Entrapment

Oil film entrapment and oil film breakup are largely side effects of rapid device operation and have no known applications. However it appears often in electrowetting devices, because in most electrowetting systems oil is used to reduce contact angle hysteresis and/or minimize the effects of gravity. For reducing hysteresis the oil film need be no thicker than surface roughness (nanometers). Unfortunately, in practice fluid dynamics prevents such thin films. Upon fast motion of an advancing conducting fluid, oil films are entrapped and become hundreds of nanometers or even a μm thick (see Section 6.2, "Entrainment of Liquid Films"). At this point there is no choice but to deal with these oil layers, including their instability under an applied electric field. Such oil films impose nonideal effects on applications such as lab-on-chip [36]. Electrowetting-induced breakup of an oil film has also proven to be a novel scientific tool to understand dewetting phenomena in applications such as electrowetting displays [21, 32].

8.10.1 Fundamentals

A rapidly advancing contact line can entrap a thin film of insulating fluid (oil) underneath the electrowetted conductive fluid. Analogous to classical dip-coating thickness theory, oil entrapment is enhanced by the speed of the advancing contact line, and the local curvature of the contact line (e.g. locally always Young's angle; Figure 5.7) [49]. As shown in Figure 8.16a, oil entrapment is easily observable in lab-on-chip devices like those of Figure 8.1 [36]. The plot in Figure 8.16a is a measure of electrical capacitance on the destination electrode

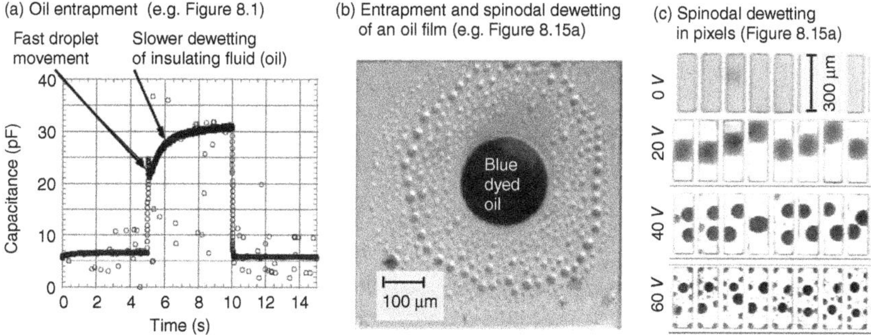

Figure 8.16 Oil film entrapment and dewetting: (a) capacitance measured at the destination electrode in lab-on-chip and (b,c) as related to electrowetting displays. (Source: (a) [36]. Adapted with permission of Springer. (b) [32]. Adapted with permission of IOP.)

(Figure 8.1). As the droplet moves quickly onto the destination electrode, it experiences a rapid increase in capacitance with the destination electrode (increase area of contact). This is the steep rise in capacitance marked *fast droplet movement* in Figure 8.16a. However, the advancing droplet also entraps an oil film that is insulating and therefore forms a series capacitance with the electrowetting dielectric (lower total capacitance). Not until this thin oil dewets over a period of seconds does the capacitance start to approach that of only the electrowetting dielectric stack (see Figure 8.16a, *slower dewetting of insulating fluid*). The capacitance again abruptly decreases as the droplet is moved onto the next electrode and leaves the electrode on which electrical capacitance is being measured. Images of this oil film breakup and theory specific to lab-on-chip applications can be found in other literature [13].

The dewetting of an oil film is an extension of same instabilities that result in nondeterministic breakup of a capillary (Chapter 1). Here instead, the starting geometry is not a perfectly round cylinder but rather a perfectly flat film. However, even in the absence of external stimuli, at room temperature such perfect flatness does not exist due to thermally induced fluctuations at the surface (refer back to Figure 6.10 and related discussion).

The physics of film breakup is even more interesting under the influence of electrowetting that provides an externally applied and stronger driving force for the breakup [49]. Just like traditional capillary or film breakup, under the influence of electrowetting, an oil film will exhibit a periodic breakup into a pattern of oil droplets, and the droplet spacing can be predicted by the fastest-growing surface wavelength (thermally induced fluctuations). This periodic oil dewetting can be referred to as form of spinodal dewetting, i.e. a linear instability, that is dominated by electrostatic forces (not intermolecular van der Waals forces). Under electromechanical pressure the many possible dewetting wavelengths for a flat oil film grow at differing exponential rates. During the dewetting, a single mode, or surface wavelength (λ), can dominate according to the following approximated equation [32]:

$$\lambda = 2\pi/k \tag{8.13}$$

$$k^2 \cong \frac{A}{4\pi \cdot d_{oil}^4 \cdot \gamma_{ci}} + \frac{\varepsilon \cdot U^2}{2 \cdot d^3 \cdot \gamma_{ci}} \tag{8.14}$$

where v is the wavenumber of the fastest-growing oil film dewetting mode, A is the Hamaker constant for the oil film, U is the sum of applied potential and the work function difference between the electrode and water, and d and ε are the total equivalent thickness and permittivity of the series capacitance of the oil film (ε_{oil}, d_{oil}) and the electrowetting dielectric (ε_{hyd}, d_{hyd}). For typical electrowetting systems $[A/(4\pi \cdot d_{oil}^4 \cdot \gamma_{ci})]$can be neglected in calculating λ since the portion of $[\varepsilon \cdot U^2/(2 \cdot d^3 \cdot \gamma_{ci})]$ dominates the dewetting (i.e. electromechanical forces far exceed intermolecular van der Waals forces). Equations (8.13) and (8.14) assume that the entrapped oil film limits the initial oil/dielectric series capacitance.

Of the many applications for electrowetting, further examination of spinodal dewetting could be suggested as most important for display devices (Figure 8.15a). In electrowetting displays, an electrostatically induced breakup of a colored oil film is utilized to alter the perceived coloration of a reflective or transmissive pixel.

This oil entrapment and dewetting is easily observable in Figure 8.16b, where an abrupt potential was applied to dewet an oil film similar to that shown at the top of Figure 8.15a. The photo in Figure 8.16b is after entrapment of the blue-dyed oil and after the entrapped oil film is broken up. The largest droplet sizes exist at a middle radius, most likely because the advancing contact line speed reached its maximum at this point and entrapped the thickest oil film (Eq. (8.14)).

Oil film breakup also tracks with theory inside electrowetting pixels, so long as the dominant wavelength for dewetting is of course smaller than the pixel dimension. As shown in Figure 8.16c, the higher the abrupt potential applied to an electrowetting pixel, the smaller the oil droplets generated (Eqs. (8.13) and (8.14)). Smaller droplets are less desirable, as they occupy more visible area due to decreasing ratio of droplet volume to area.

8.11 1D, 2D, and 3D Control of Rigid Objects

This section pertains utilizing electrowetting to provide motion or geometrical control onto solid objects, with applications including microgripping, micromotors, and tiltable mirrors for beam steering [11, 17, 33, 35]. This section will not focus on carrying along particles or solid items dispersed or stabilized inside a moving electrowetting droplet.

8.11.1 Fundamentals

Electrowetting imparts a force, which also in some environments such as channels can be interpreted as a change in pressure. Connecting an electrowetting fluid to a solid object can then be used to translate the electrowetting force to the solid object, so long as such force does not exceed the force per unit length (surface tension) of the electrowetted fluid.

For example, as shown in Figure 8.17A, a microgripper utilizes liquid droplet to pick up and release a micro object [35]. Lifting force is evolved from a liquid bridge formed between the gripper surface and an object. Electrowetting is used to dynamically change the capillary lifting forces and enable object release. The capillary force can be predicted as

$$F_{capillary} = \frac{\gamma_{ci}(\cos\theta_S + \cos\theta_{object})A}{d} \tag{8.15}$$

where A is the shared area between the object (a bead in Figure 8.17A) and the electrowetting substrate, d is the smallest separation distance between object and substrate, θ_S is the contact angle on the substrate that can be reduced by electrowetting, and θ_{object} is the contact angle on the object. In Figure 8.17A various potentials and resulting maximum gripping forces are shown for holding of a 13.9 mg (136 μN) glass bead. The 138 μN photo was the limit observed for holding the bead. If the potential were further reduced or removed, the glass bead would

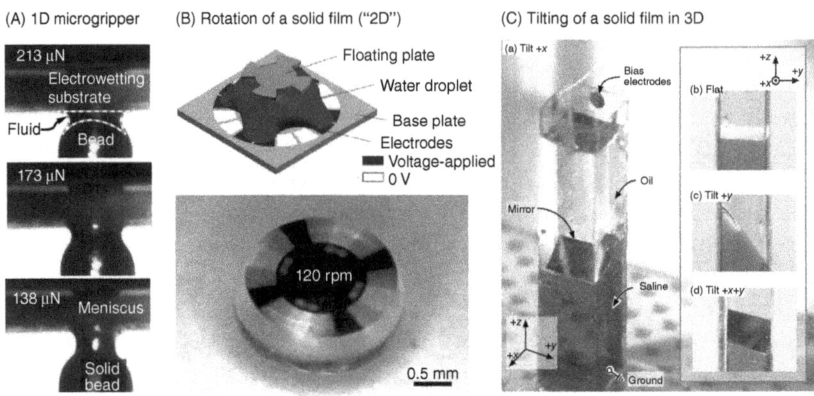

Figure 8.17 Manipulation of solid objects in (A) 1D, (B) 2D, and (C) 3D using electrowetting control [11, 17, 33, 35].

be dropped (not shown). Maximum lifting forces in this particular experiment could be as high as 213 μN at a driving potential of 28 V.

Extending electrowetting control to a 2D array of electrodes, a micromotor can be demonstrated [17, 33]. As shown in Figure 8.17B, the droplet's boundary conditions at the bottom plate and the floating plate are noncircular, and therefore capillary torque is exerted on a film suspended on the droplet. By changing the boundary conditions continuously with electrowetting, rotational motion of a plate was achieved. A 3.0 mL water droplet between two plates could achieve up to a maximum of 720 rpm.

Lastly, objects can be tilted/moved in 3D as well, as shown in Figure 8.17C [11]. A square channel was constructed with four independently controllable electrowetting sidewalls (using transparent electrodes also). A mirror film was suspended between the oil/saline meniscus by coating one side of the mirror film hydrophobic (wetted by the oil) and plasma-treating the other side of the film hydrophilic (wetted by the water). Electrowetting at each sidewall produced a saline contact angle change of 35° to 170°, which tilted the mirror and was used to achieve >200° of laser-beam deflection.

These are just several examples of the possible ways in which electrowetting can be used to control or move rigid objects. Further examples are not provided (but do exist), and the examples shown in Figure 8.17 are adequate to stimulate creative design of new concepts.

8.12 Reverse Electrowetting and Energy Harvesting

In a conventional electrowetting system, external potential is applied, which results in current flowing into the conducting fluid in order to provide the required net electrical charge accumulation near the electrowetting surface. Imagine a system instead where a conducting fluid was electrically charged and, instead using mechanical pressure or other suitable techniques, the liquid was forced to dewet from the electrowetting surface. Where is the charge to go? Could it be harvested by an external circuit?

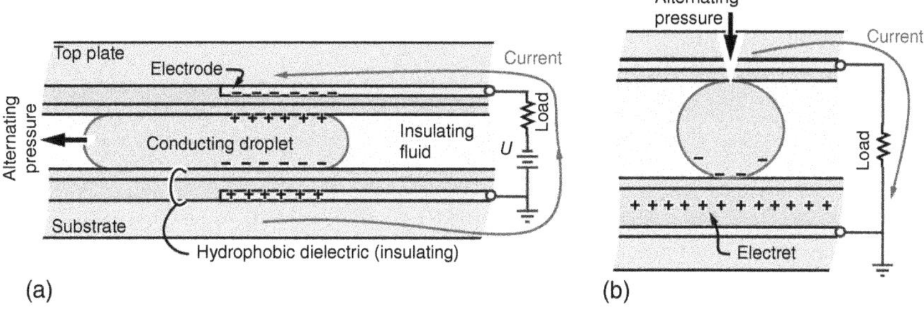

Figure 8.18 Energy harvesting using (a) a moving conducting fluid and an external potential and (b) a conducting fluid with increasing contact area and an internal potential (electret).

The above described scenario can be used for energy harvesting and is referred to as reverse electrowetting in a most recent demonstration by Krupenkin and Taylor [14]. For those who may explore this concept and publish on it, to mitigate missed references in Krupenkin's paper, it should be noted that only the term *reverse electrowetting* is new, as there are numerous previous publications showing the same fundamental concept [6, 37]. In fact, in Boland's work they even refer to electrowetting in the manuscript.

A *reverse electrowetting* system is shown in Figure 8.18. In Figure 8.18a, an external potential provides an initial charge to an electrowetting system. As the conducting fluid is removed, this charge is returned to the battery, but not without first traveling through a load (resistor) and therefore performing work (dissipating energy). The first-order calculations for the amount of energy harvested in a single cycle are quite simple, as the initial charge provided can be calculated by $Q = cU$. As the conducting fluid is moved, the area of the capacitance c decreases, and eventually, nearly the entire charge Q must flow back through the resistor to the battery. This process can be repeated by reversing the flow of the conducting fluid periodically or by passing multiple conducting fluid droplets through the channel [14]. If the battery were perfectly able to reabsorb the charge without loss (like a simple charged capacitor), then the only power dissipation occurs in the resistance of the system. If the load is the largest resistance, it will dominate the power dissipation.

It is also possible to operate the system with not an external battery as shown in Figure 8.18a, but using an internal battery (potential) as shown in Figure 8.18b. Such internal batteries are often referred to as electrets. Electrets are any insulating material that exhibits a net electrical charge (e.g. charged trapped inside an insulator) or dipole moment (e.g. piezoelectric materials under strain). It is quite interesting but not surprising that those exploring electret materials use many of the same insulating Parylene and fluoropolymer materials used in electrowetting [19], and one might recall the significant Teflon charging in water results of Figure 7.2. For a typical electret, once the charge is stored by ion implantation, by electrical discharge, or by controlled dielectric breakdown, you want the charge to stay in place, similar to how you want charge to stay outside a dielectric in an electrowetting system. For the device shown in Figure 8.18b, the top substrate is lowered with pressure, which will increase the charge Q as the electrical capacitance c increases due to increase contact area between the droplet and the dielectric surface. Again, this system could be periodically reversed, passing charge through the load.

Problems

8.1 Consider a 1 mm diameter $70\,\text{mN}\,\text{m}^{-1}$ droplet in a ~100 μm high microchannel. (a) Calculate the total Laplace pressure inside the droplet for the case of Young's angles of 110° on the top and bottom plates. You should note how the Laplace pressure is dominated by the vertically oriented radius of curvature. (b) Assume one edge of the droplet is adjacent to an electrowetting surface (similar to Figure 8.1) and potential is applied and reduces the contact angle on the bottom plate to 70°. Calculate the pressure created by electrowetting to drive the droplet onto the electrowetting surface. (c) Lastly, assume a contact angle hysteresis of 10° (as droplet is moved, the advancing contact angle *sticks* to 5° less than Young's angle and the receding contact angle *sticks* to 5° more than Young's angle). Calculate the minimum potential needed to move the droplet in this case by electrowetting.

8.2 The world-chip interface is always a challenge for lab-on-chip applications. Getting fluid into the chip itself can be a challenge. In most chips, only the bottom plate is electrowetting (like Figure 8.1) and oil is used (180° Young's angle). Describe what is needed to get fluid into the channel from the outside.

8.3 A water droplet on an electrowetting surface has a 90° Young's angle and a 1 mm base radius (similar to the experiments of Figure 8.7e). Calculate the theoretical resonant frequencies between 50 and 500 Hz, and identify which of your calculations are likely those shown in Figure 8.7e.

8.4 An easy problem... For Figure 8.10a, assume oil (180° Young's angle) and the taller channel height is 100 μm and the smaller channel height is 50 μm. Determine the electrowetted contact angle that is at the threshold for moving the conducting fluid into the smaller channel. No calculations are allowed, and you must draw and visually determine the radii of curvatures needed and therefore the contact angle needed.

8.5 For the pump of Figure 8.10, assume the maximum speed at which the conducting fluid can be moved is $2\,\text{cm}\,\text{s}^{-1}$ (both advancing and retracting). Design a device layout and calculate the time-averaged pumping rate ($\mu\text{l}\,\text{min}^{-1}$) that can be achieved for your pump. In your design, assume a channel height that is 100 μm and that for any conducting fluid the maximum horizontal radius of curvature no more than 10× greater than the maximum vertical radius of curvature. This radius of curvature requirement is generally a good design requirement if the fluid is to maintain its geometrical integrity during actuation (if the horizontal of radius is too large, its contribution to Laplace pressure can be so weak that the conducting fluid can split if it is rapidly retracted).

8.6 Design a system to pump the conducting fluid, not the oil (in Figure 8.10, only oil is pumped). Note that oil is a *continuous* phase in Figure 8.4 and it is never completely split even when the valves are pinched off (the conducting fluid will not wet into all the corners of the channel, try to visualize this). In your design, the water will need to be split/merged.

8.7 Electrowetting lenses and prisms are limited in their ability to steer light. Assume a laser beam is incident on the bottom of the electrowetting prism shown in Figure 8.14 (incident perpendicular to the lower surface of the prism device). Assume the conducting fluid is water (refractive index $n \sim 1.33$) and can be electrowetted down to 60° minimum. Assume oil (180° Young's angle) and that it is a silicone oil with $n \sim 1.58$. Using Snell's law (twice) calculate (a) the maximum refracted angle possible as light moves from the oil into the water and (b) the maximum refracted angle as that same refracted beam of light then exits from the water into air outside the device.

8.8 How much more optically efficient is an electrowetting light valve (display) than a liquid crystal display? Liquid crystal displays utilize polarizers, which limits the maximum transmission of a pixel to 40% at best. With reference to Figure 8.15a, assume an oil film is dyed black and is spread 4 μm thick (180° contact angle for the conducting fluid), a thickness adequate to block light transmission. Next assume you electrowet a conducting fluid reliably to 90°. Using simple geometrical formulas, calculate the change in coverage of black oil, and determine if an electrowetting light valve is more or less efficient than the maximum efficiency of a liquid crystal light valve.

References

1 Banerjee, A., Kreit, E., Liu, Y. et al. (2012). Reconfigurable virtual electrowetting channels. *Lab Chip* 12 (4): 758–764. http://xlink.rsc.org/?DOI=C2LC20842C.

2 Banerjee, A., Liu, Y., Heikenfeld, J., and Papautsky, I. (2012). Deterministic splitting of fluid volumes in electrowetting microfluidics. *Lab Chip* 12 (24): 5138–5141. http://xlink.rsc.org/?DOI=c2lc40723j.

3 Berge, B. and Peseux, J. (2000). Variable focal lens controlled by an external voltage: an application of electrowetting. *Eur. Phys. J. E* 3 (2): 159–163. http://link.springer.com/10.1007/s101890070029.

4 Berry S, Kedzierski J. New methods to transport fluids in micro-sized devices. *Lincoln Lab. J.* 2008;17(2). http://scholar.google.com/scholar?hl=en&btnG=Search&q=intitle:New+Methods+to+Transport+Fluids+in+Micro-Sized+Devices#0

5 Berthier J. *Microdrops and digital microfluidics*. William Andrew, Norwich, NY. 2008. 434 p. http://scholar.google.com/scholar?hl=en&btnG=Search&q=intitle:MICRODROPS+AND+DIGITAL+MICROFLUIDICS#0

6 Boland, J. S., Messenger, J. D. M., Lo, H. W., and Tai, Y. C. (2005). Arrayed liquid rotor electret power generator systems. In: *18th IEEE International Conference on Micro Electro Mechanical Systems, 2005*, 618–621. MEMS.

7 Cho, S. K., Moon, H., Fowler, J., and Kim, C.-J. (2001). Splitting a liquid droplet for electrowetting-based microfluidics. In: *Proceedings of the ASME International Mechanical Engineering Congress and Exposition*, 2781–2787. ASME https://www.scopus.com/inward/record.uri?eid=2-s2.0-1542411515&partnerID=40&md5=a278f825df090f7c1df70bfdb2b82a6c.

8 Dhindsa, M., Heikenfeld, J., Kwon, S. et al. (2010). Virtual electrowetting channels: electronic liquid transport with continuous channel functionality. *Lab Chip* 10 (7): 832–836. http://xlink.rsc.org/?DOI=b925278a.

9 Gu, H., Duits, M. H. G., and Mugele, F. (2010). A hybrid microfluidic chip with electrowetting functionality using ultraviolet (UV)-curable polymer. *Lab Chip* 10 (12): 1550–1556. http://xlink.rsc.org/?DOI=c001524e.

10 Hagedon, M., Yang, S., Russell, A., and Heikenfeld, J. (2012). Bright e-Paper by transport of ink through a white electrofluidic imaging film. *Nat. Commun.* 3.

11 Hou, L., Smith, N. R., and Heikenfeld, J. (2007). Electrowetting manipulation of any optical film. *Appl. Phys. Lett.* 90 (25): 251114.

12 Kedzierski, J., Berry, S., and Abedian, B. (2009). New generation of digital microfluidic devices. *J. Microelectromech. Syst.* 18 (4): 845–851.

13 Kleinert, J., Srinivasan, V., Rival, A. et al. (2015). The dynamics and stability of lubricating oil films during droplet transport by electrowetting in microfluidic devices. *Biomicrofluidics* 9 (3): 034104.

14 Krupenkin, T. and Taylor, J. A. (2011). Reverse electrowetting as a new approach to high-power energy harvesting. *Nat. Commun.* 2 (1).

15 Kuiper, S. and Hendriks, B. H. W. (2004). Variable-focus liquid lens for miniature cameras. *Appl. Phys. Lett.* 85 (7): 1128–1130.

16 Lee, H., Yun, S., Ko, S. H., and Kang, K. H. (2009). An electrohydrodynamic flow in ac electrowetting. *Biomicrofluidics* 3 (4): 044113.

17 Lee, J. L. J. and Kim, C.-J. C. (1998). Liquid micromotor driven by continuous electrowetting (Cat No98CH36176). In: *Proceeding of MEMS 98 IEEE Elev Annu Int Work Micro Electro Mech Syst An Investig Micro Struct Sensors, Actuators, Mach Syst*, 538–543.

18 Li, X., Tian, H., Wang, C. et al. (2014). Electrowetting assisted air detrapping in transfer micromolding for difficult-to-mold microstructures. *ACS Appl. Mater. Interfaces* 6 (15): 12737–12743.

19 Lo, H. W. and Tai, Y. C. (2008). Parylene-based electret power generators. *J. Micromech. Microeng.* 18 (10): 104006.

20 Lu, H.-W., Bottausci, F., Fowler, J. D. et al. (2008). A study of EWOD-driven droplets by PIV investigation. *Lab Chip* 8 (3): 456–461. http://xlink.rsc.org/?DOI=b717141b.

21 Mugele, F., Baret, J.-C., and Steinhauser, D. (2006). Microfluidic mixing through electrowetting-induced droplet oscillations. *Appl. Phys. Lett.* 88 (20): 204106.

22 Mugele, F., Staicu, A., Bakker, R., and van den Ende, D. (2011). Capillary Stokes drift: a new driving mechanism for mixing in AC-electrowetting. *Lab Chip* 11 (12): 2011. http://xlink.rsc.org/?DOI=c0lc00702a.

23 Ni, Q., Capecci, D. E., and Crane, N. B. (2015). Electrowetting force and velocity dependence on fluid surface energy. *Microfluid. Nanofluid.* 19 (1): 181–189. http://link.springer.com/10.1007/s10404-015-1563-7.

24 Oh, J. M., Legendre, D., and Mugele, F. (2012). Shaken not stirred – on internal flow patterns in oscillating sessile drops. *Europhys. Lett.* 98 (3): 34003.

25 Oh, J. M., Ko, S. H., and Kang, K. H. (2008). Shape oscillation of a drop in ac electrowetting. *Langmuir* 24 (15): 8379–8386.

26 Paik, P., Pamula, V. K., Pollack, M. G., and Fair, R. B. (2003). Electrowetting-based droplet mixers for microfluidic systems. *Lab Chip* 3 (1): 28–33.

27 Paik, P., Pamula, V. K., and Fair, R. B. (2003). Rapid droplet mixers for digital microfluidic systems. *Lab Chip* 3 (4): 253–259. http://xlink.rsc.org/?DOI=B307628H.

28 Pollack, M. G., Shenderov, A. D., and Fair, R. B. (2002). Electrowetting-based actuation of droplets for integrated microfluidics. *Lab Chip* 2 (2): 96.

29 Pollack, M. G., Fair, R. B., and Shenderov, A. D. (2000). Electrowetting-based actuation of liquid droplets for microfluidic applications. *Appl. Phys. Lett.* 77 (11): 1725–1726.

30 Song, H., Bringer, M. R., Tice, J. D. et al. (2003). Experimental test of scaling of mixing by chaotic advection in droplets moving through microfluidic channels. *Appl. Phys. Lett.* 83 (22): 4664–4666.

31 Song, J. H., Evans, R., Lin, Y.-Y. et al. (2009). A scaling model for electrowetting-on-dielectric microfluidic actuators. *Microfluid. Nanofluid.* 7 (1): 75–89.

32 Sun, B. and Heikenfeld, J. (2008). Observation and optical implications of oil dewetting patterns in electrowetting displays. *J. Micromech. Microeng.* 18 (2): 025027.

33 Takei, A., Matsumoto, K., and Shomoyama, I. (2010). Capillary motor driven by electrowetting. *Lab Chip* 10 (14): 1781–1786. http://xlink.rsc.org/?DOI=c001211d.

34 Tan, Y. C., Cristini, V., and Lee, A. P. (2006). Monodispersed microfluidic droplet generation by shear focusing microfluidic device. *Sens. Actuators, B* 114 (1): 350–356.

35 Vasudev, A. and Zhe, J. (2009). A low voltage capillary microgripper using electrowetting. In: *TRANSDUCERS 2009 - 15th International Conference on Solid-State Sensors, Actuators and Microsystems*, 825–828.

36 Fair, R. B. (2007). Digital microfluidics: is a true lab-on-a-chip possible? *Microfluid. Nanofluid.* 3: 245–281.

37 Boland, J., Messenger, J.D.M., and Tai, Y.C. (2004). Alternative designs of liquid rotor electret power generator systems. In: The Fourth International

Workshop on Micro and Nanotechnology for Power Generation and Energy Conversion Applications, PowerMEMS, November 28–30 2004, Kyoto, Japan.

38 Hayes, R. A. and Feenstra, B. J. (2003). Video-speed electronic paper based on electrowetting. *Nature* 425: 383–385.

39 Li, F. and Mugele, F. (2008). How to make sticky surfaces slippery: Contact angle hysteresis in electrowetting with alternating voltage. *Appl. Phys. Lett.* 92: 244108.

40 Pollack, M.G. (2001). Electrowetting-based microactuation of droplets for digital microfluidics. PhD Thesis, Duke University.

41 Berthier, J. (2013). *Micro-Drops and Digital Microfluidics*, 2e. William Andrew, p. 560.

42 Sung Kwon Cho, Hyejin Moon, and Chang-Jin Kim (2003). Creating, transporting, cutting, and merging liquid droplets by electrowetting-based actuation for digital microfluidic circuits. *J. Microelectromech. Syst.* 12 (1): 70–80.

43 Jean-Christophe Baret and Michel M.J. Decré (2005). Transport dynamics in open microfluidic grooves. *Langmuir* 23 (9): 5200–5204.

44 Kreit, E., Dhindsa, M., Yang, S., Hagedon, M., Zhou, K., Papautsky, I. and Heikenfeld, J. (2011). Laplace barriers for electrowetting thresholding and virtual fluid confinement. *Langmuir* 26 (23): 18550–18556.

45 Neil R. Smith, Don C. Abeysinghe, Joseph W. Haus and Jason Heikenfeld (2006). Agile wide-angle beam steering with electrowetting microprisms. *Opt. Express* 14 (14): 6557–6563.

46 Heikenfeld, J., Zhou, K., Kreit, E., Raj, B., Yang, S., Sun, B., Milarcik, A., Clapp, L. and Schwartz, R. (2009). Electrofluidic displays using Young-Laplace transposition of brilliant pigment dispersions. *Nat. Photon.* 3: 292–296.

47 Kreit, E., Mognetti, B.M., Yeomans, J.M. and Heikenfeld, J. (2011). Partial-post laplace barriers for virtual confinement, stable displacement, and $>5\,\text{cm}\,\text{s}^{-1}$ electrowetting transport. *Lab Chip* 11: 4221–4227.

48 Hou, L., Zhang, J., Smith, N., Yang, J. and Heikenfeld, J. (2009). A full description of a scalable microfabrication process for arrayed electrowetting microprisms. *J. Micromech. Microeng.* 20 (1).

49 Adrian Staicu and Frieder Mugele (2006). Electrowetting-induced oil film entrapment and instability. *Phys. Rev. Lett.* 97: 167801.

9

Related and Emerging Topics

9.1 Introduction and Scope

This chapter briefly reviews several emergent topics. Several of these emergent topics are classical topics such as electrocapillarity and dielectrophoresis, but are being revisited in a new and interesting context. The last topic is new and involves geometrical control of the interface between conducting and insulating fluids by electronic feedback control, without any contact angle change or even contact angle (wetting) formation on a substrate. This chapter will not attempt to provide an exhaustive theoretical treatment of such topics or provide full details on fabrication and construction. Rather, this chapter aims to provide the reader gain an appreciation of broader physics and applications that are a natural extension from conventional electrowetting. Lastly, these topics do not represent a complete set of all possible newly emergent topics and are just a small sampling.

9.2 Dielectrophoresis and Dielectrowetting

Dielectrowetting is a special subtype of dielectrophoresis, which has a voltage-squared response similar to electrowetting (hence the name, coined by McHale and Brown; [7, 8]). Therefore, a brief description of dielectrophoresis should be provided first. Dielectrophoresis of a fluid occurs when the liquid is adequately insulating such that a significant nonuniform electric field exists across the body of the fluid and therefore exerts a body force toward regions of higher electrical field. As discussed in Chapter 2 (see end of Section 2.2), perfect dielectric response and perfect conductive response are two limits of the general *leaky dielectric* behavior of fluids in which direct ohmic and polarization currents occur in parallel. Pure electrowetting corresponds to the extreme case of perfect conduction where screening charges lead to a perfect localization of electromechanical forces at the liquid surface.

9.2.1 Basic Dielectrophoresis

Shown in Figure 9.1a is an object that is polarized in a uniform electric field. As with any dielectric material, polarization gives rise to opposite but equal

Electrowetting: Fundamental Principles and Practical Applications,
First Edition. Frieder Mugele and Jason Heikenfeld.

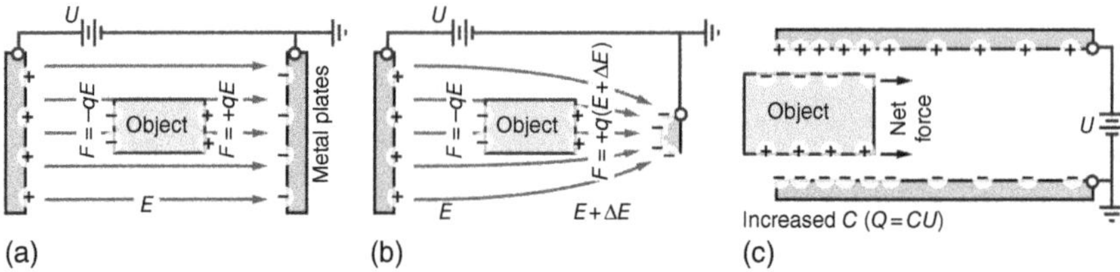

Figure 9.1 Basic models of dielectrophoresis. (a) An object with no net force in uniform electric field, (b) an object with a net force in a nonuniform electric field, and (c) an object that creates its own nonuniform field and resulting force as it enters into a uniform electric field.

charges on opposite ends of the object. The object could be a molecule, a solid, a fluid, etc. At equilibrium, the object is polarized, but there is no net force on object. We assume the object is a dielectric, but this case is accurate also for an object that is only semi-insulating, and the applied potential is alternating. For semi-insulating materials, at adequately high frequency, the electrical impedance of semi-insulating material would appear to be dominantly capacitive (an insulator) because charge flow within the material is not fast enough to respond to the electric field (too resistive), and instead the instantaneous effect of dielectric polarization dominates. Next, consider the case shown in Figure 9.1b, where the electric field is nonuniform and is stronger by a factor ΔE at the right side. Although the charges on the object remain opposite but equal, the portion of the object in the area of higher electric field obviously experiences a stronger Coulombic force $Q(E+\Delta E)$. The resulting force imbalance $Q\Delta E$ will move the object toward the smaller electrode plate. Now, the object also has higher permittivity than vacuum or air, and therefore as it moves, the total electrical capacitance of the system increases (more of the E-field will be through a higher electrical permittivity material). Therefore the work done on moving the object is powered by additional electrical charge (current) brought into the system due to the increased capacitance, according to $Q = CU$. The amount of energy required to move the object could be easily calculated from the increased capacitance and therefore increased electrical energy stored in the system ($CU^2/2$).

Now, consider the case presented in Figure 9.1c (note that for simplicity the E-field lines are not shown). In between the metal plates of Figure 9.1c the E-field was uniform before the object entered the space in between the plates. The object moves into space between the plates, which can be interpreted in two ways. First, at the advancing edge of the object, the charges will clearly have a partially horizontal electric field and force component to them, a force that does not exist or is far weaker at the trailing edge of the object. This results in a net force, which pulls the object in between the electrodes. Figure 9.1c can also be understood in terms of potential energy. As the object moves in between the plates, it increases the electric capacitance, and like Figure 9.1b, the mechanical energy expended is equal to the increased electrical energy supplied to the system ($CU^2/2$).

The cases presented in Figure 9.1b,c are readily extendable to a fluid object surrounded by another immiscible fluid. The primary requirement for a similar direction for the force imbalance would be that the fluid object has a higher permittivity than the surrounding fluid.

In general dielectrophoretic movement and dielectrophoretic wetting of fluids follow the basic principles taught for Figure 9.1. Liquid dielectrophoresis is a well-established field and will not be further reviewed here. As discussed in detail in Section 5.3, however, dielectrophoresis and electrowetting theories seamless hand off to each other. As shown by Jones et al. [4], a semi-insulating fluid will follow electrowetting theory at low alternating potential frequencies (f), because the fluid acts as resistive (R) conductor according to electrical impedance $Z = R + 1/(2\pi fC)$. At higher frequencies the semi-insulating fluid is dominated by capacitive coupling of the applied potential (again, as impedance changes with frequency). Most importantly, Jones' theory and experiment both show continuity between the electrowetting and dielectrophoretic regimes, and their respective contributions to the wetting simply change with the frequency [4].

Of course exact geometry of the object and electrodes is needed to arrive at a quantified dielectrophoretic force. For example, the dielectrophoretic force on sphere of relative permittivity ε_2 radius r inside an insulating fluid of permittivity ε_1 can be calculated according to

$$F = 2\pi r^3 \varepsilon_1(\varepsilon_2 - \varepsilon_1)/(\varepsilon_2 + 2\varepsilon_1)\nabla E^2, \quad E = \nabla U \tag{9.1}$$

The higher the relative permittivity of the sphere, the stronger the dielectric force, and the force vanishes if the relative permittivity sphere approaches that of the surrounding fluid [13]. Importantly, following our previous discussion for Figure 9.1, Eq. (9.1) clearly shows the importance of the E-field gradient. The stronger the E-field gradient, the stronger the dielectrophoretic force. We will next look a particular case of dielectrowetting, which obtains an electrowetting voltage response, but which is unique from electrowetting as it exhibits no saturation of the apparent contact angle.

9.2.2 Dielectrowetting

The term dielectrowetting was recently introduced by McHale and Brown [7, 8] for a subset of fluid dielectrophoresis, but where the cosine of the apparent contact angle follows a voltage-squared wetting dependence similar to conventional electrowetting. The best way to describe dielectrowetting may be to simply jump right to discussion of the commonly utilized device structure shown in Figure 9.2. The metal film electrodes are interdigitated, flat, and in plane with each other. Therefore the applied electric field will be strongest adjacent to the surface of the substrate (obviously electric field lines decay as you move away from the electrodes; see discussion of Figure 2.4c). Initially the fluid rests at its Young's angle on the hydrophobic device surface. As the magnitude of the alternating potential is increased, the dielectrophoretic forces further bring the fluid toward the direction of the increase in electric field. Note, of course, that the actual microscopic contact angle at the contact line remains invariant in dielectrowetting. Like in the case of conventional electrowetting, only the apparent contact angle beyond the *localization length scale* of the electric field decreases. For conventional electrowetting, this length scale is given by the thickness d of the insulating layer, as we discussed in Section 5.1.3. In the case of dielectrowetting, this length scale is given by the pitch $2w$ of the interdigitated

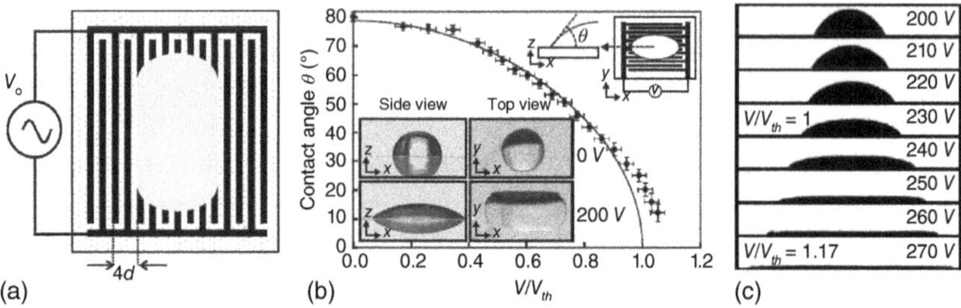

Figure 9.2 Dielectrowetting with 10 kHz sine wave and propylene glycol. (a) Top view of device and experimental setup where $d = 40\,\mu\text{m}$ is both the spacing between electrodes and the width of the electrodes. The device includes ~1 μm thickness of a hydrophobic dielectric over the electrodes. (b) Experiment versus theory (solid line) plotted versus the applied potential (V) divided by the theoretical voltage for superspreading ($U = 240$ V). (c) Side profile of the fluid including the experimental case for complete wetting (superspreading). Source: Adapted from [7, 8].

electrode pattern, i.e. it is typically much larger, often of the order of 100 μm. Initially before pseudo-complete wetting, the fluid preferentially wets strongest in between the electrodes (where electric field is strongest). If adequately high potential is applied, the fluid forms a completely wetting film (Young's angle of zero). The potential utilized in Figure 9.2 is a 10 kHz sine wave, which is adequately high in frequency such that the semi-insulating propylene glycol fluid responds as an electrically insulating fluid.

As further taught by McHale and Brown, dielectrowetting behavior below the threshold voltage for pseudo-complete wetting (superspreading) in air is predicted by

$$\cos\theta = \cos\theta_Y + \frac{\pi\varepsilon_0(\varepsilon_i - 1)}{8w\gamma_{lv}}(CU)^2$$

$$C = \left(1 + \frac{\varepsilon_d - \varepsilon_i}{\varepsilon_d + \varepsilon_i}\right) e^{-\pi t_d/2W} \Bigg/ \left(1 + \left(\frac{\varepsilon - \varepsilon_i}{\varepsilon_d + \varepsilon_i}\right) e^{-\pi t_d/W}\right) \tag{9.2}$$

where ε_i is the dielectric constant of the fluid, ε_d is the dielectric constant of the dielectric layer, t_d is the thickness of the dielectric layer, w denotes the electrode width and gap spacing, and γ_{lv} is the liquid–vapor (air) interfacial surface tension of the fluid. Equation (9.2) is only accurate well below the threshold for superspreading and represents only the first of three total wetting regimes. The second wetting regime is near the threshold for superspreading and the third beyond the threshold. The interested reader directed to read [7, 8] for further details and detailed derivation of Eq. (9.2).

An interesting applied demonstration of dielectrowetting was performed by coauthor Heikenfeld's group [3, 11, 15]. A device much like that shown in Figure 9.2 was utilized but with two significant modifications to support use as a high-transparency and very-large-area optical shutter. Firstly, the fluid chosen was a very high permittivity $\varepsilon \sim 35$ liquid crystal fluid colored with black dyes like those used in electrowetting displays [9, 10]. Because liquid crystal fluids are highly purified and designed to be electrically insulating, dielectrowetting superspreading was demonstrated all the way down to 0.1 Hz (only tens of microwatts per square centimeter power consumption). Secondly,

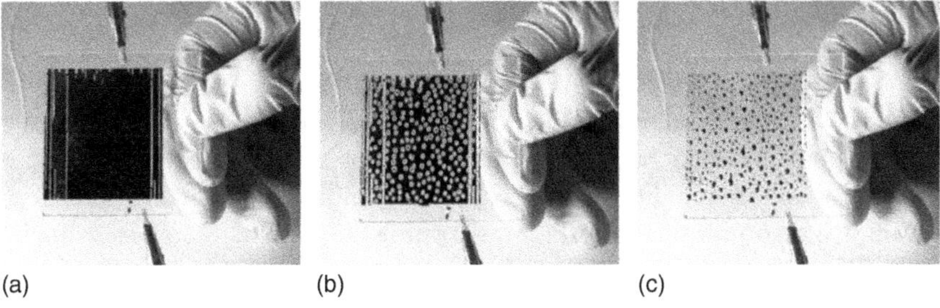

Figure 9.3 Large-area dielectrowetting optical shutter. Source: [15, 16]. Reproduced with permission of Elsevier.

to enable dewetting into smaller droplets that is faster and that is less visible, the surface was patterned with polymer bumps, which act as wetting defects in the superspread film. The dewetting around these polymer bumps is visually obvious in the center panel of Figure 9.3.

9.3 Innovations in Liquid Metal Electrowetting and Electrocapillarity

Liquid metals and electrowetting are not a new topic, especially considering the Hg electrocapillarity work of Lippmann (Chapter 4) and C. J. Kim group's work on Hg electrowetting in the 1990s. Kim and colleagues even demonstrated continuous electrowetting in a channel up to at least an impressive 10^7 cycles [6]. Liquid metals are typically of interest for one or more of the following reasons: They exhibit ultrahigh surface tension (>500 mN m^{-1}), which can lead to extremely rapid actuation speeds and/or high fluidic pressures; they are highly electrically conductive, making them suitable for applications in electronics and strong interaction with electromagnetic waves; they are completely optically opaque and exhibit mirror like optical reflection; and lastly, as will be presented at the end of this section, they have very interesting and powerful electrochemical interfacial behavior. Regarding wetting speeds, with such high surface tension and with scaling to nanometer dimensions, even nanosecond wetting speeds are postulated [1]. Recently, there has been a renewed interest in liquid metals, particularly because the new investigations are not with toxic Hg but rather with GaInSn alloys.

9.3.1 Electrowetting of GaInSn Liquid Metal Alloys

Only recently, during the time of writing of this book, has reliable and robust electrowetting been achieved with non-Hg liquid metals. Alloys such as GaInSn have long been available to electrowetting researchers but are very difficult to work with as they readily form an immobilizing oxide on their surface. The level of an oxygen-free environment needed for testing is ppm, and even at ppm oxidation will slowly occur, which has historically made general experimentation and device development completely impractical. One might first attempt to solve this by operating in a slight acidic vapor, such as HCl, which continually removes any surface oxide that forms. However, HCl and the solutes produced by the oxide-etching process adhere to the droplet and form an annulus around the base of the droplet contact on the electrowetting plate. When potential is applied, only this acid-solution annulus electrowets, and the liquid metal itself does not change in contact angle. Therefore an alternate approach is needed.

Recently, the Heikenfeld group [17] has solved this at least partially by deoxygenation of a common silicone oil (Dow Corning OS-20) and then dissolving in the oil ~0.6–6.28 wt% HCl or HBr by weight. Silicone oils are unique in their very high miscibilities with HCl, as miscibility in other oils such as alkanes is extremely low. The acid's associative incorporation into the oil does not require ionization and therefore retains the electrical resistance of the oil. This *acidic oil* slows the rate at which oxygen reaches the liquid metal and furthermore etches any oxide that does form. Currently, electrowetting of GaInSn in this acidic oil on a 3.8 μm Parylene C dielectric with an AC potential of 330 V and 1 kHz provides a fairly reliable and reversible contact angle reduction from 180° down to 109°.

However, this acidic oil approach does not resolve all the challenges for liquid metal electrowetting, as the high surface tension of the liquid metals requires very high electric fields for electrowetting (with a similar dielectric, ~10× greater than for water), which are near the electrical breakdown strength for many dielectrics. To make matters even more challenging, no surfactant systems have been developed that can appropriately reduce interfacial tension to the point where lower potential operation is possible. This is an excellent segue into the next section, where interfacial surface tension on a liquid metal surface can be electrochemically switched over a surprisingly large range.

9.3.2 Giant Electrochemical Changes in Liquid Metal Interfacial Surface Tensions

One of the unique attributes of a gallium-based liquid metal allow is that the liquid metal surface can be rapidly oxidized or reduced when the liquid metal under electric potential is placed in an electrolyte. The Dickey group has utilized this effect to achieve what is arguably the largest ever change in electrically controlled interfacial tension between two liquids [2, 5]. A liquid metal surface that is oxide-free (electrically reduced) still typically has an interfacial tension of 500 mN m^{-1} or greater, with any fluid it is brought into contact with. This is illustrated in Figure 9.4a where a eutectic GaIn droplet (EGaIn) is immersed in an aqueous solution of NaOH. Now, instead of reducing the liquid metal surface to ensure there is no oxide, electrochemically oxidize it. As shown in Figure 9.4b, simply reducing the applied potential to the point at which redox no longer occurs (>−0.5 V) allows the solution to oxide the droplet. The resulting oxide–solution interfacial tension is near zero (was not reliably measurable), resulting in a giant change in interfacial tension of >500 mN m^{-1}. The process is fully reversible (however, the maximum number of cycles has not yet been determined). The large change in interfacial tension enables extreme changes in the geometry of the liquid metal droplet as shown in the top-view diagrams of Figure 9.4a,b. Furthermore, the process can be utilized to pull the liquid metal

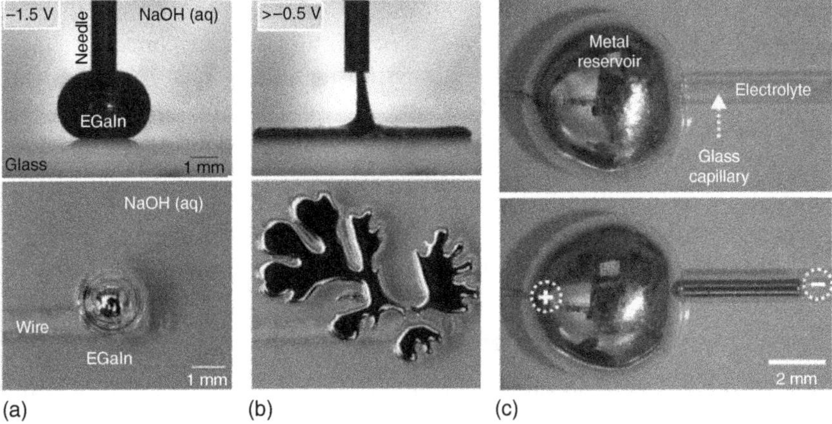

Figure 9.4 Demonstrations of electrochemical control of interfacial surface tensions with eutectic GaIn liquid metal alloy. Source: [2]. Reproduced with permission of ACS.

into a capillary (Figure 9.4c) by applying reducing potential outside the capillary and oxidative potential within the capillary. Imagine pulling the liquid metal into a very small capillary in this manner and then reducing the liquid metal. Such actuation could result in enormously large Laplace pressures and stored potential energy/pressure.

9.4 Nonequilibrium Electrical Control Without Contact Angle Modulation

The concept of an equilibrium limit in electrowetting systems is well known. For example, equilibrium limits appear in superhydrophobic surfaces [14] at the limit of stability between the Cassie–Baxter and Wenzel states and in electrowetting pixels [12] at the limit of stability (threshold potential) for beginning of dewetting of the insulating fluid. The latter of these topics leads us to the primary topic of this section: nonequilibrium electrical control without contact angle modulation.

9.4.1 Some Limitations of Conventional Electrowetting

This book has taught several techniques for electrowetting control of the interface between two immiscible fluids, such as simple periodic geometries (symmetric waves; Section 8.4, Figure 8.7) or energy-minimized geometries (only two principal radii of curvature; Section 8.8, Figure 8.14). The energy-minimized requirement for the meniscus to have only two principal radii of curvature is simple to understand. Energy minimization *at equilibrium* requires that a meniscus minimize its surface area and therefore its surface energy or interfacial energy, which can only occur if the surface is defined by two radii (e.g. no dimples, corners, inflection points, etc. are allowed on the surface). You might then ask: can more complex meniscus geometries be created?

Another limit of conventional electrowetting is that it in all cases, it is accompanied by contact angle modulation. Why is this of concern? Electrowetting contact angle modulation always involves a divergent electric field and therefore enhanced electrical field near the contact line (Figure 5.1). As taught in Chapter 7, the electrical field stress on the insulating layers forming the electrowetting capacitor will over time cause electrical degradation. With infinite time, of course, electrical breakdown will inevitably occur. Can a method be developed that does not require contact of the conducting fluid on an electrowetting capacitor (e.g. not require contact angle modulation)?

9.4.2 Electrowetting Without Wetting

Coauthor Heikenfeld's group has recently proposed and modeled a feedback control method that can impart complex nonequilibrium geometries onto the meniscus between a conducting fluid and insulating fluid [3]. The approach is fundamentally simple: (i) A potential is applied between at least one electrode and the conducting fluid with an oil film in between; (ii) the electrical capacitance is measured between the electrode and the conducting fluid; and (iii) the applied potential is increased or decreased in magnitude using feedback control to achieve a desired measured capacitance, and this capacitance represents a specific thickness desired for the oil. Importantly, because of feedback control, the system can go well beyond the point of stable equilibrium. The basic principles of operation are better taught in the electrohydrodynamic modeling work shown in Figure 9.5.

The small rectangles in Figure 9.5 are electrodes, and the black shape is an oil layer (dodecane). The white space above the black oil layer is an electrically conducting fluid (water) that is electrically grounded. At the bottom of Figure 9.5a, three of five repeating electrodes are used in this example (V_{e1}, V_{e4}, V_{e5}). The electrodes are allowed to provide a potential of 5 or 70 V. Seventy volts would cause the oil to completely dewet, and 5 V allows the oil to relax (move back toward a planar geometry). The series of photographs in Figure 9.5a show time evolution for feedback control of applying these potentials to achieve the *stable* geometry shown at the bottom of Figure 9.5a (e.g. not truly stable, only appears *stable* because of the fast speed of continuous feedback control). The feedback control algorithm used in this example switches the potential and measures capacitance until the desired oil thicknesses are achieved above each electrode, as plotted versus time in Figure 9.5b. In this case, for an oil layer that was initially 10 μm thick dodecane covered by water, the target oil thicknesses are achieved in only ~300 μs. Importantly, the final geometry achieved is nonsymmetrical and not representing an equilibrium state (e.g. if the 70 V were constantly applied, the oil would be completely dewetted by the water). Figure 9.5c,d show the results of thin versus thicker oil layers, where thicker oil layers allow closer approximation of a sawtooth waveform geometry for the oil–water meniscus.

As a reminder, two major outcomes are achieved: (1) Nonequilibrium shapes with more than two principal radii of curvature can be achieved, and (2) the conducting fluid never needs to touch (wet) the solid surface beneath the oil, which eliminates the need for creating a high quality dielectric. Furthermore, geometrical control is achieved *without* contact angle change. This work

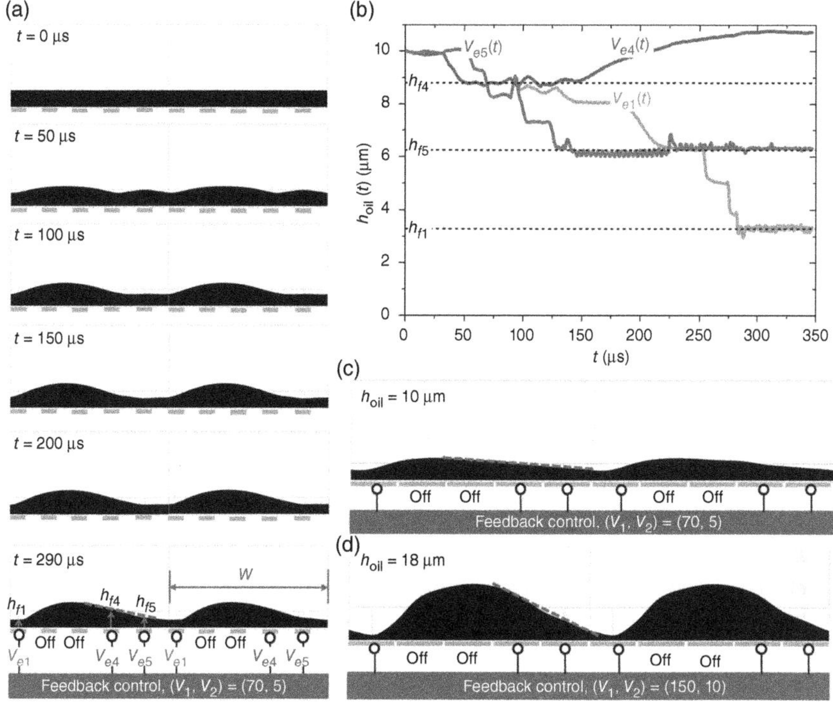

Figure 9.5 Using feedback control to build up triangular profiles. (a) An asymmetric triangular profile ($t = 290\,\mu s$) is created from the initially flat oil film ($t = 0\,\mu s$) based on the feedback method. (b) The plot of the oil film thickness $h_{oil}(t)$ above each of the three actuated electrodes as a function of time. (c,d) Plots of the fluid film profile with $h_{oil} = 10$ and $18\,\mu m$ and a reduced gap width between electrodes. Source: [3]. Reproduced with permission of Royal Society of Chemistry.

could serve numerous applications including particle or fluid transport (e.g. lab-on-chip) or adaptive optical surfaces (e.g. liquid prism arrays). Importantly, the results can be achieved using conventional materials, and the fluids respond with speeds that are adequately slow (ms-μs) such that even conventional control electronics (μs-ns) are likely more than adequately fast for the feedback control. However, this work is only modeling work, and physical experimental validation has not yet been performed.

Problems

9.1 Let us do a comparison of dielectrowetting to electrowetting. Answer as electrowetting (EW), dielectrowetting (DEW), or both (B).

a) Cosine of the contact angle increases with a potential squared (U^2) relationship.
b) Requires a solid dielectric layer over which in the ideal case all the electric field appears (low frequency, no DC current).

c) Has lower electrical power requirements as the electrical resistivity is increased for the fluid to be wetted electrically.
d) Is driven by a force that acts on the body of the fluid.
e) Is driven by a force that acts primarily on the meniscus of the fluid near the contact line.
f) With proper materials, it has been experimentally shown to wet down to zero degrees as voltage is increased.
g) Lower voltage can be achieved by increasing the dielectric constant of the fluid to be wetted compared with the surrounding fluid or gas.
h) Can be achieved with two parallel plate electrodes (careful, think about this!).
i) Can be achieved with two interdigitated electrodes like those shown in Figure 6.2.
(Hint: This can be done for electrowetting too, but why would it be less effective using the exact geometry of Figure 9.2, and how would you modify it?)

9.2 Assume a liquid metal with a surface tension of 500 mN m^{-1}. If you want to electrowet down to a contact angle of 60°, calculate the dielectric thickness needed and voltage if you are the very limit of electrical breakdown. Assume a polymer dielectric like Parylene with a dielectric constant of 2.5 and a breakdown field of 200 V µm^{-1}.
(Hint: First solve for the value of electrowetting number required, and then substitute in the dielectric thickness as a function of potential at electrical breakdown.)

9.3 Wow, the voltage is really low in the results shown in Figure 9.4! Show, however, that the energy required to wet from 180° to 0° is the same as it would be in theory for electrowetting as well (of course, it is not possible with electrowetting due to saturation). You can do a theoretical derivation to prove this or do a simple calculation for both techniques for a droplet that is 10 µl in volume.
(Hint: For the electrochemical approach of 9.4, assume the surface energy is changed by the entire 500 mN m^{-1}, roughly.)

9.4 Prove to yourself that the geometries like that shown in Figure 9.5d are energetically unfavorable (e.g. not at equilibrium). Do this by comparing the surface energy of a flat oil meniscus that is 100 µm wide against an oil meniscus that is a right triangle with a 100 µm base and 40 µm height (the geometries of Figure 9.5d are not a perfect triangle, but this calculation should be illustrative enough).

References

1 Chen, J. Y., Kutana, A., Collier, C. P., and Giapis, K. P. (2005). Electrowetting in carbon nanotubes. *Science* 310 (5753): 1480–1483. http://www.sciencemag.org/cgi/doi/10.1126/science.1120385.

2 Dickey, M. D. (2014). Emerging applications of liquid metals featuring surface oxides. *ACS Appl. Mater. Interfaces* 6 (21): 18369–18379.
3 Hsieh, W.-L., Chen, K.-C., and Heikenfeld, J. (2015). Sophisticated oil film geometries through incomplete electrical dewetting by feedback control and Fourier construction. *Lab Chip* 15 (12): 2615–2624. http://xlink.rsc.org/?DOI=C5LC00274E.
4 Jones, T. B., Fowler, J. D., Chang, Y. S., and Kim, C. J. (2003). Frequency-based relationship of electrowetting and dielectrophoretic liquid microactuation. *Langmuir* 19 (18): 7646–7651.
5 Khan, M. R., Eaker, C. B., Bowden, E. F., and Dickey, M. D. (2014). Giant and switchable surface activity of liquid metal via surface oxidation. *Proc. Natl. Acad. Sci. U.S.A.* 111 (39): 14047–14051. http://www.pnas.org/lookup/doi/10.1073/pnas.1412227111.
6 Lee, J. and Kim, C. J. (2000). Surface-tension-driven microactuation based on continuous electrowetting. *J. Microelectromech. Syst.* 9 (2): 171–180.
7 McHale, G., Brown, C. V., Newton, M. I. et al. (2011). Dielectrowetting driven spreading of droplets. *Phys. Rev. Lett.* 107 (18): 186101.
8 McHale, G., Brown, C. V., and Sampara, N. (2013). Voltage-induced spreading and superspreading of liquids. *Nat. Commun.* 4: 1605.
9 Ren, H., Xu, S., and Wu, S.-T. (2011). Voltage-expandable liquid crystal surface. *Lab Chip* 11 (20): 3426–3430. http://www.ncbi.nlm.nih.gov/pubmed/21901206.
10 Xu, S., Ren, H., Liu, Y., and Wu, S. T. (2012). Color displays based on voltage-stretchable liquid crystal droplet. *IEEE/OSA J. Disp. Technol.* 8 (6): 336–340.
11 Schultz, A., Chevalliot, S., Kuiper, S., and Heikenfeld, J. (2013). Detailed analysis of defect reduction in electrowetting dielectrics through a two-layer "barrier" approach. *Thin Solid Films* 534: 348–355.
12 Sun, B. and Heikenfeld, J. (2008). Observation and optical implications of oil dewetting patterns in electrowetting displays. *J. Micromech. Microeng.* 18: 025027.
13 Jones, T. B. (2003). Basic theory of dielectrophoresis and electrorotation. *IEEE Eng. Med. Biol. Mag.* 22 (6): 33–42.
14 Manukyan, G., Oh, J. M., van den Ende, D. et al. (2011). Electrical switching of wetting states on superhydrophobic surfaces: a route towards reversible Cassie-to-Wenzel transitions. *Phys. Rev. Lett.* 106: 014501.
15 Zhao, R., Cumby, B., Russell, A., and Heikenfeld, J. (2014). Large area and low power dielectrowetting optical shutter with local deterministic fluid film breakup. *Appl. Phys. Lett.* 104: 019901.
16 Russell, A.C., Hsieh, W.L., Chen, K.C., and Heikenfeld, J. (2015). Experimental and numerical insights into isotropic spreading and deterministic dewetting of dielectrowetted films. *Langmuir* 31 (1): 637–642.
17 Sarah Holcomb, Michael Brothers, Aaron Diebold, William Thatcher, David Mast, Christopher Tabor, and Jason Heikenfeld (2016). Oxide-free actuation of gallium liquid metal alloys enabled by novel acidified siloxane oils. *Langmuir* 32 (48): 12656–12663.

Appendix

Historical Perspective of Modern Electrowetting: Individual Testimonials

Introduction and Scope

This appendix serves as a first documented history of modern electrowetting. The chapter traces the involvement of several leading individuals in electrowetting science and technology, primarily those who have remained committed to the field's development up until the time of writing the first edition of this book. Here we also recognize the many organizers of the international meetings on electrowetting, which has created the vibrant sense of community that currently is existing during the writing of this book in 2014. Rather than attempt to "distil" the inputs found here into a cohesive story, the contributions are left here in their raw form, as provided. This allows the contributions to be truly seen through the eyes of the individual, an individual possibly like you, the reader, who is just now starting to move into electrowetting. Our work is not nearly done yet. Additional commercialization success is needed, and our theoretical understanding is still incomplete in aspects such as saturation. Now, here are the testimonials, which are provided in no particular order.

"CJ" Kim

When I joined UCLA in 1993, I wanted to move liquids without pump. Electrostatic actuation of droplets was the first option (extending the MEMS comb drive to liquids), but the resistance was too large (Teflon coating unavailable at the time) except for mercury. So we focused on electrostatic actuation of mercury droplets (my 1997 NSF CAREER and a few publications). For water, we instead explored thermal-bubble-based pumping mechanisms. I knew of electrowetting because of the attached paper in a MEMS conference (Colgate, JVST 1990, electrowetting-based microactuation), but its implementation was not possible, either.

After developing new fabrication techniques and finding a super student (Junghoon Lee), in 1996 I asked him to develop a continuous electrowetting device using mercury. After a great success, in 1997 he started to explore an electrowetting (on metal)-based pumping of water. We got some success but struggled until later that year, when he found electrowetting was possible with a dielectric layer (from the Philips website for their lens). With the subsequent

Electrowetting: Fundamental Principles and Practical Applications,
First Edition. Frieder Mugele and Jason Heikenfeld.

success, I submitted a proposal ("Integrated Digital Microfluidic Circuits Operated by Electrowetting-on-Dielectrics (EWODs) Principle") to DARPA in late 1997. We demonstrated various droplet movements in DARPA PI meetings in 1998 and 1999 but didn't consider it a success because we could not create droplets from a reservoir on chip. In the meantime, my DARPA program manager told me he found that someone was working on electrowetting (Duke) although in oil. Robin Garrell was strongly against oil for bio-applications, so we did everything in air. This is the way I followed before starting the EWOD and digital microfluidics (which I called digital microfluidic circuits in the DARPA proposal). I am attaching the DARPA proposal for your eyes; you will see what I was thinking in 1997.

Actually, the story I was trying to tell my students was a bit different in focus. As Junghoon Lee was graduating, in summer 2000 he submitted his up-to-date results to the IEEE MEMS Conference (only conference because no droplet creation yet). But soon we found Duke's APL 2000 paper. So, he expanded his MEMS 2001 results and submitted it to Sensor and Actuators after starting at Northwestern (printed in February 2002). Junghoon Lee lost his chance to publish the droplet movements well before 2000 because my goal was too high. So, I wanted to advise my students: don't try to be too perfect!

Authors Note from Heikenfeld

CJ's students have gone onto great success themselves. Any of them would have an exciting story to tell as well! Also, CJ has an MS student who did some unpublished work for which I later found was similar to my own concept for electrofluidic displays.

Johan Feenstra

When I joined Philips Research Laboratories on 1 January 1999, they still had the habit for (most) new employees to get a choice out of a number of projects to work on. In my case, I was offered a choice of three potential project teams to join. The first, a semiconductor project ended up being transferred to a foreign location within six months after I joined. The second, a display project, was terminated within the same six months… Lucky me, I actually chose the third, which was a project that focused on fundamental understanding and potential applications of a technology that I had never heard of: electrowetting.

At pretty much the same time, another new employee joined Philips Research: Rob Hayes, who actually had a lot of experience with electrowetting. We ended working together from day 1. On the fundamental side, we started the first electrowetting workshop, organized in Mons, with participation from the University of Mons (Joel de Coninck a.o.), Kodak (Terry Blake a.o.), University of Lyon (Bruno Berge a.o.), and our side. I think this meeting sets the tone for the subsequent meetings where a relatively small group of people engaged on great technical discussions as well as excellent personal interactions during lunches, dinners, and other off-conference activities.

As part of delivering on the other goal of the project (*For what applications can we use this technology?*), we came up with a large number of applications. In fact, we had the practical joke within the lab that any problem people had could be solved with electrowetting.

The most interesting application we ended up pursuing from about 2001 was that of an electronic display. Initially together, but later, by strengthening the team with over 20 students in a period of four years, we worked on developing this application from scratch. After a number of years of internal, secretive development (and passing the stage where we had to explain to people what the on- and the off-state was), we wanted to open up the work for external partners and decided to publish it. As we were very excited about the work and its prospects, we submitted the paper to *Nature*. The rest is history. *Nature* picked it up and chose our paper for the cover and their own PR, and this is now (by far) the best-cited paper of my career to date [1]. Needless to say that we succeeded on the goal of creating external interest (including that of Jason and Andrew Steckl). I'm very proud of the fact that our electrowetting work at Philips Research has strongly contributed to expanding the interest and activity in the field, as can be seen from a significant increase in the number of publications and patents around the same time we went public.

Since then, the work of my group has been spun off from Philips in a separate company: Liquavista, focusing on the development of electronic displays, based on electrowetting. While we now leave the fundamental understanding of electrowetting to others (but clearly keep abreast of the latest developments), I do think we still strongly contribute to the field by giving it a trajectory into commercial applications, thereby providing a strong viability statement for the technology.

Tom Jones

My own research in microfluidics got its start in 2000 while I was on sabbatical leave at Kyoto University in Japan. My host and long-time colleague at Kyoto, Prof. Masao Washizu, was good enough to assign his best graduate student to work with me for the eight months of my stay. This student, Masahide Gunji, was a remarkably talented experimentalist and managed to provide a first demonstration of the scheme I had in mind even before I arrived in Kyoto. It was a method for transporting aqueous liquids using simple coplanar electrodes fabricated on a glass substrate. The required AC voltage was 700 V-rms (!) at a frequency of ~100 kHz (!). This was *not* electrowetting but rather an example of liquid dielectrophoresis, performed with DI water instead of the usual insulating liquids. I presented this work at the 2001 IEEE MEMS Meeting in Interlaken, Switzerland, and after my talk, C.J. Kim raised his hand and asked a question, the specifics of which now escape me. I am fairly sure that, at that time, CJ did not know what dielectrophoresis was and I am absolutely certain that I was clueless about electrowetting. Whatever answer I gave to his question could not have been particularly illuminating. After returning home from the trip, I played catch-up, reading his papers, all of which were unknown to me up to that point.

Eventually I worked out the relationship between EWOD and liquid DEP and – as many of you know – have been sermonizing about it ever since. To wit, EWOD and liquid DEP are, respectively, the low and high frequency limits of the electromechanical force exerted on a liquid by a nonuniform electric field.

The essentials of my contribution to the understanding of the EWOD and liquid DEP are adequately represented by five papers, which are listed below with some companion notes:

- Made distinction between translational force (that enables droplet-based microfluidic applications) and apparent contact angle change (voltage-controlled lenses) [2]. Showed that lumped parameter electromechanical model (supported by Maxwell stress tensor approach) correctly predicts the force on droplets without any reference to contact angle change. Theory only, no experiments.
- Used simple RC circuit and lumped parameter electromechanics to show frequency-based relationship of EWOD (low frequency) and liquid DEP (high frequency) [3]. Showed that saturation effect influences the observed pressure drop across an interface. Experimental data obtained with Quincke bubble method. Paper contains an acknowledgement of K. H. Kang. Unfortunately, I no longer have copies of our correspondence.
- Used the Pellat apparatus to obtain much more reliable data for the frequency dependence of the electromechanical force [4]. Used lumped parameter analysis and RC circuit models. Stated more clearly the view that the translational force and the contact angle effect are distinct observables and backed-up statement with a thought experiment.
- The dynamic Pellat experiments provide further support for the electromechanical model with RC circuit [5]. Used Blake/Haynes model for dynamic contact line friction to correlate data successfully. Experiments revealed some evidence that the saturation effect is time and/or motion dependent.
- This paper distilled and refined the essential arguments about the independence of contact angle change and the observable translational force, relying on the fact that the lumped parameter method (and the Maxwell stress method) of force calculation does not depend at all on the liquid–air profile [6].

Frieder Mugele

I joined the University of Ulm in late 1999 to set up my own little research group as a young assistant within the team of Stephan Herminghaus. I had worked on confined liquids before and Stephan was very active in various aspects of wetting ranging from the molecular scale understanding to mesoscopic phenomenology on structured surfaces. Stephan had heard about this new thing "electrowetting," and he suggested to me to go and have a look at it. At the time, I was interested in the effect of the disjoining pressure on the equilibrium shape of liquid surfaces in the vicinity of the contact line. This was a tough problem to study because the effects were so small and local. Electrowetting offered the potential of a long-range force that should deform the interfaces on a much larger scale, which sounded very interesting to me. So, in 2000 I joined what we call now second International Workshop on Electrowetting, which was in fact a meeting with 15–20 people in a small seminar room at Philips Research in Eindhoven. Menno Prins was present at the time, as well as Rob Hayes, Johan Feenstra, and a few other people who had already attended the first electrowetting workshop in Mons.

Our focus early on was to try to understand the fundamental mechanisms that govern electrowetting. At the time, it was not clear in which sense the then modern EWOD was different from the classical experiments by Frumkin with mercury–electrolyte interfaces that made direct use of Lippmann's electrocapillary effect. Together with Jürgen Bührle, a very talented master student at the time, we performed numerical calculations to determine self-consistently the distribution of the electric field and the equilibrium configuration of the liquid surface in the vicinity of the three-phase contact line. The interpretation of the results was relatively straightforward for us because we were familiar with the deformations of the liquid surface that the disjoining pressure causes on the nanometer scale. Electrowetting turned out to be very similar – except that the molecular interaction forces had to be replaced by the long-range electrostatic forces that would also depend on the shape of the liquid surface. Together with Tom Jones in Rochester and Kwan Kang in Pohang, we argued that EWOD is primarily an electromechanical effect and that the variation of the macroscopic contact angle is the specific response of the fluid to this primary driving force. This interpretation finally became widely accepted when we demonstrated experimentally that the local contact angle at the three-phase contact line indeed remains constant as predicted by the theory.

I continued to have the pleasure of working together with a number of excellent students and senior coworkers, including Jean-Christophe Baret, now a professor in Bordeaux, France, with whom I published the first broad review article on electrowetting in 2005. After moving to the Netherlands in 2004, electrowetting continued to be a focus of my group. We tried to understand various fluid dynamics aspects of electrowetting that had been mentioned here in there but lacked quantitative physical understanding, such as the effect of confined oil films into drops, the origin of mixing flows in oscillating drops, contact angle hysteresis and the electrically driven Cassie-to-Wenzel transition on superhydrophobic surfaces. More than once, it turned out that we were working at the same time on the same problems as Kwan Kang did – our scientific counterpart on the other side of the planet who had to leave us way too early.

Throughout all the years, it has always been an inspiring pleasure to meet the community at the regular Electrowetting Meetings that by now include a stretch of eleven meetings (Mons 1999, Eindhoven 2000, Grenoble 2002, Blaubeuren 2004, Rochester 2006, UCLA 2008, Pohang 2010, Athens 2012, Cincinnati 2014, Taipeh 2016, Twente 2018). It was at the 2012 meeting that Jason Heikenfeld approached and asked me whether I would be interested in writing this book on electrowetting together with him. Since I had been approached by Wiley with the same question just a few weeks earlier, I accepted the challenge. It turned out to be a tough challenge that requested a lot of time. Yet, I certainly learned a lot during the writing process, and I hope that you as a reader enjoy the beauty, the simplicity, and the flexibility of the physical principles of electrowetting as much as I came to do over the years.

Richard Fair

I have attached three docs. The first is the 2000 APL paper we published that described the first electrowetting array actuator. The second is a 2007 paper that gives credit to the contributions of many who played a role in exploring electrowetting devices and characteristics (see pp. 247–248). The third is a recent recount of work in our lab from the late 1990s (see pp. 5–6). This piece is available on the Duke ECE website (http://dukeececorneroffice.wordpress.com/) and recounts the translation of electrowetting to the successful acquisition of ALL.

Michael Pollack coined the term "digital microfluidics" in 1998. He used the term in a DARPA research review, and I still have his power point slides. In August of 2000 I attended a DARPA program review at Park City, Utah. I set up my laptop and poster and showed videos of droplets moving on 2D arrays. Our APL paper came out in September 2000, so no one knew about the electrowetting array result. Suffice it to say, I had not heard about CJ nor had he heard about me. CJ and his group were also there and had a poster that proposed electrowetting actuation of droplets. Their previous work was on continuous electrowetting of mercury. When I stopped by their poster, I made the smart ass comment: "Would you like to see how your proposal turned out?" There were some hard feelings afterward. There is also an interesting history in the pursuit of patents, which is described in the third pdf file. It's clear that Duke/ALL won the patent wars, with over 100 issued patents. In summary, the key contributions to electrowetting devices were made by Berge and Shenderov. We built the first planar droplet devices, but the success of the field was built on the contributions of many around the world.

Author's Note from Heikenfeld

Like CJ, Richard's former students/postdocs went onto and founded Advanced Liquid Logic, which has resulted in some of the highest levels of commercial success for electrowetting. Their road to profitability was long, and we appreciate the dedication it must have taken along the way.

Bruno Berge

In 1989–1991, I was a visiting scientist at the University of Chicago in Albert Libchaber's group. My research subject was to investigate 2D solid–liquid line instabilities during situations of solid growth. My favorite toy at this time was a monolayer of amphiphilic molecules deposited on a water surface. In order to visualize the solidification, I was using a classical trick: you would deposit along with the molecule of interest a zest of fluorescing molecules, also amphiphilic in nature: the partition ratio between solid and liquid phase gives a very good contrast that is observable with a standard fluorescence microscope.

The overall goal of this physics study was to investigate chaos theory: how a regular or periodic system, once driven harder, could be pushed into a chaotic dynamical behavior? Libchaber had showed that in brilliant experiments in helium convection cells, which were basically one degree of freedom (0-dimensional system). We were hoping that a solid–liquid line in those true 2D systems would give additional interesting clues on chaos! At the end, it didn't give any, but it gave me the chance of a (re)discovery of EWOD.

Playing with a monolayer of molecules on top of a water surface is really easy! Take a clean glass of water, and deposit a drop of dodecanol on the surface. After a shaky time, the drop will rest somewhere, surrounded by a monolayer of dodecanol that exhibits a solid–liquid transition about 20 °C below the bulk one. At the time I arrived at the University of Chicago in Libchaber's team, I visited another professor of the chemistry department (third floor) who had made all his career in monolayer research: after keeping me waiting during one hour in his corridor, he told me I was too young and way too inexperienced to even think of starting a serious project on monolayers physics.[1] I was a little shaken, but I took it as an additional data point to what I could feel during discussions with experienced physical chemists in France: experiments that were challenges in the 1950s and 1960s, needing brilliant and smart tricks, could be done routinely in the 1980s–1990s due to laboratory technique evolution.

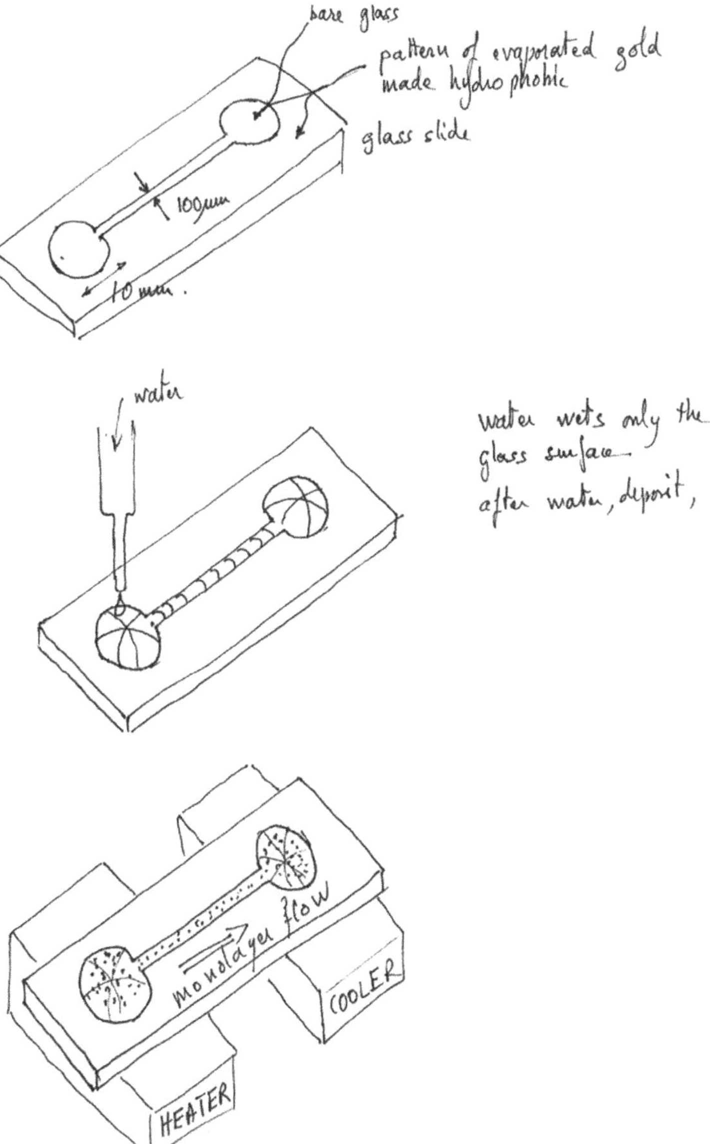

1 I realize now that my visit to this professor, with no appointment, might have been seen as arrogant by him!

After about one year in playing with these systems, my experiments came to a roadblock: I wanted to reproduce in two dimensions an in-capillary directional solidification (see attached figure). For that purpose, I designed a hydrophilic pattern on a glass planar surface, on which I would deposit a water phase. At first this worked very well, leaving a long thin wire of water on the planar surface. The monolayer spread nicely on this surface, and I could apply a temperature gradient between extreme reservoirs. I could observe fascinating behaviors, numerous patterns linked to the flowing 2D structure. Nevertheless, every time I started an experiment, after a few hours the water thin line would break: irreversibly! That was a severe problem and I fought like mad to solve it: better cleaning of the glass plate, better hydrophobization of the gold surface, etc. It took me several months with no solution at the end, thinking that the third floor professor might have been right!

At this time, I had explored most of "passive" ways of constructing a hydrophobic/hydrophilic pattern on a surface, and I thought I should try to achieve that same goal with forcing the wetting with an external field that could be as high as needed: "active wetting!" As soon as I had this idea in mind, I arrived very quickly to the electrowetting nowadays standard configuration: metal/hydrophobic insulator/water. I derived the formula for electrowetting – $\Delta \cos\theta = \frac{\varepsilon\varepsilon_0}{2e\gamma}V^2$ – and then started to think how to make an experimental test.

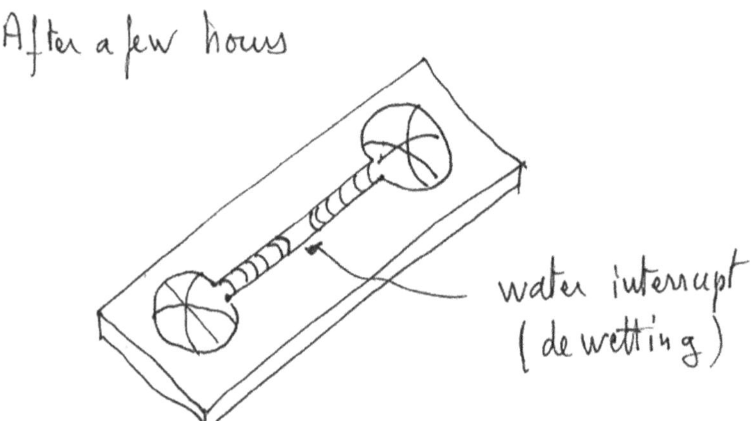

While I was bringing all the parts of a simple first experiment, I went to the theory department, just across the corridor. At this time there was an extraordinary emulation in the physics department. I already mentioned Libchaber and his numerous bright experiments on convections, turbulence, chaos, etc. Close to our team was also Leo Kadanoff and his group of bright students and postdocs. I visited one of them and discussed about my calculation of electrowetting, and he immediately turned my idea down: "it will never work," he said. He had quickly made a calculation of the E-field, compared with tabulated dielectric breakdown for standard polymers! I was very disappointed! But not to the point I would stop the experiment![2]

2 Later we realized that the breakdown E-field is increasing very much when thinning the insulating material, and that's the reason why EW works so well.

The very first experiment was done using a sheet of Saran as hydrophobic insulator (12.5 μm thick) stretched on a brass cylinder flat extremity using a small drop of salted water to fill the gap below the film. On top of that, I deposited a drop of tap water slightly salted, and a simple electric wire dipped into the drop served as electrode. As a DC source, I used a photomultiplier power supply, capable of 1 kV maximum voltage. The very first time I was extremely excited: after thinking, again and again, here was the time of seeing if this could work! I believe many experimentalists have experienced this excitement, at the exact moment of turning *on* the main power button! The experiment was ready in the evening, but I waited till the morning after, at a time where the physics lab was more quiet.[3] After a few experiments, I realized that even when applying high voltages (it's amazing that the Saran film can sustain easily 1 kV!), there was no effect: only a small transient effect when the DC was applied. I was about to quit, but when I turned the voltage *off*, I was very surprised to see the same effect. This made me realize that there was a DC filtering by some charge leak or some parallel resistance, which made the capacitance not perfect, and very quickly after, I could observe a response that is close to modern work on the subject using AC voltage.

At this time I had to return to France, and my electrowetting work was paused for several months. Before that I could benefit from the fantastic resources of the library of the University of Chicago: coming from France, this was a positive shock to have access 24/24 to an extremely well-documented library, sometimes finding French articles written by Lippmann that I could hardly find in France! I also dipped delightedly in early papers about applications of electrowetting, electrometers, optical switches or coolers using mercury as a liquid, etc. Later I could find also a very short paper by Nicholas Sheridan (disclosures from Xerox 1978), who designed what is probably the very first EWOD system. That's why I consider this personal contribution as the rediscovery of EWODs.

Writing this short paper has revived my memory of this fantastic period of postdoc at the University of Chicago. For me that was the starting of electrowetting as a wide playground, pioneering the theory, and developing applications – this is another story.

Glen McHale

In the 1990s I had an interest in simple and inexpensive but elegant science. At that time research funding was scarce and equipment basic, and what we did was as much about creating undergraduate student projects as staff research. Studying droplets on surfaces seemed interesting and was certainly accessible. Having published work on *dynamic wetting/spreading*, *evaporation*, and *contact angles*, I came across the *electrowetting* work of Bruno Berge in Grenoble, Terry

3 As I already had a family at that time, during evenings I was not working as late as many students were. Then I usually arrive first in the lab in the morning. This was a quiet time, and I had often the privilege of face-to-face discussions with top class scientists, Nobel-winning physicists or chemists, which were wandering in the lab looking for some interesting physics to see before their plenary talk later in the day!

Blake and Andrew Clarke at Kodak UK, Mike Pollack and colleagues at Duke University, and Altti Torkelli at VTT, and so I attended the first Electrowetting Meeting in Grenoble (2002). It was a fantastic informal meeting of enthusiasts where I was allowed to talk about "wetting" rather than "electrowetting" as my only electrowetting work was from undergraduate student projects. By the time of Blaubeuren (2004), my research had been pulled into superhydrophobicity, but I had also gained funding from the UK EPSRC for electrowetting and was able to talk about the possibilities of electrowetting on superhydrophobic surfaces. It took a couple of years more before myself and Mike Newton published on combining electrowetting and superhydrophobicity, and by that time we were also able to report about how electrowetting could be combined with ideas of liquid marbles.

I had listened over the years to discussions at the electrowetting workshops between Tom Jones, Frieder Mugele, and many others about the relationship between electrowetting and *liquid dielectrophoresis* and seen the commercial optical and display applications by Bruno Berge, Stein Kuiper, Rob Hayes, and Johan Feenstra. We had also been fortunate to be joined by a new lecturer, Carl Brown, in 2003, who had an interest in liquid crystal displays. Eventually chats over coffee led us to filing a UK patent application in 2005 on "switchable phase grating" and initiating a PhD project with Kodak in the United Kingdom on liquid dielectrophoresis and electrowetting. That work proved far harder for the student (Gary Wells) than we initially thought it would, and we dropped the patent, but eventually it led us to our *Nature Photonics* and *Nature Communications* articles on voltage programmable liquid diffraction gratings and voltage-induced superspreading, which can use the ability of our approach to force spreading into films. More generally, it led us into the concepts of *interface localized liquid dielectrophoresis* and, in its droplet form, what we have called *dielectrowetting* and which Jason Heikenfeld has shown has potential for optical shutters. Whether these concepts will take hold in the academic community and industry in quite the same way as electrowetting remains an open question.

In reading my testimonial I am conscious of the significant contributions of all the people I have not mentioned by name – some of whom are mentioned elsewhere here Thanasis, "CJ," and Kwan, and some who are not – but all of whom influenced my thinking through their papers and discussions.

Stein Kuiper

My first encounter with electrowetting was in June 2000, while applying for a position at Philips Research. They offered me twelve different projects to choose from, and one of them was "electrowetting-based liquid lenses." I was at the end of my PhD work on microfiltration in which I had been struggling with wetting on hydrophobic surfaces. And now they showed me a method for controlling hydrophobicity with a voltage! The elegance of the technology, combined with the enthusiasm of researcher Menno Prins who demonstrated it, made the choice easy.

When I started at Philips, there was a large group working on a dynamic beam attenuator. This was a pixelated X-ray absorber, consisting of an array of channels that could be filled with an X-ray absorbing liquid using electrowetting. Menno Prins had started the electrowetting work at Philips in 1994, one year after Bruno Berge made electrowetting accessible with his discovery of EWOD. Since then a huge amount of fundamental research had been done at Philips, all captured in confidential internal reports. What an intellectual treasure to start my work with! I joined the subgroup electrowetting fundamentals and was confronted with a poorly understood phenomenon called saturation. It seemed a simple phenomenon, asking for a simple explanation. I intended to find this explanation and surprise my colleagues. However, I underestimated the problem or overestimated myself. Saturation remained a mystery.

A few months after my start, the X-ray project was stopped. The reason was that the hydrophobic fluoropolymer turned hydrophilic during exposure to X-rays. It was a show stopper. However, the X-ray researchers had noticed that the water/oil interface in the channels deformed from convex to concave when the channels were filled and thought this could work as a lens. My task would be to develop this lens. They had also seen that the transparent side walls of the channels would turn translucent during electrical actuation, because tiny oil droplets formed. This effect could possibly be used for display purposes. One of the main inventors of these ideas, my roommate Johan Feenstra, changed the group electrowetting fundamentals into optofluidics, and we started developing the ideas. Johan and Rob Hayes focused on displays and I focused on lenses. We both made appealing demonstrators and together won the Philips Management Award in 2004. However, soon after that, Philips reorganized, which made funding difficult. In 2006, Johan and Rob spun out in Liquavista. I could continue at Philips on a low budget, concentrating again on more fundamental work and collaborating with that very enthusiastic American researcher Jason Heikenfeld. He came up with an elegant hypothesis for explaining saturation, comparing it with a Taylor cone. I am convinced this hypothesis is a large step in the right direction, but the last Electrowetting Meeting in Cincinnati in 2014 showed that more work needs to be done to convince the community.

In December 2000, the second Electrowetting Meeting took place at Philips. I had just made my first lens demonstrator. During the meeting, Bruno Berge presented his work on liquid lenses. I was shocked. He had already made and patented what I was doing! I thought of stopping the project, but Johan convinced me to continue and work on technical improvements. There was a lot to gain in developing the right set of liquids and dielectrics. In 2002, optical expert Benno Hendriks joined the lens project. His experience in generating patents for optical recording led to an avalanche of patents in the largely unexplored area of liquid lenses. Bruno had founded Varioptic and we had to work around his patents. I met Bruno on the Electrowetting Meetings. We talked about his lenses, but I could not tell him that I was working on the same subject. It was an awkward situation, especially because he was so open about his work. I proposed at Philips to join forces with him, because together we would cover the whole liquid lens field. However, Philips did not want to. When I finally managed to convince Philips, it was too late: Varioptic and Samsung had just signed a collaboration agreement.

In 2011, Philips stopped the liquid lens work. Addicted to electrowetting, I left Philips to be able to remain active in the field. As an independent consultant, I am now in an ideal position to share my enthusiasm for this charming technology.

Jason Heikenfeld

My personal story has some parallels with CJ's and involves a project I was working on with another electrowetting practitioner Andrew Steckl (who was my PhD advisor). In 2003, my formal training was mainly in electroluminescent devices, which meant high-temperature semiconductor growth and dielectric optimization, basically a lot of thin film work. I had just come up for an exciting new emissive display architecture called "light wave coupling," but as far as I could tell, I needed MEMS-type switching for the pixels. I was working at a startup company that Andrew and I had co-founded after I completed my PhD in 2001. I knew in my gut that MEMS switching would be too difficult to economically manufacture anytime soon (to this day, the display industry is still proving me correct). Also in 2003, Hayes and Feenstra at Philips published their breakthrough *Nature* cover-feature paper on reflective electrowetting displays. As I did every day, I was checking the news features on the website for the Society for Information when they highlighted Johan's work showing the famous "four-square" image of four colored pixels in operation. Instantly, I realized electrowetting could be a far more compelling architecture for making the type of transparent and emissive display we were after. I got hold of Jim Brown at Cytonix to secure some FluoroPel (FYI, Jim did some very early electrowetting devices too!) and made a dip-coating station with a mixer pump that wound up string pulling substrates out of FluoroPel solution, and we were off and running. We developed light wave coupling displays based on electrowetting, and our first publication made the cover of *Applied Physics Letters*. At that point I leaped entirely from an expert "solid-state, hard materials" research into a very eager and naïve "fluids, soft-matter" researcher.

I joined the faculty at my alma mater, the University of Cincinnati, in 2005. Presenting our display work-up at the University of Dayton, Joseph Haus prodded us to consider methods to utilize electrowetting for optical beam steering. My first PhD student and I shortly thereafter published for the first time electrowetting microprisms and gained our first large research grant on the subject (an NSF CAREER award). At that point, I went "all in" with electrowetting, betting my tenure success entirely upon it. Our rapidly growing volume work spread to cover significant contributions to electrowetting optics, displays, and lab-on-chip. I vividly remember my first Electrowetting Meeting in 2005 in Rochester and an informal pre-meeting gathering Tom Jones scheduled near his beautiful garden, where I was able to meet for the first time many of the giants in the field. To say the least, I felt completely miniscule and out of place at that informal gather for drinks but eager and ready to make my mark over the coming years.

I am very thankful to the electrowetting community, because my present day success would have been impossible without them. I also am thankful to Johan for running a company (Liquavista) that kept secret all new developments in electrowetting displays, which made my lab the one-stop shop in the world where industry and academia could come to learn about the technology. It really helped pay the lab bills! Because we were focusing on commercialization of several types of displays also (in my second startup company on electrofluidic displays), we were forced to really focus down on materials development for reliable electrowetting. This and a new collaboration with Stein Kuiper at Philips helped us publish many of the leading papers for going from "10 second lifetime" electrowetting experiments that many in the community would not admit to toward devices that could last the entirety of testing needed for a PhD dissertation! At the 2012 Electrowetting Meeting in Greece, on the bus during the excursion to the Temple of Neptune, I suggested to Frieder that it is time to write a book that he should lead (an easy call to make), and I noted that I would like to contribute the chapters related to the applied aspects of electrowetting. The rest is history! "Now, as I finish writing this, if I can just get Frieder to complete his chapters..."

Kwan Hyung Kang: An Appreciation by T. B. Jones

Prof. Kwan Kang's career was cut short by cancer, not long after he hosted the international Electrowetting Meeting in Pohang, Korea.

Some years ago, Kwan Kang asked me to write a letter supporting his candidacy for a regular faculty appointment in the Department of Mechanical Engineering at Pohang University of Science and Technology (POSTECH). Excerpts from that letter, found below, capture well enough my views of the significance of his earlier work on the origins of electrowetting.

> My interactions with him were initiated about ten months ago after I read his paper entitled "How electrostatic fields change contact angle in electrowetting" [7]. In this contribution, Kang argued convincingly that it is an electrostatic force, and not a "change in surface tension," that reduces the contact angle of a sessile droplet under electric stress on a substrate. He specifically addressed the issue of how the electric surface charge and its associated electric field concentrate near the contact line where liquid, solid, and gas come together. What impressed me immediately about the paper was that Kang used the same classical methods of electromechanics that I have been using, principally the Maxwell stress tensor, to determine the forces that control contact angle and droplet shape. This approach is in contradistinction to the very general thermodynamic methods used by certain electrochemists, who, in my view, unnecessarily complicate and confuse the matter by lumping together electromechanical and electrochemical phenomena in situations where they are best kept separate.

Kang and I, not being aware of each other's work, had published complementary work that employed the Maxwell stress tensor to evaluate forces of electrical origin. It can be said that we were kindred spirits. It is nice to be able to say such a thing on this, the 150th anniversary of the original publication of the famous equations of James Clerk Maxwell. As an aside, the Maxwell stress tensor formulation resulted from his effort in the 1860s to understand electricity and magnetism using a mechanical analogy. That cumbersome analogy is long forgotten but the Maxwell stress tensor reigns.

My recommendation letter continued:

> For several years, I have been studying the relationship of the ponderomotive force on liquid dielectrics (which is responsible for the DEP effect) to electrowetting on dielectric-coated conductors. The conventional view is that the observed actuation exploited in electrowetting systems results from the change in the liquid/solid contact angle as voltage is applied. There are serious questions about this view that I posed and considered in my own paper, "On the relationship of dielectrophoresis and electrowetting" [2]. Despite never having met him, my interactions with Dr. Kang have been quite valuable. By email, we regularly exchange papers and comment extensively on each other's work. He asks tough questions that force me to think hard about my work. At Rochester, we are doing variable frequency measurements of the electromechanical response of liquids that are yielding some surprises. Kang's theory might help us to understand our data, and so I look forward to meeting him someday and perhaps to collaborating with him in the area of surface microfluidics.

Kang and I of course did meet at a couple of the Electrowetting Meetings and enjoyed some interesting discussions. Though we never did collaborate in a direct way, we were strongly influenced and supported by one another. This is a great example of how science should work.

Author's Note from Mugele

Like Tom Jones, I interacted very intensively with Kwan Kang. Initially, this interaction was mostly at a distance; later, Jung Min Oh, Kwan's former student, joined my lab and was a very successful and well-trained postdoc in my group who brought a lot of Kwan's spirit. From the early 2000s until his premature death in 2011, it often felt like Kwan and I were reading each other's minds. Not knowing from each other in the beginning, we published a series of papers throughout the years, in which we used very similar approaches to address the exactly same question. Sometimes, Kwan published a little bit ahead of us and we went a little bit deeper; sometimes it was the other way round. Examples include the basis of the electrical stress distribution around the contact line [7, 8], internal flows in oscillating droplets shaken by AC electrowetting [9, 10], and the electrothermal flows induced in very high frequency electrowetting [11, 12]. The community lost a creative mind and a dear colleague when he passed away.

References

1 Hayes, R. A. and Feenstra, B. J. (2003). Video speed electronic paper based on electrowetting. *Nature* 425: 383.

2 Jones, T. B. (2002). On the relationship of dielectrophoresis and electrowetting. *Langmuir* 18: 4437–4443.

3 Jones, T. B., Fowler, J. D., Chang, Y. S., and Kim, C.-J. (2003). Frequency-based relationship of electrowetting and dielectrophoretic liquid microactuation. *Langmuir* 19: 7646–7651.

4 Jones, T. B., Wang, K.-L., and Yao, D.-J. (2004). Frequency-dependent electromechanics of aqueous liquids: electrowetting and dielectrophoresis. *Langmuir* 20: 2813–2818.

5 Wang, K.-L. and Jones, T. B. (2005). Electrowetting dynamics of microfluidic actuation. *Langmuir* 21: 4211–4217.

6 Jones, T. B. (2005). An electromechanical interpretation of electrowetting. *J. Micromech. Microengrg.* 15: 1184–1187.

7 Kang, K. H. (2002). How electrostatic fields change contact angle in electrowetting. *Langmuir* 18: 10318–10322.

8 Buehrle, Herminghaus, S., and Mugele, F. (2003). Interface profiles near three-phase contact lines in electric fields. *Phys. Rev. Lett.* 91: 086101.

9 Ko, S. H., Lee, H., and Kang, K. H. (2008). Hydrodynamic flows in electrowetting. *Langmuir* 24: 1094.

10 Mugele, F., Baret, J.-C., and Steinhauser, D. (2006). Microfluidic mixing through electrowetting-induced droplet oscillations. *Appl. Phys. Lett.* 88, 204106.

11 Lee, H., Yun, S., Ko, S. H., and Kang, K. H. (2009). An electrohydrodynamic flow in ac electrowetting. *Biomicrofluidics* 3, 044113.

12 García-Sánchez, P., Ramos, A., and Mugele, F. (2010). Electrothermally driven flows in ac electrowetting. *Phys. Rev. E* 81, 015303(R).

Index

Electrowetting: Fundamental Principles and Practical Applications,
First Edition. Frieder Mugele and Jason Heikenfeld.